THE LIBRARY
ST. MARY'S COLLEGE OF MARYLAND
ST. MARY'S CITY, MARYLAND 20686

SEED AGING

SEED AGING

*Implications for Seed Storage
and Persistence in the Soil*

By DAVID A. PRIESTLEY

Boyce Thompson Institute for Plant Research

COMSTOCK PUBLISHING ASSOCIATES
A division of Cornell University Press
Ithaca and London

Library of Congress Cataloging-in-Publication Data
Priestley, David A.
　　Seed aging.

　　Bibliography: p.
　　Includes index.
　　1. Seeds—Aging.　2. Seeds—Viability.　3. Seeds—Storage.　I. Title.
QK661.P74　1986　　　631.5′21　　　85-21334
ISBN 0-8014-1865-8　(alk. paper)

Copyright © 1986 by Cornell University

All rights reserved. Except for brief quotations in a review, this book, or parts thereof, must not be reproduced in any form without permission in writing from the publisher. For information, address Cornell University Press, 124 Roberts Place, Ithaca, New York 14850.

First published 1986 by Cornell University Press.

Printed in the United States of America

The paper in this book is acid-free and meets the guidelines for permanence and durability of the Committee on Production Guidelines for Book Longevity of the Council on Library Resources.

Contents

	Preface	9
Chapter 1	**Introduction**	13

    1.1 The historical background *13*
    1.2 Concepts and definitions *17*

Chapter 2	**Seeds as Dry Systems**	21

    2.1 The state of water in seeds *22*
    2.2 Enzymic and metabolic activity in dry seeds *29*
    2.3 Cellular organization and the state of membranes in seeds *34*

Chapter 3	**Loss of Seed Quality in Storage**	39

    3.1 The consequences of seed deterioration *40*
        3.1.1 Loss of viability *40*
        3.1.2 Loss of vigor *43*
        3.1.3 Changes in development and yield *45*
    3.2 The effects of humidity and temperature on seed deterioration *51*
        3.2.1 Deterioration equations *51*
        3.2.2 Climatic effects *56*
        3.2.3 Aging under extreme conditions *57*
    3.3 Sealed and gaseous storage *59*
    3.4 Preharvest conditions and other factors that limit longevity *62*

3.4.1 Mechanical injury 62
3.4.2 Maturity, size, and specific gravity 64
3.4.3 Hard seeds 66
3.5 Species differences in storability 67
3.6 Genetic effects on storability 71
　3.6.1 Varietal differences in susceptibility to aging 71
　3.6.2 Genetic studies of susceptibility to aging 73

Chapter 4 Surveys of the Longevity of Seeds Stored Dry in Collections 76

4.1 Systematic surveys 77
4.2 Records of extreme longevity 84

Chapter 5 The Longevity of Seeds in the Soil 88

5.1 Evidence from studies of seed banks 89
5.2 Buried seed experiments 91
5.3 Factors influencing persistence of seeds in the soil 98
5.4 Seed longevity in the soil: taxonomy and ecology 102

Chapter 6 Ancient Seeds 105

6.1 Seeds recovered from dry archeological sites 106
　6.1.1 Seeds of ancient Egypt 107
　6.1.2 Other ancient seeds 111
　6.1.3 The structure and chemistry of ancient dry seeds 114
6.2 Ancient seeds recovered from moist or frozen ground 117

Chapter 7 Morphological, Structural, and Biochemical Changes Associated with Seed Aging 125

7.1 Morphological changes 128
7.2 Ultrastructural changes 130
7.3 Cell membranes and permeability 132
7.4 Lipids 137
　7.4.1 Changes in lipid content 137
　7.4.2 Fat acidity 140
　7.4.3 Lipid peroxidation 142
　7.4.4 The free radical status of seeds 151
7.5 Changes in the structure and chemistry of proteins 156
7.6 Enzymes 158
　7.6.1 Changes in enzyme activity induced by aging 158
　7.6.2 The use of enzyme assays in testing for vigor and viability 162

Contents

    7.7 Respiration   *165*
        7.7.1 Quantitative and qualitative changes in metabolic reserves   *165*
        7.7.2 Alterations in respiratory pathways   *166*
        7.7.3 ATP content and energy charge   *169*
    7.8 Chromosomal aberrations and deterioration of DNA   *171*
        7.8.1 Chromosomal aberrations   *172*
        7.8.2 Other genetic irregularities   *175*
        7.8.3 Changes in the integrity of DNA   *177*
    7.9 RNA and protein synthesis   *180*
        7.9.1 Deficiencies in messenger RNA   *180*
        7.9.2 Deficiencies in translation   *181*
    7.10 Hormones   *183*
        7.10.1 Hormonal deficiencies   *183*
        7.10.2 Growth inhibitors and seed aging   *185*
    7.11 Repair and reinvigoration   *186*
        7.11.1 Cellular repair   *187*
        7.11.2 Reinvigoration by chemical and physical treatments   *192*
    7.12 Synopsis   *193*

Appendix **Selected Tabulations of Seed Longevity**   197

Bibliography   209

Species Index   289

Subject Index   299

Preface

The literature on seed deterioration is extensive and continues to grow rapidly. In attempting to organize this knowledge into a single coherent account, I have been surprised to find that much that has been learned over the years has simply disappeared from view. Several good reviews and even books have been produced on the subject of seed aging, but the range of sources consulted has usually been quite narrow; hundreds of contributions have been passed over, apparently unread and unassessed by workers in the field. In writing this book I have attempted to incorporate all accounts, no matter how obscure, that seem to offer material of value.

My approach to the subject of seed aging has been largely that of a plant physiologist and biochemist. I have devoted little attention, for example, to practical procedures for efficient storage of seeds, as others more capable of addressing such questions have already done so. For similar reasons I have not attempted to consider in great detail the effects of fungi and bacteria on stored seed. My overriding concern has been to interpret seed aging as a biochemical and physiological phenomenon, without losing sight of agronomic, ecological, and taxonomic considerations. For the most part, I have restricted my treatment to so-called orthodox seeds, which usually deteriorate in a relatively umimbibed state; the specialized problems associated with ''recalcitrant'' seeds—those incapable of desiccation—have received very little analysis at the biochemical level, and they are not considered at length here.

I am indebted to many individuals for their help and advice. Carl Leopold, in particular, provided a generous flow of ideas and encouragement. Several others have also been kind enough to read and comment on various sections of the manuscript: Derek Bewley, Martin Caffrey, Kees Karssen, Eric Kueneman, Daphne Osborne, and Eric Roos. No one could hope to survey a literature spanning more than two dozen languages without considerable assistance, and I am grateful to many friends and colleagues for contributing their linguistic talents; in particular I thank Sherman Hsu, Tazuko Kawazu, Vlado Macko, Mary Musgrave, Sandor Pongor, Aladar Szalay, and Takefumi Tezuka. Roger Lew deserves special mention for enthusiastically expounding on some of the more obscure Japanese literature relating to ancient lotus seeds. In addition, I am grateful to Greta Colavito and Brenda Werner for their occasional bibliographic assistance, and I am also conscious of the debt I owe to the efficiency and professionalism of the staff of the Cornell University Libraries in Ithaca and the National Agricultural Library in Beltsville. To my wife, Angela, I am grateful for much help, both professional and personal; among her many other talents, she organized the figures and helped regularize the taxonomy. Finally, to the two younger members of the Priestley family, Matthew and Joanna, I give my thanks for their curiosity and enthusiasm.

DAVID PRIESTLEY

Ithaca, New York

SEED AGING

Chapter 1

Introduction

1.1 The historical background

Farming has always provided the foundation for the development of complex societies and civilizations because stable social structures develop only when future resources—especially food supplies—are assured. In Neolithic times the shift from food gathering to food producing introduced the prospect of a permanent supply of nourishment for an increasing population without the need to move away from depleted areas every few years. The keystone for this "Neolithic revolution" was the development of appropriate foresight and thrift: a portion of every crop was set aside for the following year's seed (Childe, 1936). Archeological investigation has yielded considerable information about bulk storage of grain in early societies, but seed for sowing may well have been subjected to privileged treatment. Prehistoric sites in the southwestern United States, for example, yield evidence that seed corn was kept separate from the store used for the food supply (Amsden, 1949). Attempts have been made to reconstruct storage pits of the European Iron Age in order to investigate deterioration in seed quality, and it has been demonstrated that a species such as barley can retain a reasonably high germinability in this way despite the creeping dampness of an English winter (Bowen and Wood, 1968; Hill et al., 1983).

Some of the oldest botanical writings still extant—those of Theophrastus (c. 372 B.C.–287 B.C.)—treat problems of seed deterioration at some length. Infestation of stored seed by pests was evidently of particular concern. In addition, Theophrastus was well aware that some kinds of seed retained vitality longer than others: onion, for example, rapidly lost viability, whereas the millets had excellent powers of preservation. In general, seed was considered best for sowing when it was a year old and useless after four years. Theophrastus also noted, however, that seed stored at high altitudes in breezy locations could remain viable and free from pests for up to forty years. He explained the malady of seed aging rather obscurely in terms of the Empedoclean doctrine of the elements championed by his close associate Artistotle: decline in vigor was thought to be due to a prolonged elemental imbalance associated with loss of the moist component. Hard seeds, such as those of the date palm, retained this element more tenaciously and consequently preserved their viability longer.

Despite the proliferation of treatises on the practice and techniques of agriculture produced by Roman authors, little of substance was added to the basic analysis of seed deterioration provided by Theophrastus. Varro (116 B.C.–27 B.C.), for example, cautioned against the "drying" of seed with age. One alarming consequence of the use of inferior seed, he noted, could be the sudden transformation of species: rape could grow from old cabbage seed and vice versa. Columella (fl. first century A.D.) recorded the commonly held view that only the largest grain should be saved for sowing, as the smaller and weaker seed was likely to succumb too rapidly during storage. Dampness, he warned, is a grievous cause of such deterioration. In China, far removed from Western influence, Fan Sheng-zhi was providing his readers with remarkably similar advice. The fragmentary remains of the first-century agricultural classic *Fan Sheng-zhi Shu* clearly demonstrate an understanding that wheat and millet seeds best retain their viability if they are kept as dry as possible, cool and free from pests. Further, only large and solid ears of wheat should be chosen for sowing, and they should be dried as thoroughly as possible by the heat of the sun before they are placed in storage (Shih, 1959).

In the twilight years of the Roman Empire, Palladius (fl. fourth century A.D.) recorded the practice of a crude form of seed testing. Before setting aside seed for fall planting, he noted, the Egyptians performed seed trials during the dog days of August. The same custom is also recorded in the tenth-century Byzantine compilation known as the *Geoponika*.[1] This strange farrago of earlier agricultural writers includes a contribution by "Zoroastres" in which the testing procedure is described in some detail. Seeds were sown a few days before the heliacal rising of Sirius (the dog star), just before that part of the year when Sirius and the sun rise simultaneously. In Classical

---

1. The *Geoponika* formed part of an encyclopedia undertaken on the orders of Constantine VII Porphyrogenitus (905–959) and is based on an older compilation of Cassianus Bassus Scholasticus of the sixth or seventh century.

times the star's warmth was thought to augment normal levels of solar radiation, thereby inducing the burning heat of a Mediterranean summer. Those seeds that failed to succumb to "star blight" were evidently from a superior lot.[2]

Further wisdom on seed storage and aging was accumulated in the greatest medieval work on agriculture, the *Kitab al-filahah*, composed in the twelfth century by the Moorish scholar Ibn al-'Awwām. His advice for the storage of fruit and vegetable seeds was quite sophisticated for its time: when the seeds were mature, they should be cleaned, dried, and packed in a suitable vase. The mouth of this vessel must then be blocked with clay and it should be removed to a dry place for storage. In those parts of the world that were not illuminated by Arab scholarship, such matters were evidently pondered only by practical men and women whose thoughts have passed largely unrecorded; but after several centuries, with the coming of the Enlightenment in Europe, scholars again became interested in problems of seed deterioration. Peter Lauremburg, whose work *Horticultura* (1654) was eventually to provide the name for an emerging discipline, listed the relative longevity of seeds of several garden species. He particularly emphasized the role of the coat and other peripheral layers in retaining the *humor foecundus* within the seed—although this was scarcely a marked improvement on the analysis provided by Theophrastus two thousand years earlier. Little attempt was made to promote seed longevity. The pessimistic advice of La Quintine (1693) to his readers was typical: "I must tell you, that generally speaking, most seeds grow naught after one or two years at most, and therefore it concerns us always to be provided with new ones if we would not run the hazard of sowing to no purpose in the spring." Fresh insights into seed deterioration were ultimately provided by studies of grain storage. Much of the treatise published by Duhamel du Monceau (1754) deals with problems associated with the heating of stored grain, considered to arise from germinative "fermentations" within the seed that were triggered by increased moisture levels. Oven-treated seed or old seed lost the "fermentative" power and so tended to survive storage better. With modern hindsight, though, it seems certain that these "fermentations" were at least partly the result of fungal activity. Duhamel also noted with some precision the differences in longevity between seeds of different species, and he attempted to prolong seed viability by

---

2. In his work on seed storage and testing, Mercer (1948) identified the *Geoponika*'s "Zoroastres" with Zoroaster (or Zarathustra), the Iranian divine, and inferred that the testing of seed quality was practiced as early as the fifth or sixth century B.C. Unfortunately, the ascription is not altogether convincing, as many of the writings formerly attributed to Zoroaster are now judged to be spurious (Hammond and Sculard, 1970). In any case, belief in the supposedly malign influence of Sirius on plants and animals seems to have been a particular proclivity of Egyptians, Greeks, and Romans rather than Iranians (Allen, 1899).

adjusting temperature and moisture (Duhamel du Monceau, 1780). During this period particular interest was shown in procedures for transporting seeds on the long sea passages between newly acquired colonies and Europe; the literature on this subject has been reviewed by Henrey (1975).

The contributions of the Candolle family usher in the modern era of studies on seed aging. Augustin de Candolle's writings on the physiology of seed deterioration bear a decidedly modern stamp. According to his *Physiologie végétale* (1832), inasmuch as oxygen, warmth, and water are all required for germination, control or exclusion of any or all of these factors should help to keep stored seed viable. His son, Alphonse de Candolle, was one of the first botanists to investigate systematically the relative longevity of the seeds of a wide variety of species; his conclusions were published in 1846 (see sec. 4.1). Gradually the effects of prolonged seed storage on the emergence of important crop species in the field began to receive rational scrutiny from agronomists. Loiseleur-Deslongchamps (1843), for example, investigated the practical limits of storage for wheat seed, beyond which sowing was inadvisable. It became common for agricultural and horticultural handbooks to list the expected longevity of the seeds of many commonly planted species (e.g., Schlipf, 1844; Thouin, 1845). Perhaps unsurprisingly in view of the many factors that affect seed longevity, the lists often failed to agree. Subsequently, objective analyses by Haberlandt (1873), Dimitrievicz (1875), Samek (1888), and others on the relative longevity of crop seeds provided a basis from which more effective techniques of seed storage could be developed in the present century. Nevertheless, viewed against the long history of agriculture, serious attempts to understand seed deterioration are very recent indeed.

Numerous reviews on seed aging have appeared. Crocker's classic contributions (1938, 1948) can still be consulted with profit, and Barton's monograph on seed preservation and longevity (1961) remains impressive for its detailed coverage. Useful summaries of much of the earlier literature in this field were published by Biasutti Owen (1956) and James (1961, 1963). More recent information on the aging of seeds in storage has been collated by Harrington (1972) and by Justice and Bass (1978). General treatments of seed aging and loss of vigor have been presented by James (1967), Abdul-Baki and Anderson (1972), Heydecker (1972), Krishtofovich and Pokrovskaya (1972), Floris et al. (1972), Roberts (1972a, 1979, 1981, 1983), Villiers (1973), Łuczyńska (1976), Lovato (1976), Sen (1977), Bass (1979, 1980a), Roos (1980), Osborne (1980), Abdul-Baki (1980), Roberts and Ellis (1982), Anderson and Baker (1983), Bewley and Black (1982), and Halmer and Bewley (1984). Several other reviews dedicated to specialized aspects of seed deterioration are considered separately in the chapters that follow.

1.2 Concepts and definitions

The study of seeds has generated an immense literature covering a broad spectrum of scientific disciplines; its vocabulary is almost equally diverse. Quite often, even relatively simple words have assumed restricted meanings in the hands of specialists. The definitions that follow are meant to be valid only within the context of the present work.

The term *seed* is used throughout this book in a very general sense and is meant to include, for example, the dry fruits of grasses and composites in which the ovary wall has become part of the normal dispersal unit. Successful *germination* entails the emergence, and more especially the development, of seed embryo structures. Seed testers have often added an extra criterion to their definition of germination: the seedling produced should be *normal*. The distinction between normal and abnormal seedlings is a useful one in the field of seed aging, although the designation of a particular seedling as abnormal must inevitably be somewhat arbitrary. One current definition lists normal seedlings as those "possessing the essential structures that are indicative of their ability to produce plants under favorable conditions" (Association of Official Seed Analysts, 1978). A *viable* seed is one that germinates under conditions that are appropriate for the species. Seeds that will not germinate because of some form of dormancy that our lack of expertise prevents us from breaking are also viable. In the absence of any universally agreed distinction between life and death, a seed rendered nongerminable by age may still contain viable tissues, cells capable of metabolism, or functional enzymes.

The decline in quality experienced by a seed lot over time is not, of course, restricted to a lowering of viability. Aged seeds that retain their capacity to germinate generally do so more slowly and with an enhanced sensitivity to external stress. Indeed, this loss of performance may be evident well before viability begins to decline. Seeds of this type have frequently been said to be low in *vigor*. Aging is only one determinant of vigor; the seed's genotype, conditions of maturation, physical integrity, and other factors may all influence vigor by determining the potential level of the seed's activity during germination and the emergence of seedlings (Perry, 1978). The definition of seed vigor has long been a contentious issue. Two recent publications have addressed the issue with admirable directness (Perry, 1981; Association of Official Seed Analysts, Seed Vigor Test Committee, 1983), outlining protocols for the laboratory analysis of vigor. For the purposes of the present work, a less stringent attitude must inevitably be adopted, inasmuch as many worthwhile data on seed aging and loss of vigor were collected long before acceptable rules for vigor testing were codified. The term *deterioration* is used throughout in a general sense to indicate declining vigor or viability rather than property that is sometimes held to be the strict reciprocal of vigor (Association of Official Seed Analysts, Seed Vigor Test Committee, 1983).

The distinction that can be drawn between the terms *aging* and *senescence*, especially when applied to plants, has given further cause for debate. The position adopted in this work is that the deteriorative changes that inevitably accrue in seeds as storage progresses are attributable to aging. Some of these perturbations lead to decreased vigor and ultimately to death; considered in this light, even the freshest seed is subject to aging. The term *senescence*, although still occasionally encountered in the seed literature, is better reserved for deteriorative changes in whole plants or plant organs that tend to occur at well-defined points in their life cycle and normally result in death. The death of a seed through aging is a purely destructive event without any positive consequence for the species. Senescense of a whole plant or organ, on the other hand, is often a process of regulated development that is likely to have favorable ecological or physiological value (Leopold, 1961).

It has sometimes been suggested that a distinction can be drawn between "physiological" age and "chronological" age; seeds stored under normal conditions over several years, for example, may be more vigorous and viable than similar material exposed to a relatively unfavorable environment for only a few days. Clearly, time alone is a poor measure of physiological deficiency in the two cases. However, once it is granted that deterioration results from a complex interaction of time, temperature, humidity, and other factors (sec. 3.2), a sharp distinction between the two types of aging can scarcely be maintained. All changes, including the deleterious physiological alterations that occur in seeds during storage, are predicated on the passage of time; chronological age is consequently only one factor that determines the level of physiological deterioration. Throughout this book, therefore, I refer simply to "aging," recognizing that aging may progress at various rates and by dissimilar means, depending on the storage environment.

Perhaps no property is so important to the understanding of seed biology as *moisture content*. Unfortunately, some confusion prevails even in the definition of this most fundamental characteristic. This book follows the prevalent preference among seed scientists to express moisture content on the basis of fresh weight.[3] The alternative practice of citing moisture levels on the basis of tissue dry weight is common in papers that deal with the chemistry and biology of dry systems.[4] Such values, whenever evident, have been transformed to a fresh weight basis for the purposes of discussion.

3. Percentage moisture content (fresh weight basis) = weight of water/fresh weight of tissue ($\times$ 100).
4. Percentage moisture content (dry weight basis) = weight of water/dry weight of tissue ($\times$ 100).

Below about 15% hydration, the difference between the two systems is fairly small. Moisture content is usually derived quite routinely by drying seed material at temperatures in excess of 100°C, but even this deceptively simple statistic may conceal hidden complexity. It has been concluded from nuclear magnetic resonance data, for example, that seeds dried at 105°C may still retain 1 to 2.5% water (Askochenskaya, 1978). The data on moisture contents given in this book are derived from oven-dried seeds, and the true level of seed hydration may therefore have been slightly underestimated.

Throughout this work, the discussion of aging is limited chiefly to seeds that experience physiologically realistic conditions. It seems almost self-evident that unnaturally destructive influences that are far removed from the biological norm should not be identified with seed aging. Unfortunately, the distinction may be difficult to maintain in all cases. Surface soil temperatures at a particular location in Gabon, for example, have been reported to reach 80°C; yet after many months of burial, seeds of annual plants begin to emerge with the coming of the rainy season (Kerner, 1890). If seeds fail to germinate after prolonged exposure to such high temperatures, it may be argued that the deficiency can be ascribed to something other than seed aging in the usual sense. In storehouses and elevators in Siberia the temperature within the grain pile may remain well below 0°C throughout the winter months (Makarov and Prokhorova, 1964). Grain with elevated moisture content rapidly loses viability when it is stored in such conditions. If we choose to regard this second case as a straightforward example of freezing injury, it may be dismissed quite readily from our study of seed aging. Nevertheless, it should be clearly recognized that those phenomena we elect to identify with the aging process gain their status to some extent from the exercise of subjective judgment.

Because the book has been written primarily from a physiological and biochemical standpoint, the role of the microflora is considered quite briefly, despite its evident involvement in many types of seed deterioration. According to extensive studies pursued by Christensen and his associates (e.g., Christensen, 1972), growth of *Aspergillus* and *Penicillium* species can play a significant role in the deteriorative process if seed hydration is in excess of 13 to 14% in cereals and above about 12% in oilier seeds, such as soybean.[5] Christensen and Kaufmann (1969) suggest that "under conditions that prevail in grain bins, storage fungi must in many cases be the primary cause of loss of germinability of the seeds." The present work, in contrast, considers primarily those changes that occur in relatively small lots of seeds under conditions that do not favor fungal invasion and growth. Even so, it should

---

5. Following the terminology adopted in sec. 2.1, the effects of the microflora become significant in sorption zone III and are relatively trivial or absent in zones I and II.

be borne in mind that seeds are almost never entirely free of microorganisms. In consequence, even if innate physiological lesions can legitimately be associated with aging, they may frequently be amplified by complex interactions with the seeds' associated microflora (McGee, 1983).

Chapter 2

Seeds as Dry Systems

One of the most striking characteristics of seeds is their ability to endure severe desiccation. Even more remarkably, once they have attained the dry state, seeds are capable of withstanding environmental stresses that would prove rapidly lethal to the parent plant. The resistance attained by the desiccated organism can often be quite extraordinary; Becquerel (1950), for example, cooled seeds of several species to within 0.05°K without compromising their viability, and Pouchet (1866) noted that hard medick seeds (*Medicago* sp.) could be boiled for several hours without imbibing and still remained germinable. In addition to conferring extreme resistance to environmental stresses, the lapse into metabolic quiescence resulting from desiccation permits broad dispersal through time and space, and it is in this way that the immobilizing constraints associated with the autotrophic habit of higher plants are largely overcome.

This book is concerned primarily with the aging of seeds that are well adapted to desiccation, the so-called orthodox seeds (King and Roberts, 1979), which include the great majority of agronomically important species in the temperate zones. Not all seeds, though, are adapted to withstand dehydration. Many seeds of commercially significant tropical trees—such as coffee, cacao, and rubber—are unable to dry out without loss of viability (King and Roberts, 1979; Chin and Roberts, 1980; Roberts et al., 1984). In tropical climates, conditions for germination and seedling development are evidently sufficiently favorable for such species to obviate the need for a seed capable of withstanding inclement conditions. Some temperate plants, notably many hardwoods and various aquatics, also produce seeds that either are destroyed by drying or at least store poorly in the dry state (Barton, 1961). Acorns (*Quercus* spp.), for example, require a minimum of about 35–40% water, but in this case careful management of temperature and humidity can permit storage of viable material for several years (Suszka and Tylkowski, 1982). The term *recalcitrant* has been applied to seeds that cannot be successfully desiccated, a label that emphasizes the difficulties inherent in their storage and handling. The physiological background to the aging of recalcitrant seeds is still largely unexplored and will not be considered further here. Indeed, the phenomenon of recalcitrance itself is poorly understood; several species once thought to be destroyed by drying have since been shown to survive desiccation under certain conditions (Nakajima, 1926; Mumford and Panggabean, 1982).

Water exerts a profound influence on orthodox as well as recalcitrant seeds. When levels of hydration are very low, conventional metabolism ceases and there is little or no outward sign of life; yet a fairly small increase in water content will often serve to reinstate many of the functions and activities that are normally diagnostic of living organisms. In practice, most orthodox seeds are stored at moisture levels that lie somewhere between these two conditions, neither truly quiescent nor fully active. In systems so poised, a small shift in water content may have profound consequences for cellular and metabolic organization. Clearly, without some understanding of the response of seeds to hydration, little light can be shed on the physiology of aging.

A diverse array of organisms other than seeds are capable of withstanding extreme desiccation. In the plant kingdom, pollen, spores, and even whole plants may routinely achieve extreme dryness. In addition, almost every invertebrate phylum includes species that can withstand desiccation with impunity at some stage in the life cycle (Crowe and Clegg, 1973). There has been a gradual realization that the physiological similarities between such dry biological systems far outweigh their taxonomic diversity. This view has been fostered by such contributions as the pioneering monograph by Shmidt (1955) and the synthesis of earlier work in the field of anhydrobiosis presented by Crowe and Clegg (1973). Considerable advances have been made in recent years toward the understanding of dry organisms, and much of this work has also appeared collectively (Crowe and Clegg, 1978). For the purposes of the present chapter, occasional reference will be made to other desiccated systems whenever our knowledge of the situation in seeds is lacking.

2.1 The state of water in seeds

"Dry" seeds are rarely devoid of all water. In open storage, certainly, levels of hydration are generally in the range of about 4 to 16% of the fresh weight of the seed, and as the relative humidity of the storage environment

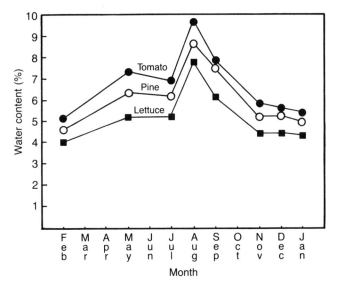

Figure 1. Moisture content of seeds of tomato, longleaf pine (*Pinus palustris*), and lettuce during one year of open storage in Yonkers, N.Y. Adapted from L. V. Barton, "Relation of Certain Air Temperatures and Humidities to Viability of Seeds," *Contributions from the Boyce Thompson Institute* 12 (1941): 95.

fluctuates, the moisture within the seed alters in response. Hydration data for three species of seed throughout one year of open storage in Yonkers, New York, are shown in Figure 1. It is evident that the level of hydration attained in the warm and humid summer months can be almost twice that of the winter minimum. Although all three types of seed respond to the changing environment in similar fashion, each species attains its own characteristic moisture content under a given set of conditions; lettuce has a consistently lower degree of hydration than pine, and pine less than tomato. Differences in equilibrium moisture content between species are due primarily to variations in the quantity of storage lipid, a component that interacts very feebly with water (Sijbring, 1963; Cromarty, 1984). Seeds with large amounts of storage lipid (such as lettuce) tend to equilibrate at relatively lower moisture levels. Consequently, if equilibrium moisture content is expressed on the basis of nonlipidic seed material rather than total seed weight, differences between species are considerably reduced (Fig. 2).[1]

The interaction between water and seeds is most easily summarized by a

---

1. Mercado and Helmer (1971) have published regression equations for peas, maize, cotton, rape, and soybean, relating equilibrium moisture content to the percentage of lipids, proteins, and carbohydrates within the seed. The lipid content, however, is by far the most significant determinant of differences of response between species.

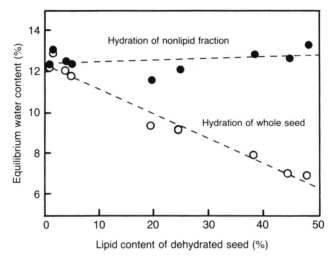

Figure 2. The relationship between seed lipid content and equilibrium moisture content attained at 60% relative humidity. Each point represents a single species. In increasing order of lipid content, they are barley, rice, sorghum, oat, soybean, cotton, flax, peanut, and rape. Information on seed lipid content is from Spector (1956). Data on moisture content for seeds held at 60% relative humidity and temperatures of 12–25°C are from the various sources tabulated by Justice and Bass (1978).

consideration of sorption isotherms, which are obtained by measuring equilibrium moisture content as a function of relative humidity at constant temperature. Typical results are shown in Figure 3. Seed sorption isotherms, like those for most biological material, have a characteristic negatively sigmoidal shape, permitting the definition of three distinct zones or regions of hydration. Tissue moisture content increases fairly rapidly with rising humidity at the driest end of the range (zone I), then slows (zone II), then rises rapidly again (zone III). The sorption isotherm obtained for the whole seed represents the integrated hygroscopic properties of its various constituents (Fig. 4). Starch, cellulose, proteins, and other seed components give negatively sigmoidal curves when they are studied in isolation (Iglesias and Chirife, 1982). In a comparative study of sorption isotherms for a number of starchy and oily seeds made by Pixton and Warburton (1971), the most evident difference between the two groups was the amount of vertical displacement of the sorption curves on the ordinate. Seeds with well-developed starch reserves tended to attain a higher equilibrium moisture content at a given humidity than did oil-containing species. Nevertheless, the relative humidities

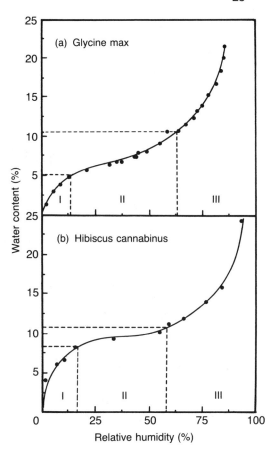

Figure 3. Sorption isotherms for seeds of soybean and kenaf (*Hibiscus cannabinus*) at 20°C. Both sorption isotherms can be divided, somewhat arbitrarily, into three zones. Each zone is characterized by a predominant form of interaction between water and seed material: Zone I, water tightly bound to ionic groups; Zone II, hydrogen bonding; Zone III, free or capillary water. Adapted from C. W. Vertucci and A. C. Leopold, "Bound Water in Soybean Seed and Its Relation to Respiration and Imbibitional Damage," *Plant Physiology* 75 (1984): 115; A. Mahama and A. Silvy, "Influence de la teneur en eau sur la radiosensibilité des semences d'*Hibiscus cannabinus* L. I. Rôle des différents états de l'eau," *Environmental and Experimental Botany* 22 (1982): 235.

at which the two principal inflection points occurred (between zones I and II and between zones II and III) varied little between the two groups.

Each sorption zone corresponds to a state of hydration in which a particular type of water binding tends to predominate. A precise definition of "bound" water has never been universally adopted, but there is a fairly broad consensus that "bound" water is that water which occurs in the vicinity of large molecules, and whose structural or dynamic properties therefore differ from those of the "bulk" water in the same system (Kuntz and Kauzmann, 1974). Zone I is dominated by the very strong interactions between water molecules and the charged carboxyl and amino groups of proteins, lipids, cell walls, and other constituents. Less tightly bound water predominates in zone II, where hydrogen bonding of water to the hydroxyl groups of starch and to the hydroxyls and amides of proteins tends to be particularly evident. Zone III

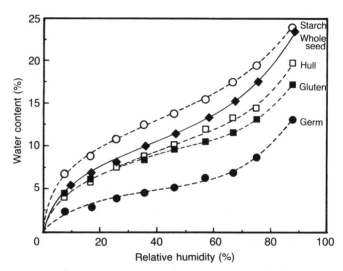

Figure 4. Sorption isotherms for whole maize seeds and their starch, hull, gluten, and germ fractions at 22–25°C. Adapted from D. S. Chung and H. B. Pfost, "Adsorption and Desorption of Water Vapor by Cereal Grains and Their Products. Pt. I. Heat and Free Energy Changes of Adsorption and Desorption," *Transactions of the American Society of Agricultural Engineers* 10 (1967): 549, by permission of the American Society of Agricultural Engineers.

can be considered a region of predominantly "mobile" or "free" water, in which capillary and osmotic forces become significant (Rockland, 1969; Acker, 1969).[2]

One evident difficulty with sorption isotherm data is the relative imprecision with which zone boundaries can be assigned. Although "free" water becomes predominant only in zone III, for example, it will occur to a minor extent in zone II. Some uncertainty is also generated by the frequently observed phenomenon of hysteresis; values may vary depending on whether sorption or desorption curves are consulted (Chung and Pfost, 1967b). Nevertheless, sorption isotherm analysis provides useful insights into the effects of hydration, as the absolute water content of seed tissue is less significant than its pattern of binding. As chapters 3 and 7 will make apparent, the mode of seed deterioration in storage depends principally on which zone of hydration is applicable to the particular storage conditions.

Some of the most detailed analysis of the state of water in seeds has come

---

2. The enthalpy of hydration as a function of seed moisture content can be calculated from sorption data collected at different temperatures. Studies of maize (Chung and Pfost, 1967a) and soybean (Vertucci and Leopold, 1984) indicate that the strength of water binding in seeds decreases rapidly as hydration progresses.

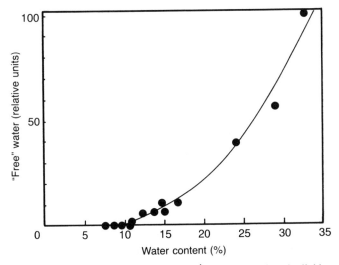

Figure 5. The ratio of the "free" water ^{1}H-NMR signal to the lipid signal in hydrating soybean cotyledons. Seed material was hydrated by air imbibition at room temperature. Spectra were recorded at 31°C. "Free" water is expressed in arbitrary units relative to the constant triacylglycerol signal. Adapted from V. Seewaldt et al., "Membrane Organization in Soybean Seeds during Hydration," *Planta* 152 (1981): 21.

from nuclear magnetic resonance (NMR) studies. ^{1}H-NMR has been used to examine a wide variety of species and it is evident that "free" water first becomes detectable in hydrating seeds when the moisture content is in the order of 5 to 15%, depending on the species (Aksenov et al., 1969; Askochenskaya et al., 1970; Mishin et al., 1973; Samuilov et al., 1976a; Samuilov et al., 1976b; Seewaldt et al., 1981). Here too, though, the method of experiment may to some extent influence the result obtained. The characteristics of the signals from the water protons may vary, for example, depending on whether the seeds used for analysis are imbibed in a saturated atmosphere or directly in water (Askochenskaya and Aksenov, 1970). Despite this complication, it is obvious from the wealth of data available that most unimbibed stored seeds possess little or no "free" water. Thus it is possible to demonstrate that for soybean cotyledons, "free" water detectable by ^{1}H-NMR is absent below about 11% moisture content (Seewaldt et al., 1981), a value that is close to the onset of zone III as it has been defined by sorption isotherm work (Fig. 5). The designation of water as "free" in this context, though, is entirely relative; the mobility of the free water in an imbibing seed may be considerably less than that of fully hydrated plant tissues or the pure bulk fluid (Aksenov et al., 1969; Callaghan et al., 1979).

A number of studies have employed pulsed NMR techniques in an attempt to define better the state of water in hydrating seeds. In these investigations, a spin echo protocol has been used to define nuclear relaxation parameters. Under conditions that satisfy nuclear resonance, a short series of radio frequency pulses can be used to provoke constructive interference from precessing nuclear spins. Observation of the decay characteristics of the resulting induction signals provides a measure of nuclear relaxation processes, and these processes in turn offer useful indications of the environment within which the protons are located (Hahn, 1950; Lechert, 1981). The most frequently used parameter in seed studies has been spin–spin relaxation (usually designated T_2), the time constant measuring the rate at which the precessing nuclei lose phase with each other. Relaxation decay for a homogeneous system is exponential, so that a plot of the logarithm of echo amplitude against time should be linear. This relationship is often approached in well-imbibed seeds with an excess of "free" water. However, for seeds at fairly low degrees of hydration, in which bound water may contribute significantly to the decay signal, markedly bi- or multiphasic echo decay is evident (Aksenov et al., 1969; Askochenskaya et al., 1970; Ratković et al., 1982). This finding is consistent with a heterogeneous system in which the exchange of material between water phases is slow compared to the relaxation times of the nuclei within the phases. A recent review by Askochenskaya (1982) suggests that at least three phases can be identified in hydrating seeds before the signal from "free" water becomes overwhelmingly predominant. Characteristic of many seeds, especially cereals, is a rapidly relaxing component (T_2 less than 1 ms) that can be equated with protons that have relatively little mobility. This component seems readily ascribable to water hydrating starch (Kuz'mina et al., 1980; Askochenskaya, 1982). The composition of a more slowly relaxing component (with T_2 greater than 10 ms, corresponding to protons of greater mobility) is less assured. In oily seeds, lipid protons apparently contribute significantly to this fraction (Aksenov et al., 1977), but for cereal grains, at least, the assertion has been made that water protons associated with nonstorage structures of the cell may predominate (Askochenskaya, 1982). A third component has been postulated with a very short relaxation time (T_2 less than 100 μs). It has been suggested that this component corresponds to water that is tightly bound to proteins (Askochenskaya and Golovina, 1981). Relaxation in this case is so rapid as to escape instrumental detection, but the presence of the fraction is postulated on the basis of discrepancies between the water known to be present in the seed and the initial amplitude of the spin echo. The difference is larger for seeds containing substantial amounts of storage protein than it is for cereal grains (Askochenskaya, 1982).

The overall conclusion drawn from ^{1}H-NMR studies of seeds at low hydration (below zone III) is that water binding is a highly complex heterophasic phenomenon. As the seed hydrates, "free" water becomes detectable at a fairly well-defined threshold near the onset of zone III. It is also evident that different seeds and different parts of the same seed interact with water in quite diverse ways, depending on chemical composition. Further, on the basis of very limited evidence it appears that these characteristics change little with seed age. According to Buchvarov and Nikolaev (1984), who used spin echo NMR to investigate soybean embryonic axes and cotyledons, both accelerated aging at high humidity and deterioration in long-term storage have fairly minimal effects on water-binding properties.

2.2 Enzymic and metabolic activity in dry seeds

Seed germination is characterized by considerable protein synthesis and the development or reactivation of enzymic functions. Nevertheless, the unimbibed seed contains many enzymes that are readily detectable immediately upon hydration. The extent to which such activity may be expressed during the normal storage of dry seeds and its relationship to seed aging are questions of considerable importance.

The structural configuration and activity of most enzymes are undoubtedly sensitive to extreme desiccation. Several studies indicate, though, that even enzymes that normally function in "soluble" fractions of fully hydrated cells are capable of limited activity at fairly low degrees of hydration. Skujins and McLaren (1967), for example, reported the urease from jack beans (*Canavalia ensiformis*) slowly decarboxylated urea at sorption levels corresponding approximately to the boundary between zones II and III. Activity increased greatly with further increments in hydration. Rupley et al. (1983) similarly found that enzymic activity of egg lysozyme became detectable at moisture levels corresponding to zone II of its sorption isotherm. Stevens and Stevens (1977) noted that glucose-6-phosphate dehydrogenase from yeast (*Torula* sp.) was capable of activity at surprisingly low levels of hydration (less than 2%), although in the absence of sorption data for the system, the significance of this result is difficult to assess.

Studies of isolated enzymes have been helpful in suggesting that many may still be functional under conditions of restricted hydration, although the situation *in vivo* must be considerably more complex. Little is known, for example, of the effects of ionic strength on enzymes within seeds, yet this factor must presumably exert a considerable influence when the quantity of "free" water is very limited. In addition, substrate mobility and product diffusion away from the reaction site undoubtedly form important limitations on enzymic rate in partially hydrated systems (Acker, 1962), and these two factors in turn are highly dependent on the nature of the cellular environment in which the enzyme is located. The problem of molecular diffusion in seeds has not been directly addressed, although Duckworth (1962) investigated the dispersal of labeled glucose through other dried plant materials. Over a three-month period he found no evidence of movement of the molecule through the tissue when hydration levels corresponded to sorption zone I. Above the zone I–zone II boundary, diffusion became gradually more evident as hydration levels rose. Limited information on the molecular mobility of potential substrates has also been provided by NMR studies. Analysis of soybeans by ^{13}C-NMR suggests that the only freely mobile components in unimbibed seeds are the triacylglycerols, neutral lipids that are apparently as undisturbed by the absence of water as they are by its presence (Schaefer and Stejskal, 1975; Anderson et al., 1977; Colnago and Seidl, 1983). Proteins and carbohydrates, on the other hand, are held relatively rigid and immobile. The moisture content of the seeds used in these studies was imprecisely defined, but evidently coincided with sorption zone II (5–10% moisture content). Priestley and de Kruijff (1982) analyzed seeds of the same species by ^{31}P-NMR and noted the appearance upon hydration of an isotropic resonance, apparently attributable to small molecular weight phosphorus-containing compounds in solution. This component was quite marked in seeds maintained at 15.1% moisture content. Only a trivial amount of isotropic motion could be detected at 8.7% water content, and even this disappeared when seed hydration dropped to 8.1%. At this latter value a very broad line shape was observed with the character of a solid-state "powder" spectrum, consistent with all the phosphorus in the sample being rigidly immobilized in randomly orientated positions. This finding suggests that such compounds as sugar phosphates and nucleotides are relatively motionless within soybeans at levels of hydration corresponding to sorption zone I or the drier part of zone II (Fig. 3).

Much information on enzymic activity at low hydration has come from model systems that correspond in varying degrees to the situation *in vivo*. Studies of this nature have generally involved the intermixing of enzyme and substrate, usually in the presence of an inert carrier material, followed by a determination of reaction rate as a function of hydration. Ground seeds have also been employed in order to evaluate storage characteristics of the meal at different moisture levels. Grinding undoubtedly accelerates reactions in comparison with the rate in the whole seed (Acker, 1962), but data obtained from such experiments still serve to illustrate the ability of enzymes to function under conditions of restricted hydration. Toda et al. (1966), for example, studied the persistence of ribonucleotides (inosine and guanosine 5'-monophosphates) added to wheat flour as a function of hydration and found that destruction by acid phosphatase became significant only in sorption zone II. Kiermeier and Codura (1955) similarly demonstrated that hydrolysis of starch by rye amylase occurred at somewhat higher sorption levels. Some insight into the activities of enzymes in ground wheat germ was given by Linko and Milner (1959a), who found evidence for alanine-glutamic acid transamination over a three-hour incubation period at a moisture level of 15%. It seems likely, though, that this figure is fairly imprecise, as hydration of the experimental material was accomplished by "dampening" with buffer; presumably some parts of the tissue were at least temporarily hydrated in excess of the value stated. More recently, Stevens and Stevens (1977) investigated the activity of glucose-6-phosphate dehydrogenase in freeze-dried barley embryo extracts at different levels of moisture content. Reduction of $NADP^+$ in the presence of glucose-6-phosphate was detectable at 7.5% hydration and increased rapidly above 10%. Taken together, these findings are broadly consistent with the view that enzymic activities normally associated with "soluble" phases of the cell are probably negligible or nonexistent in sorption zone I, are very low in zone II, and increase rapidly with the onset of zone III.

Investigations of dry model systems have indicated that lipolytic activities may be expressed at extremely low hydration levels, the rate-limiting factor being the mobility of the lipid substrate. Acker and Wiese (1972), for example, were able to demonstrate that the activity of oat seed lipase in dehydrated model systems was directly controlled by the fluidity of its triacylglycerol substrate. Free water is evidently not an absolute requirement for hydrolytic activity in desiccated material (Potthast, 1978). In similar studies with lipoxygenase of soybean seeds, Brockmann and Acker (1977) observed slow enzymic oxidation at moisture contents well within sorption zone I. Degradation of amphipathic lipids by phospholipases, on the other hand, seems to occur only at higher levels of hydration; the effects of barley phospholipases B and D were evident at moisture contents corresponding to sorption zones II and III (Acker and Kaiser, 1959).

A number of investigators have used radioactive tracers in attempts to describe metabolism in dry systems. Some of the most convincing studies have been performed on organisms other than seeds. Cowan et al. (1979), for example, incubated lichens in humid air over tritiated water. They were able to demonstrate that in sorption zone II some enzymes of the tricarboxylic acid cycle and related transamination reactions were clearly active; they discounted nonmetabolic incorporation by performing parallel experiments on dead tissue. There was little evidence for lipid metabolism in the lichen at this level of hydration or for macromolecular synthesis. Similar experiments were performed on *Pinus ponderosa* pollen dried to less than 8.6% moisture (Wilson et al., 1979). Alanine, aspartic acid, γ-amino butyric acid, glutamic acid, and malic acid became labeled; but there was no indication of incorporation into proteins or other soluble macromolecules, and no labeling was noted in steam-killed pollen.

The evidence that has accrued for some form of metabolism in dry seeds is rather less assured because of methodological uncertainties. The few studies that have been pursued with seeds have used ^{14}C-labeled compounds, and much ingenuity has been expended in attempts to introduce these compounds into the tissue with minimal disturbance. Edwards (1976) employed gaseous ^{14}C-carbon dioxide and detected incorporation into several loosely defined cellular fractions of charlock seeds (*Brassica kaber*) at levels of hydration corresponding to sorption zone II.[3] It is not altogether clear, though, by what pathways seeds should be able actively to fix and metabolize atmospheric carbon dioxide. The reactions of dead material were not examined. In further experiments (Edwards, 1976), 2-^{14}C-acetate or ^{14}C-leucine infiltrated aqueously into charlock seeds were both apparently assimilated when moisture levels were in sorption zone II. Chen (1972, 1978) has pursued similar studies with wild oat seeds (*Avena fatua*) at about 10–11% water content. Following exposure to ^{14}C-ethanol for three days, amino acids, organic acids, and sugars were all labeled. Evidence of incorporation in protein was also presented. The labeling was assumed to be metabolic in every case, although no killed controls were employed. Sorption isotherm data for wild oat were not given, but according to information provided by Justice and Bass (1978) for the cultivated oat, a hydration of 10–11% corresponds to the central region of sorption zone II.

Before the advent of radioactive tracer techniques, attempts to identify metabolism in "dry" seeds rested almost exclusively on respiratory measurements, but such observations can scarcely be considered definitive. All workers agree that if respiratory gas exchange occurs in dry seeds, it must do so at a very low level indeed. Conventional manometric techniques indicate little or no respiration in sorption zone II (compare Fig. 6 with Fig. 3). Even if we assume the existence of feeble respiratory activity at this level of hydration, however, a number of factors are apt to confound its measurement. It has long been held that in stored seeds maintained at hydration levels near the zone II–zone III boundary, fungi in or under the seed coat are the principal source of respiratory gas exchange rather than the seed material itself (Oxley and Jones, 1944; Christensen, 1972). At lower water contents, at which fungal metabolism is usually assumed to be far less significant, nonrespiratory reactions within the seed may give rise to significant gas exchange. Enzymic decarboxylations of amino acids or other substrates could lead to a slow release of carbon dioxide from stored seed held at hydration levels below that needed for integrated metabolism (Linko and Milner, 1959b). In certain cases this gaseous evolution might arise from an even simpler source: seeds

---

3. The sorption data provided in this study give no indication of the existence of zone I. Zones II and III are apparently well defined.

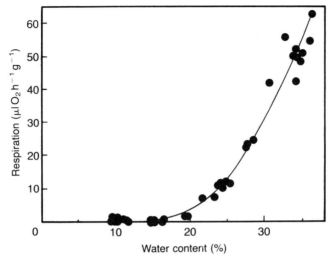

Figure 6. The respiratory rate of soybeans as a function of hydration. Oxygen consumption is expressed on the basis of dry weight of seed tissue. Adapted from C. W. Vertucci and A. C. Leopold, "Bound Water in Soybean Seed and Its Relation to Respiration and Imbibitional Damage," *Plant Physiology* 75 (1984): 115.

contain considerable amounts of trapped carbon dioxide, some of which is normally dissolved in the lipids (Sankara Rao and Achaya, 1969) and some adsorbed onto the seed coat (Côme, 1971). Oxygen uptake may also be accounted for by nonrespiratory processes. Becquerel (1907), for example, indicated that the oxygen taken up by "dry" legume seeds during storage was almost entirely consumed by the seed coat rather than by the cotyledons or embryo. More recently, Marbach and Mayer (1974) have shown that legume seed coats have a highly developed capacity for oxidative reactions. Even if this latter contribution is discounted, the oxidation of labile compounds within the seed itself could presumably obscure a weak respiratory activity. Consequently, studies that have demonstrated a minor amount of oxygen uptake or carbon dioxide release from dry seeds (e.g., Ching, 1961; Bryant, 1972) still fall short of establishing the existence of respiratory activity. In view of these problems of interpretation, definitive evidence for respiration in desiccated seeds has remained distinctly elusive (Bartholemew and Loomis, 1967). If we may extrapolate from studies of dry and rehydrating chloroplast membranes (Matorin et al., 1982), electron transport in seed mitochondria probably demands the presence of "free" water. The limited data currently available therefore make it seem probable that although some enzymic reactions can occur in sorption zone II, complete respiratory functioning may be dependent on the existence of less tightly bound water.

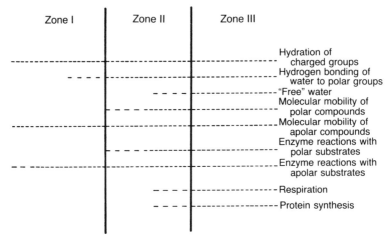

Figure 7. The state of metabolic organization in seeds as a function of hydration. Zones I, II, and III are defined from sorption isotherms (see Figure 1). The summary given here lists the feasibility of a particular state, reaction, or process; it does not necessarily indicate that it will occur in intact seeds to a significant extent.

A diagrammatic summary of the evidence reviewed in this section is provided in Figure 7.

2.3 Cellular organization and the state of membranes in seeds

There is no dearth of studies detailing the structural disposition of cellular components in "dry" seeds. Unfortunately, to an even greater extent than for most other plant tissues, the usefulness of ultrastructural analyses of seeds is highly dependent on the mode of fixation employed. The three types of investigation that have been attempted with seeds include conventional fixation in the presence of water, nonaqueous fixation, and freeze fracture. The conclusions that can be drawn from each type of study are rather rigidly constrained by the method of tissue preparation employed.

Most studies of the fine structure of dry seeds have used conventional fixatives (glutaraldehyde, paraformaldehyde, and lithium or potassium permanganate) infiltrated into the tissues as aqueous solutions (e.g., Yatsu, 1965; Paulson and Srivastava, 1968; Yoo, 1970; Hallam, 1972; Öpik, 1972; Swift and O'Brien, 1972; Webster and Leopold, 1977; Baird et al., 1979). A comparison of the results achieved in these studies with those obtained in the absence of water (reviewed below) suggests that seed material responds more rapidly to hydration than to fixation. Nevertheless, the incidence and general disposition of organelles observed in aqueously fixed cells are probably faithful representations of the situation in the desiccated system, despite the undeniable suspicion that much of the image may be artifactual. The studies cited tend to concur in their descriptions of cells; oil bodies are likely to be aligned on the plasmalemma and around protein bodies, for example; dictyosomes are commonly absent and the endoplasmic reticulum is frequently, but not always, reported to be poorly developed; the structural definition of mitochondria has often been found to be deficient.

Perner (1965) attempted to avoid the shortcomings of aqueous preparation by fixing the radicles of pea seeds (10.6% water content) in osmium vapor over several weeks. Although the levels of contrast were poor, the appearance of cells differed noticeably from that of aqueously fixed material: the cell walls and organelles were tightly crumpled. Similar techniques applied by Öpik (1980) to dry rice coleoptile cells yielded comparable results. As in Perner's study, organelles were compressed and irregular in shape although cellular membranes appeared to have a high degree of organization and integrity. The lipid bodies, in particular, did not display the smoothly rounded outline seen in most aqueously fixed preparations. An alternative to the osmium vapor technique suggested by Thomson (1979) involved a preliminary fixation by formaldehyde dissolved in glycerol that had been adulterated with a small amount of water. This study laid considerable emphasis on the ill-defined appearance of the membranes, and suggested that some specialized configuration could be assumed as a direct consequence of the dry state. It seems quite possible, though, that this image may have been an artifact of the interaction of glycerol with the amphipathic membrane lipids (Smith, 1980). Chabot and Leopold's (1982) application of this technique to soybeans led to rather better defined membrane profiles. Attempts to fix unimbibed seed tissue by exposure to paraformaldehyde and acrolein vapors have also been briefly noted by Smith (1979).

Freeze fracturing has provided useful insights into seed structure and the manner in which it responds to the entry of water. Buttrose (1973), for example, has indicated that seed tissue frozen after a few seconds of hydration may differ markedly from the same tissue in the unimbibed state. The images obtained by freeze fracture of dry seed tissue tend to parallel those achieved by osmium vapor fixation: walls are strongly convoluted, oil bodies are polyhedral rather than smoothly rounded, and organelles are somewhat indented (Swift and Buttrose, 1972; Buttrose, 1973). When water content is low (less than 20%), fine detail is frequently lacking. Several freeze-fracture studies of seed membranes, for example, have yielded rather poorly defined images that defy simple interpretation (Buttrose and Soeffky, 1973; Swift and Buttrose, 1973; Chabot and Leopold, 1982). Investigations of the cowpea radicle (about 8–9% moisture content), however, have demonstrated extensive membranous areas that differ little in appearance from the hydrated norm (Thomson and Platt-Aloia, 1982; Bliss et al., 1984). Crèvecoeur et al. (1982)

published similar images of the plasmalemma of maize radicle cells (12% hydration), and Vigil et al. (1984) also claimed to identify a relatively intact plasmalemma in a study of cotton radicles at 12% hydration. That quite profound changes in membrane structure may be provoked as a consequence of desiccation seems probable, but the details have obviously yet to be clarified. There is at present no compelling evidence from ultrastructural studies to suggest that the characteristic bilayer organization need necessarily be lost upon dehydration.

The integrity of membranes in seeds is of crucial importance to the maintenance of viability. In the presence of water, membranes perform functions that are essential for the survival of any hydrated organism. By regulating the flow of materials between subcellular compartments, membranes supply a means for the control of intermediary metabolism; they provide a dynamic framework for numerous enzymic activities and maintain separations between mutually incompatible cellular compartments. Any undue disruption of the dried membrane system is likely to be of immediate consequence once the seed imbibes.

Upon undergoing hydration, many dry organisms tend to release cellular constituents in a manner suggestive of incomplete membrane integrity. Yeast cells (Herrera et al., 1956) and pollen (Hoekstra, 1984), for example, characteristically display such effects during imbibition. Seeds, too, are noted for their tendency to leak on hydration, although large-scale cracking of the seed surface as swelling progresses can undoubtedly contribute to this effect, especially if the seed material is very dry (Cohn and Obendorf, 1978). Attempts have been made to relate the leakage kinetics from seeds during imbibition to some form of configurational transformation in the membranes (Simon and Raja Harun, 1972; Parrish and Leopold, 1977). Although the earlier interpretation of such kinetics has now been questioned (Powell and Matthews, 1981a; McKersie and Senaratna, 1983), it is nevertheless true that stressed or aged seeds generally leak considerably more than their healthy counterparts (see sec. 7.3).

A great deal of conjecture has arisen concerning the state of membranes in dry systems, although data directly pertinent to seeds are scarce. Extrapolating from the lipid mixtures studied by Luzzatti (1968), which were atypical of plants, Simon (1974) advanced the interesting speculation that seed membranes may undergo a mesomorphic phase transition from a bilayer to a "hexagonal II" configuration at low levels of hydration. In this latter conformation, groups of phospholipids aggregate to form porous or tubular structures, in which water-filled areas are contained within a lining derived from the hydrophilic head groups of the lipids (Cullis and de Kruijff, 1979). In an extension of this model, Simon (1978) suggested that mesomorphic rearrangement of the membrane lipids would provoke displacement of protein from the bilayer, so that considerable reconfiguration would have to take place upon hydration. On the admittedly rather limited basis of studies with dehydrated erythrocytes, however, it seems probable that membrane proteins are comparatively unaffected by desiccation. In this system protein structure is reported to be highly stable during drying (Schneider and Schneider, 1972; Schneider, 1981). That hexagonal-type lipid domains may occur to a minor extent even in fully hydrated biological membranes is quite probable (de Kruijff et al., 1980), but evidence for its occurrence as a generalized response to dehydration has not been overwhelming. Toivio-Kinnucan and Stushnoff (1981) reported the presence of hexagonal phase in freeze-fractured lettuce seed, although alternative explanations of the images in their micrographs are possible. More compelling evidence has emerged from systems other than seeds. Ultrastructural investigations, for example, have established that hexagonal II phase arises in the plasmalemma of nonacclimated rye protoplasts as a consequence of freezing-induced dehydration (Gordon-Kamm and Steponkus, 1984). Crowe and his coworkers have also identified the presence of hexagonal phase by ^{31}P-NMR and freeze-fracture electron microscopy in dried sarcoplasmic reticulum isolated from lobster abdominal muscle (Crowe and Crowe, 1982; Crowe et al., 1983), but this is an unusual system that does not normally undergo desiccation. Similar considerations apply to the dried bovine retinal rod membranes studied by Gruner et al. (1982), which formed a binary phase of protein-containing lamellae interspersed with microdomains of hexagonal II configuration.

It seems likely that membranes within anhydrobiotic organisms are protected from such changes by high levels of stabilizing carbohydrates associated with the bilayer (Crowe et al., 1984). X-ray diffraction patterns of isolated seed membrane lipids suggest a tendency to remain in the bilayer configuration even when dried (McKersie and Stinson, 1980; Seewaldt et al., 1981), although it can be argued that these systems, too, are artificial. However, a study in which ^{31}P-NMR was used to investigate phospholipids of *Typha latifolia* pollen *in vivo* demonstrated no tendency for the formation of hexagonal phase at levels of hydration at least as low as 11% (Priestley and de Kruijff, 1982), and this conclusion has been reinforced by a similar investigation of the desiccation-tolerant moss *Tortula ruralis* (Singh et al., 1984). On the basis of these scant data it seems likely that although some systems display a tendency toward hexagonal mesomorphism upon drying *in vitro*, membranes within anhydrobiotic organisms are protected against such alterations.

Even if the changes incurred by seed membranes on dehydration fall short of a shift to the hexagonal II conformation, other forms of perturbation are readily conceivable. It is known, for example, that the packing of phospholipid molecules becomes tighter (i.e., the area of membrane surface per molecule decreases) as model layers of egg phosphatidylcholine dry (Levine and Wilkins, 1971), and Finean (1969) suggested that some molecular reorganization in membranes occurred when hydration dropped below 20%, possibly as a result of the lateral separation of lipid phases. Alternatively, Beker (1977) has developed a model in which the membrane components are displaced from the bilayer and randomized in their orientation as water bound to the polar head groups is lost by dehydration. Whether loss of water would provide sufficient stimulus to provoke dissolution of the bilayer is uncertain, however. Sorption studies of phosphatidylcholine have shown it to bind about 10 moles of water per mole of lipid, principally by hydrogen bonding to the ester and phosphate regions of the polar head group (Lundberg et al., 1978). Although this hydration sphere would be depleted in a desiccated seed, it is doubtful that it could ever be completely removed under normal conditions of seed storage. The influence of dipole–dipole interactions between head groups and van der Waals forces among the fatty acyl chains would also be significant in preventing the complete displacement and disorientation of lipids (Hauser, 1975). Clearly, much work remains to be done in this area. The only certainty is that in the imbibing seed considerable stress is placed on the membranes, as changes in cellular volume are both rapid and large (Buttrose, 1973; Leopold, 1983; Simon and Mills, 1983).

Without prompt and efficient reestablishment of functioning membranes when they are challenged by a swiftly changing hydration environment, cells would inevitably be condemned to degeneration and death. It is unsurprising, therefore, that deficiencies in membranes have frequently been regarded as highly pertinent to explanations of seed aging (chap. 7).

Chapter 3

Loss of Seed Quality in Storage

Chapter 2 has emphasized that the physiological state of seeds in storage covers a broad spectrum, from the almost total quiescence of sorption zone I to the active metabolism found in sorption zone III. For the purposes of normal agronomic practice, the best form of storage is usually one that minimizes change within the seed by efficient management of moisture and temperature. Most orthodox seeds will maintain viability at high levels for many years if they are kept suitably dry and cool. In practice, the principal constraint is economic: large quantities of seed are of necessity stored at temperatures and moisture levels at which deterioration is kept barely within acceptable limits.

For the biologist, the investigation of seeds stored at exceptionally low temperatures or at greatly reduced water content leads to important speculations on the nature of the living process. It is now accepted that in some extremely cold conditions, virtually all change within the seed can be suspended indefinitely without compromising the viability of the material. A consideration of such anabiotic systems leads inevitably to the conclusion that "death is not the mere cessation of vital processes . . . [but] is conditional upon the *disruption* of the mechanisms of life due to irreversible changes in the living material" (Shmidt, 1955). Although probably not strictly accurate for seed material at normal temperatures, Gautier's (1896) graphic image of the quiescent seed as a clockwork device, wound but with its action temporarily suspended until an external stimulus sets it going again, conveys quite adequately the basic notion involved. According to this model, aging proceeds under storage conditions that tend to degrade structural organization, preventing the mechanism from functioning effectively when it is called upon at a later stage.

Our present knowledge of seed physiology is sufficient to indicate that these simple analogies are no longer completely tenable; the evidence to be considered in section 7.11, for example, makes it seem increasingly certain that in some circumstances seeds are capable of self-repair. Further, it seems probable that some of the debilitating changes that occur in seeds are self-generated rather than merely imposed by a hostile environment. As far as the relative merits of different storage regimes are concerned, however, it is still generally true that those conditions that minimize changes in metabolic and structural organization within the seed are likely to prove most advantageous.

3.1 The consequences of seed deterioration

3.1.1 Loss of viability

The most obvious, though not the most informative, indication of the quality of a seed lot is its germinability. The exigencies of seed testing are such that this characteristic is usually determined under unnatural conditions in a laboratory, so that a variety of debilitating stresses that are normally encountered in the field are avoided. One consequence is that the gulf between the viability of a seed in the laboratory and its performance in the field is often uncomfortably wide. Despite its shortcomings, however, the measurement of germinability under unstressed conditions has provided a fundamental yardstick in the great majority of studies on seed aging. It is unsurprising, therefore, that predictive and descriptive models of seed deterioration relate almost exclusively to the germination of seeds under unstressed conditions.

The curve that describes the loss of germinability of orthodox seeds over time in storage is generally held to be negatively sigmoidal (Fig. 8a). Roberts (1960; Roberts et al., 1967; Ellis and Roberts, 1980) has suggested that the deterioration curve is probably best considered in terms of an inverted cumulative normal distribution of seed deaths over time. Fortunately, the properties of such a function lend themselves readily to probit analysis (Finney, 1971). Transformation of germination data from percentages to probits linearizes the deteriorative trend (Fig. 8b), and when the results are expressed in this form, statistical assessment is greatly facilitated. An important quantity that emerges from such analysis is the P50 value, or half-viability period, the time taken for germination to drop to the level of 50%. There is general agreement that this descriptive model satisfactorily fits the circumstances of seed lots stored under constant controlled conditions, although Roberts et al. (1967) and Moore and Roos (1982) have described situations in which deviations from this trend may occur. Seeds exposed in the laboratory to ex-

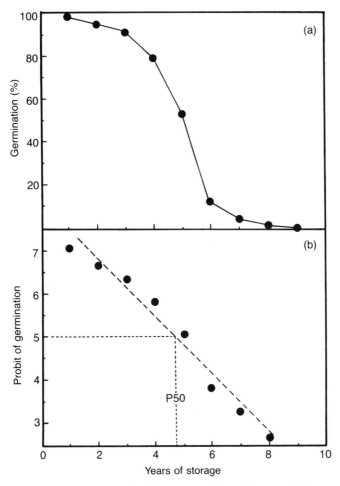

Figure 8. Loss of germinability in a seed lot of timothy (*Phleum pratense*) in open storage in Denmark. A probit value of 5 (equivalent to 50%) determines the half-viability period (P50). Calculated from data of Dorph-Petersen (1924).

ceptionally unfavorable combinations of temperature and moisture, for example, may fail to show the pseudo-asymptotic "tailing" normally associated with the later stages of storage. Deteriorative curves for seed lots in open storage, where temperature and humidity fluctuate, are particularly prone to stray from the ideal. The admirably symmetrical curve of Figure 8a is rather unusual for seed in open storage; results like those shown in Figure 9 are more frequently encountered. When such data are subjected to probit analysis, significantly nonnormal distributions are evident. The roots of this nonconformity may lie partly in external influences, such as unpre-

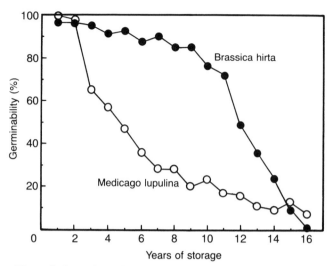

Figure 9. Loss of germinability in open storage. Seeds of black medick (*Medicago lupulina*) and white mustard (*Brassica hirta*) provide examples of strongly skewed aging patterns. Data selected from Dorph-Petersen (1924).

dictable climatic fluctuations; an unusually prolonged period of hot and humid weather, for example, obviously tends to accelerate deterioration. Factors innate to the seed probably play a more significant role in producing skewed distributions. In some species, residual dormancy may alter the pattern of germinability in storage, and in legumes especially, significant tailing of the deteriorative curve is ascribable to a small percentage of harder and more resistant seeds. Despite these reservations, the pattern of decline most commonly encountered in seed studies is of a recognizably negative sigmoidal type. Impressive data involving 51 seed lots and 26 species assembled by Hernø (1944) over a 13-year period provide graphic evidence of this tendency.

Although no physiological interpretation of the characteristic germinability-decay curve has been firmly established, one attractive hypothesis has been advanced by Roberts et al. (1967). According to this model, the germination of a seed is dependent on the proper functioning of a relatively large number of key cells in the embryo. The model assumes that some of these key cells are rendered nonfunctional during aging by some unspecified factor, and that the interaction of this debilitating process with the cellular population is describable in terms of a Poisson distribution. Roberts and his colleagues were able to demonstrate that under such conditions, assuming that failure of an individual seed to germinate was contingent on the loss of a relatively small proportion of key cells, the deterioration of a seed lot should conform to an inverted cumulative normal distribution.

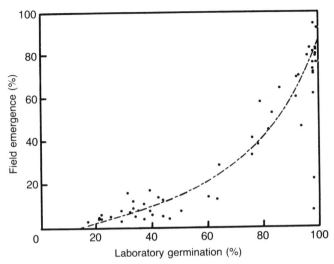

Figure 10. The relationship between laboratory germination and field emergence for maize. Each point represents one of 60 seed lots, ranging in age from 3 to 27 years. Data of Yarchuk and Leizerson (1971).

3.1.2 Loss of vigor

Except under very favorable circumstances, the percentage of a seed lot that emerges in the field is generally lower than the percentage that germinates under unstressed conditions (e.g., Sherf, 1953; Delouche, 1981). This tendency is illustrated in Figure 10 by data for maize seed. The shortfall between viability in the field and that in the laboratory becomes increasingly conspicuous as the seeds age (Laughland and Laughland, 1939; Fig. 11). Clearly, seeds that have attained the half-viability period (P50) in storage are likely to prove altogether useless for planting. For such species as maize, the half of the seed lot that still retains viability at the P50 point is likely to consist of debilitated material, low in vigor and poorly fitted to survive the rigors of the field environment. Indeed, older seeds are generally less able to withstand a wide variety of environmental stresses. In a study by Floris (1967), for example, seeds of durum wheat (*Triticum durum*) became increasingly sensitive to chilling injury during the imbibitional phase as they aged, and Ellerton and Perry (1983) similarly reported that aged barley seeds are more susceptible to the influence of anoxia. Highly artificial stresses may also serve to highlight the differences between fresh and aged seeds: data of Pocsai and Szabó (1982) indicate that older seeds are less resistant to applied methanol; according to Babayan (1971), the exposure of unimbibed wheat seeds to a temperature of 100°C for a few minutes is far more lethal to older material; and Darwin (1855b) long ago noted that aged seeds were more

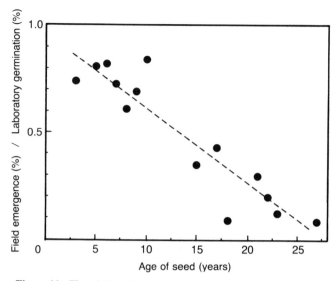

Figure 11. The relationship between laboratory germination and field emergence for maize after 3 to 27 years of open storage. Each point represents the mean of from 1 to 16 seed lots of the same age. Data of Yarchuk and Leizerson (1971).

sensitive to the stress of being soaked in sea water. The susceptibility of older seeds to chilling and their sensitivity to accelerated aging treatments of high temperature and humidity provide the bases for some current vigor tests (Byrd and Delouche, 1971).

The speed of germination, which has long been recognized as an indicator of seed vigor, is usually a more sensitive measure of seed deterioration than is loss of viability. The data presented in Figure 12 demonstrate that as a seed lot deteriorates, the remaining viable seeds are characterized by increasingly lethargic germination, a response typical of many other species (Dorph-Petersen, 1924; Filter, 1932). Indeed, it is commonly observed that the rate of radicle extension slows even before viability begins to decline (Toole et al., 1948; Associated Seed Growers, 1954; Guy, 1982). Occasionally an increase in the speed of germination with seed age has been noted, perhaps as a result of deterioration in the effectiveness of endogenous germination inhibitors (Harrison, 1966). Such effects have been particularly well documented in stored grass seeds, in which dormancy-related mechanisms may slow the rate of germination in relatively fresh material (Shaidee et al., 1969; Harty et al., 1983).

Concern over the inability of conventional germination tests to predict field performance has stimulated recent advances in the quantification of seed vigor. A formidable battery of vigor tests is now available (Perry, 1981; Association of Official Seed Analysts, Seed Vigor Test Committee, 1983),

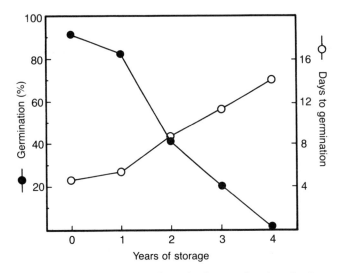

Figure 12. Decreased speed of germination as a function of aging. The average lag time to the start of germination for Norway spruce seeds (*Picea abies*) was determined in a laboratory germinator. Data of Lipkin (1927).

each procedure designed to give some practical measure of seed quality that is more expressive than a simple assessment of the capacity to germinate under favorable circumstances.

3.1.3 Changes in development and yield

Extreme developmental abnormalities in seedlings are commonly encountered following the germination of aged seeds (sec. 7.8), and several studies indicate that the morphology of relatively mature plants may also be influenced by the age of the seed from which they are grown. In *Arabidopsis thaliana*, for example, the number of leaves that appear on the main stem before flowering decreases with the age of seed sown (Napp-Zinn, 1964). Hartmann (1970) found that many plants from older *Oenothera biennis* seed had a tendency to development of the lateral axis, while the primary stem remained relatively undeveloped; other plants grown from old seed displayed an increase in the number of stem leaves. The sowing of deteriorated seed may even influence flower morphology. There have been reports that double flowers occur with diminished frequency in *Matthiola* and *Delphinium* species when plants are grown from less vigorous seed (Schmid et al., 1888; MacLeod et al., 1890); and according to Schwemmle (1952), flowers of plants grown from old *Oenothera campylocalyx* seeds have shorter hypanthia and smaller petals.

Surprisingly, opinions are sharply divided on the question of the planting quality of seed lots that have become partially deteriorated through age. (In this context "partially deteriorated" seeds are those that are still more than 50% viable in laboratory germination assays.) Some studies, but not all, have indicated that as field emergence drops with age, the quality and productivity of the plants that develop are markedly inferior. The inference to be drawn is that adjustment of seeding rate may not always be a wise practice when partially deteriorated seed is used. Investigations of the relationship between seed age and yield are deceptively complex, however, and generalizations are difficult for several reasons. First, there may be genuine differences between species; it may be naive, for example, to compare the productivity of root crops from low-vigor seeds with a similar situation in cereals. Second, disparity in vigor between seed lots is usually much more evident under stressful planting conditions, yet few studies have employed a range of sowing environments to investigate the productivity of aged seeds. Third, yield per plant can often increase if competition is reduced by lower emergence. To circumvent this problem, plots have been thinned to equal density following emergence in some studies; in others, seedlings have been started under glasshouse conditions and subsequently transplanted to the field. The latter solution was adapted by Barton and Garman (1946), for example, who contended that viable seeds in aged lots of verbena (*Verbena platensis*), tabasco pepper (*Capsicum frutescens*), and lettuce produced yields that were equal to or better than those grown from fresher seed lots of higher germinability. It remains questionable whether these conclusions would be valid for seeds planted directly in the field.

Studies with maize (*Zea mays*) generally suggest that inferior plants and yields result from partially aged seeds, even though levels of germinability are still relatively high. A result of this sort was noted for an age series of maize (up to ten years old) by Dungan and Koehler (1944), and a comparable conclusion was reached in a more extensive study of the same species (up to five years of aging) by Funk et al. (1962): planting of older maize seed led to slower and lesser field emergence, shorter ear height, and individually less productive plants. Similar trends seen in a study by Naumenko and Tkachev (1976) are illustrated in Figure 13, and further evidence for an effect on yield per plant in this species can be seen in observations of Clark (1963) and Grabe (1966, 1967).

Broadly similar conclusions arise from work on other cereals. Gelmond et al. (1978) indicated that individual yields fell in sorghum following accelerated aging (17% moisture, 30°C), even though laboratory germination values declined only slightly. Reduction in development and yield of plants as a consequence of the use of moderately aged seeds was noted in wheat by Crocioni (1934) and in extensive investigations of barley by Neoral (1923). Rajbhandary (1971/76) found a slight downward trend in yield with seed age

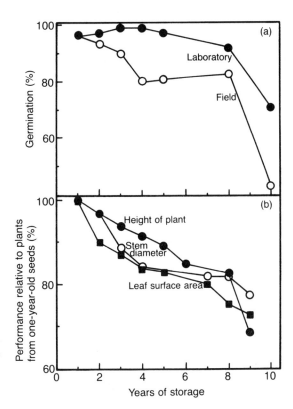

Figure 13. Decreased field performance with age of seed planted. The results shown are for maize seed (cv. Voronezhskaya 76) in open storage. Data of Naumenko and Tkachev (1976).

in wheat, even though germinability remained high; more recently, Schuch and Lin (1982) investigated the field performance of wheat that had been subjected to accelerated deterioration under laboratory conditions for up to four days (42°C and saturating humidity). Field emergence was reduced over the aging series (82.5% to 68.9%), as was the number of panicles per plant, but plant height, seed weight, and number of grains per panicle showed no significant reduction with age of seed planted. Several field trials in Poland have also indicated that partial deterioration in seeds of wheat and other cereals has relatively slight effects on plant productivity (Lonc, 1980, 1984; Wiłkojć et al., 1983). Perry and Harrison (1977) noted that individual plant yields in barley could increase with seed age under some sowing conditions, but differences in response varied from year to year depending on the degree of stress prevailing during planting. Brown (1962), working with oat seeds, reported no influence of age on yield per area, although germinability was lower in the older seed.

Among legumes grown for their seed, a variety of responses have been noted. In a growth-room investigation of partially deteriorated snapbeans

(*Phaseolus vulgaris*), Toole et al. (1957) concluded that both the number and weight of pods per plant were reduced in seed lots with inferior germinability. A plot study conducted by Vieira (1966) on seed of the same species that had aged in open storage in Brazil also indicated that yield per plant decreased with declining emergence. In contrast, Rodrigo (1939) claimed that 11- to 13-year-old seed of mung bean (*Vigna radiata*) was considerably more productive than fresher material, for reasons that remain elusive. Garner and Sanders (1935), on the other hand, noted loss of yield when older broad beans (*Vicia faba*) were planted, although the conditions prevalent in the year of the seeds' production were probably of greater significance than their age. Abdalla and Roberts (1969b) found the early stages of plant development to be slower when aged seeds of broad bean, peas, and barley were sown. The differences between plants from young and old seeds, however, were much less evident at later stages of development, a phenomenon noted by others (Hurd, 1969).

Studies on oilseed legumes similarly lead to few useful generalizations. Beattie et al. (1932) reported that although yields per plot attained with partially deteriorated peanut seeds were lower, there was some tendency for individual plant productivity to increase as a result of reduced competition. Effects of a comparable nature were evidently encountered by Burlison et al. (1940), who investigated the effect of prolonged storage of soybeans on yield per area; productivity was found to decrease only slightly with declining field emergence, provided the latter remained in excess of 50%. In a study of artificially aged soybeans (deteriorated at 40°C and 13% seed moisture), Edje and Burris (1971) found that although the rate of early development of viable seedlings was slower in seed lots with lower emergence, differences between aged and unaged seed lots diminished as plant development progressed. There was again evidence that yield per plant sometimes increased as emergence decreased as a result of reduction in population density. Torrie (1958), however, reported lower yields from two- or three-year-old soybean seed than from fresh material.

A variety of consequences resulting from seed aging have been recorded among forage legumes. Crocioni (1934) reported a decrease in yield of plants grown from older alfalfa and red clover (*Trifolium pratense*) seed. Ching and Calhoun (1968), on the other hand, investigated crimson clover (*Trifolium incarnatum*) subjected to different degrees of aging and found little or no decline in forage production per plant, even when laboratory germination values had dropped below 80%. Some improvement with age has even been suggested in clovers: in a field study of 51 samples of three clover species (*T. pratense, T. hybridum*, and *T. repens*) of various ages up to 34 years, Teräsvuori (1930) concluded that despite declining germinability, the plants produced by older seed lots tended to be as vigorous as those produced by fresh seeds, or more so.

In some investigations, the performance of root crops has seemed to be minimally affected by the use of partially deteriorated seed. In a study of beet seed, Filutowicz and Bejnar (1954) reported that although both viability and speed of germination dropped slightly over seven years of storage, the yields of beets and foliage in the surviving plants were unaffected. A subsequent investigation (Bejnar, 1958) led to similar conclusions concerning the productivity of older seed of mangels (fodder beets). In an analysis of radishes and turnips, Lazukov (1969b) indicated little linkage between germinability of seed lots and the size and weight of roots grown from them, and in the case of onions, yields per hectare were unaffected by laboratory germination values as low as 67% (Ugarynko, 1967/68). The latter experience is apparently contradicted by work of Newhall and Hoff (1960), however: when the number of bulbs per plot declined 15% with the use of older seed, weight per bulb decreased 16%. Chen et al. (1972) also noted loss of productivity per plant when they used turnip seed subjected to accelerated aging (42°C, 100% relative humidity). Similar work with radish gave less conclusive results: yields from aged seeds were reduced in one sowing, but in a second trial differences were not significant.

Other crops have, predictably perhaps, given a variety of responses. In an investigation of eggplant seeds of various ages, Kadota (1937) found that productivity per plant decreased as germination began to decline. Crocioni (1934) suggested that same to be true of field mustard, and Chirkovsii (1953) also encountered poorer yields (leaf dry weight) from tobacco plants grown from older seed. Harrison (1966) likewise found lower individual yields from lettuce plants derived from seed that had been aged in long-term storage. Relatively small declines in germinability (10–20%) entailed substantial losses in individual plant productivity among the survivors. For lettuce seed that had deteriorated rapidly over a few days of accelerated aging, however, yield per plant did not drop until germination of the seed lot had fallen below about 50%. Nowosielska and Schneider (1980), on the other hand, found that yields from lettuce seeds 10 to 12 years old (73–95% germinability) were not inferior to those from fresher material. When Ghosh and Basak (1958) investigated the effect of seed age on the yield of fiber in jute (*Corchorus* spp.), they found that although yield per plot decreased with declining emergence, yield per plant often remained acceptable; and Christidis (1954), working with cotton, suggested that differences in yield due to seed age were relatively trivial compared to the influence of cultural conditions.

It is difficult to generalize from this variety of experiences with old seed, although it seems evident that in some species a moderate amount of age-related deterioration need not seriously imperil yields, at least under favorable growth conditions. Roberts (1972c) has suggested that two categories of species are definable: those in which the viability is correlated with individual plant yield and those in which it is not. Inasmuch as a species such as lettuce may evidently belong to either of the proposed categories, depending on the mode of storage, such a classification may be difficult to maintain.

A curious postscript to this discussion of plant productivity and seed age is provided by the properties supposedly possessed by some members of the Cucurbitaceae, particularly melons (*Cucumis melo*) and cucumbers (*C. sativus*): plants grown from older, less vigorous seeds have often been held to be more desirable than those grown from fresh seed. In these monoecious crops the female flowers develop later than the male, and according to Cazzuola (1877), D'Arbaumont (1878), Troop (1914), and many others, the sowing of older, less vigorous seed results in more compact growth, with a greater number of the pistillate flowers and hence more fruits per plant. Recent scientific opinion on this venerable supposition has been divided.

Boos (1966), on the one hand, reported that plants grown from two- or three-year-old cucumber seeds produced more female flowers than plants grown from fresh seeds, so that yields of fruit were greater. He recommended that fresh seed should be heated in an oven for several hours at 50–55°C to age it slightly before planting. Similar conclusions were reached by Šmerda and Tichý (1957). Ovečka (1960) and Fröhlich and Henkel (1964), however, found no evidence to suggest that older seed was of superior quality; yields tended to drop slightly as seed age increased. Environmental influences may control the outcome of such trials. Experience gained with watermelon and melon seeds led Esitasvili (1956) to suggest that younger, fresher seed is to be preferred in a warm climate with a long growing season, whereas plants from older seed enjoy some advantage in those areas where the growing season is relatively short. In the related squashes (*Cucurbita pepo*), accelerated aging of seed at 42°C and 100% relative humidity has been reported to lead to a proportionally similar decline in the numbers of both male and female flowers and a decrease in fruit yield (Chen et al., 1972). Others, however, have found an increased proportion of female flowers on plants grown from seeds aged for one or two years in storage (Abou Hussein and El-Beltagy, 1977).

Partly aged seeds have seemed to possess desirable cultural properties in at least one other case: pea seeds harvested in extreme northerly latitudes have been reported to improve as a result of prolonged storage (Olsson, 1950). In short growing seasons in Sweden, plants from fresh seed were found to be too vigorous and likely to develop an overabundance of foliage; this deficiency was evidently remedied when the seed was aged. Instances of this sort are atypical, however; generally the planting of older, less vigorous seed involves a considerable risk of failure.

3.2. The effects of humidity and temperature on seed deterioration

Two environmental influences, relative humidity and temperature, are known to exert profound influences on seed aging. Proper control of both parameters provides the basis for all rational conservation of seeds. The effects of a third factor, the gaseous environment in which seeds are stored, is far less well understood and is discussed separately in section 3.3.

3.2.1 Deterioration equations

Detailed investigations of the effects of temperature and relative humidity on the aging of stored seed have been pursued over many decades. Much of the earlier work has been definitively summarized by Barton (1961); more recent research appears in the monograph by Justice and Bass (1978). One general tenet is that the drier and cooler the storage environment, the longer orthodox seeds retain the ability to germinate. Three riders should be attached to this statement. First, it is well established that seeds with elevated moisture levels may experience freezing damage, associated with the presence of free water (Roberts, 1972a). This problem does not seem to occur with drier seeds, although at temperatures well below 0°C structural deformations of an uncertain character can apparently compromise viability in a few species (Stanwood and Bass, 1981). Second, metabolic repair processes may be triggered at high levels of hydration, and these processes may serve to limit deterioration (see sec. 7.11). A third consideration is that many seeds, after being dried to very low moisture levels, tend to be afflicted with deleterious cracking during rehydration, especially if imbibition is relatively rapid (Phillis and Mason, 1945; Powell and Matthews, 1978).

It has sometimes been suggested that when seeds are stored in an excessively dry state, their rate of deterioration may increase. Thus Nutile (1964) intimated that for seeds of celery, eggplant, Kentucky bluegrass (*Poa pratensis*), and tabasco pepper (*Capsicum frutescens*), germinability decreased more rapidly at 1% hydration than at 4%. Comparable effects have also been noted in very dry lettuce seed (Kosar and Thompson, 1957) and in legumes and crucifers (Nakamura, 1958, 1975; Takayanagi, 1980). The question of whether the increased loss of germinability represents an acceleration of the aging process per se, as Harrington (1972) has suggested, or results from the compounding of aging effects with imbibitional stresses appears not yet to have been fully addressed. Observations on soybeans, for example, clearly suggest that aged material is more susceptible than fresh seed to injury from rapid hydration (Woodstock and Tao, 1981; Woodstock and Taylorson, 1981b; Tilden and West, 1985). Further, not all seeds show an apparently diminished longevity when stored at very low levels of hydration. According to Zentsch (1970), for example, seeds of spruce (*Picea abies*) store better at 0.5% moisture content than at higher levels of hydration, and similar considerations seem to apply to sweet pepper (*Capsicum annuum*) and onion (Woodstock et al., 1976) and to peanut seeds (Norden, 1981). Touzard (1961, 1975) also noted that extreme dryness had no deleterious influence on longevity in such species as *Callistephus chinensis*, whereas others, such as *Salvia splendens* and *Primula obconica*, were sensitive to storage under conditions of extreme desiccation.

The work of Roberts and his associates (1960; Roberts et al., 1967; Ellis and Roberts, 1980; Ellis, 1984) has led to detailed descriptions of the effects of temperature and moisture content on loss of viability in a number of species. Observations of seed aging in a wide variety of conditions of controlled temperature and humidity have provided data from which a generalized descriptive equation has been developed:

$$v = K_i - p/10^{[K_E - (C_W \times \log m) - (C_H \times t) - (C_Q \times t^2)]}$$

where v represents the probit of percentage germinability after a storage period of p days; K_i is the probit of initial germinability for the seed lot; K_E, C_W, C_H, and C_Q are species-specific constants; m is seed moisture (expressed on a fresh weight basis); and t is the storage temperature (°C).[1]

It is quite evident that different species lose viability at markedly different rates under the same storage conditions, but the pattern of deterioration in each case can be described by means of the same general equation. Roberts and his co-workers have supplied species-specific constants (see Table 1) that can be used in the deterioration equation. Given a value for initial germinability and temperature and moisture during storage, the viability of a seed lot at a later point in time can be predicted with some accuracy. The useful range of the equation is thought to extend from −20 to 90°C, and from about 5 to 18% or higher in seed moisture content. Because of freezing damage, however, combinations of high seed moisture and temperatures much below 0°C are likely to prove more destructive than the model predicts. There is good evidence that at higher hydration levels the generalized deterioration relationship begins to fail dramatically in some species; it does so in lettuce when moisture content rises above about 15–20%, a point at

1. This equation replaces a simpler version of lower overall fidelity. Application of the earlier equation to the aging of seeds of rice, wheat, broad bean, peas, and lettuce has been described by Roberts and Roberts (1972). Constants for soybean were published by Burris (1980), for the tropical tree *Agathis macrophylla* by Smith (1984), and for *Lupinus polyphyllus* by Dickie et al. (1985).

Table 1
Species-specific aging constants for use in the general seed deterioration equation. Data of Ellis and Roberts (1980, 1981) and Ellis et al. (1982).

Species	K_E	C_W	C_H	C_Q
Barley *Hordeum vulgare*	9.983	5.896	0.040	0.000428
Chickpea *Cicer arietinum*	9.070	4.829	0.045	0.000324
Cowpea *Vigna unguiculata*	8.690	4.715	0.026	0.000498
Onion *Allium cepa*	6.975	3.470	0.040	0.000428
Soybean *Glycine max*	7.748	3.979	0.053	0.000228

which metabolic repair processes become effective in inhibiting the decline in viability (Ibrahim and Roberts, 1983).

Curves generated when the deterioration equation is applied to soybean (Fig. 14) summarize in rather idealized fashion some well-established characteristics of seed in storage. The effects of increasing temperature at constant moisture content and of increasing moisture content at constant temperature are shown in Figures 14a and 14b, respectively. The higher rate of viability loss characteristic of a partially deteriorated seed lot in comparison with a fresh one is indicated in Figure 14c. In practice, a seed lot of high initial germinability is usually favored by a considerable induction period before losses in viability become evident (e.g., Agrawal and Sinha, 1980).

Comparative analyses of loss of viability in different species under identical conditions of temperature and seed moisture content are shown in Figure 15a. It should be borne in mind that the predicted deterioration curves apply to circumstances of closed or hermetic storage, as the model equation and its constants were developed from studies that employed such conditions. Some readjustment may be required in cases where gaseous exchanges are to be permitted, and the relationships shown in Figure 15a will change markedly under conditions of open storage, even when temperature and humidity are held constant, because of differences in sorption characteristics between species. If the five species were exposed to the same relative humidity rather than being maintained at an identical moisture content, the equilibrium hydration attained by each species would depend on its lipid content. More specifically, the five species would attain characteristically distinctive moisture contents, but their nonlipid fractions would be about equally hydrated (see Fig. 2, in chap. 2). When this consideration is taken

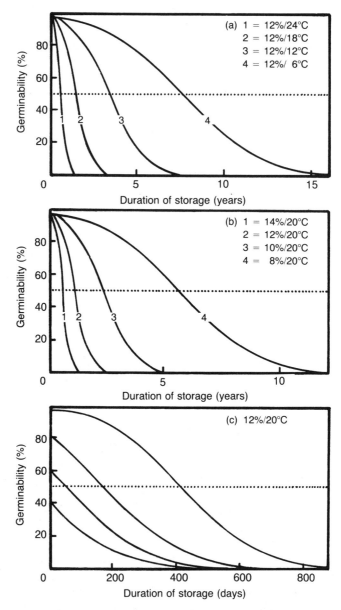

Figure 14. Theoretical deterioration curves for soybean. Curves were computed for various constant storage conditions: (a) different temperatures at 12% moisture content with initial germinability 98%, (b) different moisture contents at 20°C with initial germinability 98%, and (c) different initial germinabilities (98%, 80%, 60%, and 40%) at 12% moisture content and 20°C. The deterioration curves were generated by means of the predictive equation and constants of Ellis et al. (1982).

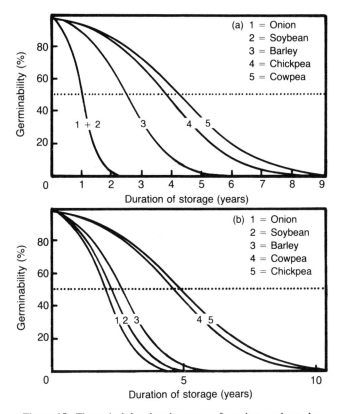

Figure 15. Theoretical deterioration curves for onion, soybean, barley, chickpea, and cowpea. Curves were computed for (a) seeds stored at 20°C and 12% moisture content, and (b) seeds stored at 20°C and a relative humidity such that the nonlipid fraction of each type of seed equilibrates to 12% moisture content. In all cases, an initial germinability of 98% is assumed. The deterioration curves were generated by means of the predictive equation and constants supplied by Ellis and Roberts (1980, 1981a) and Ellis et al. (1982). Data on lipid content are from Spector (1956) and Gvozdeva (1971).

into account, comparative deterioration curves for the five species appear rather different (compare Fig. 15a with Fig. 15b).[2]

Calculation of predicted seed longevity is fairly straightforward with the general equation and appropriate species-specific constants, or by means of specially constructed nomographs (Roberts and Roberts, 1972; Ellis and

---

2. This analysis assumes that under a given set of conditions, hygroscopic equilibria are similar for all samples of a given species. In practice, fairly small differences in varietal response can often be detected; e.g., Paricha et al. (1977), Pixton and Henderson (1981).

Roberts, 1980; Ellis et al., 1982).[3] Considerable caution must be exercised in the use of a value for K_i, the probit of initial germinability. Although the result of a simple germination assay could be used to define this property, a seemingly trivial uncertainty at this point can cause substantial variation in predicted longevity. One remedy, discussed at some length by Ellis and Roberts (1980), is to use the intercept on the ordinate of the probit of germination-versus-time deterioration line to generate a more precise value for K_i. A useful treatment of this problem has also been given by Bewley and Black (1982).

The descriptive analysis provided by Roberts and his colleagues implies that seed deterioration can be evaluated in terms of a smoothly continuous function, although it has frequently been suggested that "critical" or safe moisture contents for acceptable seed storage are definable. Lists of such critical moisture contents on a species-by-species basis (e.g., Lityński, 1957; Kuleshov, 1963; Gvozdeva, 1965) are often based on much practical experience at a particular locality, but may be misleading elsewhere if the effects of temperature are not also considered. Generally, the critical moisture content for safe storage declines with increasing temperature (Kreyger, 1963). Further, it cannot be inferred that the existence of such "threshold" values is an indication that physiological factors that influence seed aging necessarily increase in a discontinuous fashion once a particular level of hydration is exceeded. Some workers consider the critical moisture content as that level at which respiration is first readily detectable (Koz'mina et al., 1964), whereas others have closely linked it to the onset of mold-induced effects (Milner and Geddes, 1946). The two phenomena are usually closely linked, as they are associated with sorption zone III.

3.2.2 Climatic effects

It is firmly established that seeds in open storage fare better in cooler and drier localities than in warm, humid ones. The results of two early studies serve to illustrate this point: Duvel (1904) placed 12 species of seed in open storage for periods of up to nine months at various localities in the United States and Puerto Rico. Deterioration was found to be most rapid in the warm and moist climates of states bordering the Gulf of Mexico and least evident in Michigan. An extensive investigation of coniferous seed storage by Tillotson (1921) pointed to very similar conclusions, although in this study some association with altitude was also established. When data for six species in open storage at 13 locations in the United States were pooled, the

---

3. A table of probit transformations must be consulted before calculations can be carried out. Such tables can be found in advanced statistical texts, e.g., Fisher and Yates (1963), Bliss (1970), and Finney (1971).

three best stations were found to be in New Mexico (elevation 1,980 m), Colorado (2,740 m), and Idaho (1,370 m). Each of these sites was characterized by low prevailing atmospheric humidity. The three worst stations were in Kentucky (120 m), Kansas (240 m), and Illinois (210 m). The deterioration of seeds in tropical or subtropical climates poses equally evident problems. In some cases, vigor and viability are lost so rapidly that the usefulness of otherwise successful species is severely diminished. Soybeans, for example, are notorious for their inferior storage characteristics even in temperate climates; but in Trinidad, at latitude 10°N and with a very humid climate, viability may be almost extinguished in just eight months of open storage (Spencer, 1931). In areas where losses reach this magnitude, such species are unlikely to be exploited to their fullest extent (Boakye-Boateng and Hume, 1975; Ravalo et al., 1980).

As section 1.2 noted, the effects of fungal pathogens become much more significant when seeds are exposed to very humid atmospheres: the effects of the microflora must consequently loom large in any discussion of seed deterioration in tropical climates. There is evidence, nevertheless, that loss of germinability during storage in the humid tropical lowlands need not always be mediated primarily by pathogens, at least in some species. Ndimande et al. (1981) have argued that loss of soybean seed quality in these conditions is due largely to innate physiological changes, and that pathogens play a secondary role. Harrison and Perry (1976) reached similar conclusions for barley stored at relatively high moisture levels (14–24%) at 20°C, suggesting that viability and vigor began to decline before substantial invasion by fungi was evident. Other species, of course, may well differ in this respect. Data of Dharmalingam et al. (1976), for example, suggest that a benefit can be gained from the use of fungicides on mung beans (*Vigna radiata*) stored in southern India. The same may also be true of maize stored in humid environments (Moreno-Martínez et al., 1985).

3.2.3 Aging under extreme conditions

The rapid deterioration of seed lots exposed to both high temperature and high humidity under laboratory conditions has been of special significance to much recent work in the field of seed aging. Following the suggestion of Delouche and Baskin (1973), "accelerated" or artificial aging of seed lots over several days of exposure to 40°C and saturating humidity has been recognized as a good predictor of storability; those seeds that deteriorate rapidly under conditions of accelerated aging also tend to perform poorly in long-term open storage. Accelerated aging has subsequently been recognized as a useful vigor test for some species (Association of Official Seed Analysts, Seed Vigor Test Committee, 1983). Increasingly, however, rapid deterioration under unfavorable conditions has been employed in studies of the physiology and biochemistry of seed aging. Two principal justifications have been advanced for its use in this context. Some workers have attempted to investigate seed deterioration under simulated tropical conditions, which may sometimes approach those used in conventional accelerated aging experiments (Minor and Paschal, 1982). More often, though, and with much less validity, accelerated aging has been used as a means to circumvent the need for experimental analyses that would otherwise extend over many years of storage.

As chapter 7 will show more fully, there is little reason to believe that the physiology of aging in a relatively unimbibed condition (i.e., sorption zone I or II) need necessarily resemble that which occurs at higher moisture levels (e.g., sorption zone III). The use of accelerated aging protocols in the investigation of the physiology of seed deterioration is long established, however, and physiologists have been loath to abandon the practice. Over a century ago, Just (1871) studied seed aging by using a range of warm temperatures and high humidity to promote deterioration, and it is possible that the technique may be even more venerable. As section 3.1.3 indicated, it has often been maintained that the horticultural value of some seed of the Cucurbitaceae can increase as a result of a moderate loss of vigor. Reimers (1979) has reported that it was customary in Russia for vegetable growers to subject fresh cucumber seed to several weeks of accelerated aging under rather unorthodox conditions: seeds were placed in a small cloth bag held constantly under the armpit, where the elevated temperature and humidity helped to induce a suitable decline in seed vigor.

The advantages that accrue from the storage of seeds at very low temperatures are attractive for the custodians of valuable collections, although the costs incurred are unacceptable for normal agronomic practice. Storage of germ plasm in sealed containers at a temperature of about -15 or $-20°C$ is now fairly routine (Roberts, 1972a; Fraser, 1980), and experience gained with subfreezing temperatures suggests that many seeds may be stored for several decades without the need for replacement (Pack and Owen, 1950; Rincker and Maguire, 1979; Rincker 1981, 1983; Wang, 1982; Ackigoz and Knowles, 1983). An increasingly attractive option is the storage of seeds at a temperature close to that of liquid nitrogen ($-196°C$). Fairly extensive surveys (e.g., Fedosenko, 1976; Stanwood and Roos, 1979; Stanwood and Bass, 1981; Stanwood, 1985) suggest that most agronomically important species can be exposed to such temperatures with relative impunity. The longevity of seed accessions stored in or above liquid nitrogen is unknown, but it seems probable that a period of centuries—or even millennia—may pass before regrowing is required.

Several species of seed have been successfully exposed to even lower temperatures, close to absolute zero ($-273°C$) (Lipman, 1936; Becquerel,

1950). Under such conditions there can be no question of metabolism or even conventional chemical activity. In the absence of all but the faintest of atomic tremors, it is likely that the structural organization of the seed, and hence its capacity to germinate under appropriate circumstances, would endure for extraordinarily prolonged periods of time (Fedosenko, 1981). A prospect of this kind was envisioned long ago by the physical chemist Svante Arrhenius, who once remarked that "viable seeds might be preserved for ever if they could be kept in a dry state at a temperature near absolute zero" (Dorph-Petersen, 1928). Arrhenius had previously elaborated such views elsewhere, proposing an exceptional longevity for simple life forms at very low temperatures (Arrhenius, 1908); there is little reason to dispute the application of a similar thesis to seeds.

3.3 Sealed and gaseous storage

Since Haberlandt (1873) first discussed the advantages of closed over open storage more than a century ago, they have been so well documented as to need no further reinforcement here; much of this evidence has been definitively reviewed by Barton (1961). The extension of viability resulting from a closed storage strategy is due largely to the efficiency with which a low level of hydration at the time of sealing can be maintained. Success depends almost entirely on proper management of moisture content.

Repeated experience has shown that for most seeds a hydration level well below 15% is mandatory before sealing if germinability is to be maintained: the conventional wisdom has often been that seeds must be dried below a so-called critical moisture content (Heinrich, 1913). For seed lots with relatively high moisture content and significant respiratory activity, closed storage leads to a depletion of oxygen and an increase in carbon dioxide. Thus when peas were sealed in bottles at 25°C and 18.4% hydration (1.9 ml head space per g seed), oxygen fell from 21% to 1.4% in 11 weeks (Roberts and Abdulla, 1968). Carbon dioxide increased from 0.03% to 12% during the same period. At these relatively high levels of hydration the deleterious influence of the microflora can scarcely be ignored; once almost complete anaerobiosis is established, however, the vitality of both the seeds and their associated fungi will be destroyed (Nash, 1978). The debilitating lesions incurred by the sealing of seeds at these higher moisture contents may therefore be complex, depending on the type of seed, the degree of fungal contamination, and the size of the storage head space, as well as the temperature and the level of hydration. Closed storage is highly deleterious not only to overly hydrated orthodox seeds but also to recalcitrant seeds that must maintain high moisture levels (King and Roberts, 1979; Tompsett, 1983) and to imbibed seeds in a state of dormancy (Ibrahim et al., 1983).

An obvious extension of the sealed storage technique involves the use of various gaseous atmospheres or a vacuum for seed conservation. Previous reviewers (e.g., Roberts, 1972a; Bass, 1973a) have noted the considerable difficulties inherent in any assessment of the literature in this area. Many reports of experimental protocols are inadequately detailed or indicate possible deficiencies in design. Poor control of seed moisture, in particular, has probably generated many false conclusions. As Ibrahim and Roberts (1983) have noted, a drop in hydration of one percentage point can lead to a doubling of longevity in some cases.

A good example of the uncertainties found in this field is furnished by the experiments of Guillaumin (1928, 1937) on soybeans, work that is still cited in the literature. Seeds of unspecified moisture content were sealed in glass vials in oxygen-free air or in a vacuum. These treatments helped maintain germinability over 15 years, whereas seeds in open storage degenerated completely. Guillaumin concluded that exclusion of oxygen was the determining factor in their longevity, although moisture content must certainly have been better controlled in the sealed tubes. Indeed, vacuum storage probably led to some reduction in hydration as a result of evaporation of seed moisture into the head space of the container. Furthermore, sodium was used as an oxygen scavenger in the other closed treatment. As sodium also serves as an excellent moisture trap, it is quite probable that the oxygen-free air present in some vials was much drier than the ambient atmosphere at the time the tubes were sealed. Consequently, even if Guillaumin had employed an additional treatment of seeds sealed in air, it is still doubtful that his experiment would have been adequately controlled. Unfortunately, these observations are still cited as evidence for the involvement of oxidative processes in seed aging (e.g., Rabinowitch and Fridovich, 1983).

The storage of seeds *in vacuo* is generally not inimical to viability provided hydration levels are relatively low. Whether it confers significant advantages for most species over sealed storage in air is questionable. Moderate benefits have been reported in some cases: Barton (1935b, 1939b) found that seeds of American elm (*Ulmus americana*) and various coniferous species were better preserved *in vacuo*. Brison (1941a, 1941b) reported some benefits for onion seed storage, as did Bockholt et al. (1969) for cotton and sorghum. Busse (1935) likewise found some advantages for pine and aspen seeds, and more recently Lougheed et al. (1976) indicated that similar effects may be noted for onion, celery, and cabbage. Nevertheless, an impressive array of investigators deny that vacuum storage offers any great benefit over other sealed treatments (e.g., Barton, 1939c, 1960; Allen, 1962; Bass et al., 1962, 1963a, 1963b; Justice and Bass, 1978; Bass and Stanwood, 1978; Tauer, 1979; Bomme et al., 1982).

A complete or a partially enhanced atmosphere of carbon dioxide has occasionally been advocated as a storage medium. When seeds are maintained at elevated moisture levels, carbon dioxide may play some role as a metabolic repressor, and for stored seed at low hydration it has a reputation as an efficient antioxidant (Sankara Rao and Achaya, 1969). Bennici et al. (1984) have suggested that some advantage accrues from exposing durum wheat (*Triticum durum*) to carbon dioxide in storage (seeds equilibrated to 11% relative humidity), and Harrison (1966) reported that sealed lettuce seed (5–6% moisture) declined in viability more slowly in this gas than in air. Similar effects had previously been noted by Lewis (1955) for lettuce, onion, and parsnip seed of undefined moisture content, and Evans (1957) reported a relatively meager advantage for red clover seed (*Trifolium pratense*) at hydration levels below 10%. Peanuts (*Arachis hypogaea*) at 6.2% moisture stored in carbon dioxide (or nitrogen) at 38–40°C, however, were afforded considerable protection in comparison with storage in air, according to Luo et al. (1983); but Marzke et al. (1976) found negligible advantages for peanuts stored in these gases (6% hydration, 4° or 27°C) over those stored in air. Many other studies have encountered little or no benefit from carbon dioxide (Kondō and Okamura, 1934a; Sayre, 1940; Simpson, 1953; Goodsell et al., 1955; Barton, 1960; Ikeda et al., 1960; Bass et al., 1962, 1963a, 1963b; Bass and Stanwood, 1978). A complication that some investigators may have overlooked is that commercial samples of carbon dioxide have often been extensively dried, whereas laboratory air usually contains some moisture. As with vacuum storage, the physiological significance of experiments conducted with carbon dioxide must remain uncertain unless moisture levels are also precisely monitored.

The value of inert atmospheres for seed preservation (usually nitrogen, argon, or helium) has still to be firmly established. Several wide-ranging studies indicate little or no advantage to the technique (e.g., Gane 1948a, 1948b; Simpson, 1953; Goodsell et al., 1955; Isely and Bass, 1959; Barton, 1960; Bass et al., 1962, 1963a, 1963b; Justice and Bass, 1978; Bass and Stanwood, 1978), although Harrison (1966) recorded some benefits for lettuce seed at 6% moisture and for onions at 8%. As Roberts (1972a) has noted, however, there were puzzling discrepencies between the performance of seeds stored in nitrogen and those stored in argon. According to Davies (1977), nitrogen storage was found to have limited advantages for red clover seed (*Trifolium pratense*) at 0°C, but was detrimental at warmer temperatures. White clover (*T. repens*) in the same study was unaffected by a nitrogenous atmosphere. Roberts (1961) found rice seed to be better preserved in nitrogen at 12% moisture, but not at 13.5% or 14.5%, and Sampietro (1931) claimed that rice retained full germinability over eight years in nitrogen at 5% hydration, but deteriorated completely at 13% hydration. Pixton et al. (1975; Pixton, 1980) reported that an atmosphere depleted in oxygen (less than 2%)

retarded loss of viability for wheat stored at about 12% moisture content in the absence of fungal growth. Quaglia et al. (1980) reached similar conclusions in regard to wheat at approximately the same level of hydration stored in nitrogen. At relatively high moisture levels (13–18%) nitrogen may slow deterioration somewhat in cereals by suppressing fungal growth (Glass et al., 1959).

The effects of storage environments enriched by oxygen have received rather less attention. Oxygen is clearly not only beneficial to seeds with elevated moisture levels but essential for their prolonged survival (Ibrahim et al., 1983). That oxygen exerts a debilitating influence on drier seeds gains support from observations of Sayre (1940), Harrison (1966), and Roberts and Abdulla (1968), although Dju and McCay (1949) reported oxygen to be relatively harmless for stored soybeans. More recently, Ohlrogge and Kernan (1982) found that soybeans subjected to accelerated aging (at 44°C and 100% relative humidity) deteriorated as rapidly in nitrogen as in oxygen; at 17% moisture and 25°C, however, oxygen was clearly deleterious. The injurious effect of oxygen may not always be as straightforward as many of these studies suggest, however. Barton (1960), for example, found pure oxygen to be detrimental to *Lobelia cardinalis* seeds stored at room temperature, whereas at 5°C seeds declined in viability at about the same rate in air, in carbon dioxide, in nitrogen, and under vacuum as they did in oxygen.

The brief survey of the effects of gases provided here serves yet again to highlight the rather limited extent of our knowledge in this area. The conclusions to be drawn are inevitably somewhat tentative. There is reason to believe that oxygen exerts a deleterious influence on seeds that are too dry to respire, and in some species, at least, inert atmospheres may help to promote longevity. As species-specific differences may be quite great, though, broad generalizations are unwarranted. Finally, there is some question about the penetrability of storage gases into seeds. Permeability of seed coats to oxygen and carbon dioxide, for example, varies greatly, depending on state of hydration and species characteristics (Côme, 1968; Porter and Wareing, 1974). Some gases are never likely to permeate certain types of seed coat to a significant extent. Such factors must obviously be examined in greater detail before a full description of the effects of gaseous storage can be integrated into our knowledge of seed aging.

3.4 Preharvest conditions and other factors that limit longevity

3.4.1 Mechanical injury

Stresses to which seeds are exposed before or during harvest frequently have a marked influence of subsequent storability. Weathering in the field, particularly in conditions of excessive moisture or freezing temperatures,

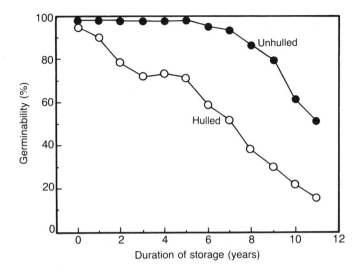

Figure 16. Longevity of hulled and unhulled seeds of timothy (*Phleum pratense*) within a single seed lot in open storage. Data of Stevens (1935).

results in a product with inferior storage prospects (e.g., Dillman and Toole, 1937; MacKay and Tonkin, 1967; Karpov, 1980). Artificial drying procedures, when improperly applied, also produce seed lots that lose their viability in storage with unusual speed (Harrington, 1972), and mechanical injury during harvesting or handling can similarly exert a depressing influence on storability (Burns et al., 1958; Metzer, 1961; Mamicpic et al., 1963; Arnold, 1963; Moore, 1972; Almeida and Falivene, 1982). Some investigators have suggested, for example, that seeds of timothy (*Phleum pratense*) and sorghum store less well after hulling (Goff, 1890; Stevens, 1935; Haferkamp et al., 1953; Kalashnik and Naumenko, 1979), implying that the hulls or chaff confer a stabilizing influence (Fig. 16).[4] The studies of Lakon (1954) on hand- and machine-hulled oats and timothy, however, suggest that much of this difference may result from threshing injury, and Esbo (1954, 1960) accepted this conclusion on the basis of his investigations of stored timothy seed. Delinting of cotton seed can also influence storability (Flores, 1938; Simpson, 1946), and seeds of Douglas fir (*Pseudotsuga menziesii*) and Scots pine (*Pinus sylvestris*) that have been damaged during dewinging show a similar increase in their rate of deterioration (Barner and Dalskov, 1954; Huss, 1956; Kamra, 1967).

Cracked material is especially susceptible to fungal invasion during storage (Oathout, 1928; Kuperman, 1950), although other deteriorative influences

---

4. Hulling entails the separation of the caryopsis from the lemma and palea.

may also be significant in this context. In a ten-year study of six cool-season grasses (*Agropyron intermedium, Phleum pratense, Bromus inermis, Alopecurus pratensis, Arrhenatherum elatius,* and *Dactylis glomerata*), Canode (1965, 1972) found that hulling influenced storability at 5°C and 40% relative humidity for all but the last two species. Since fungal and microbial influences should be trivial under these storage conditions, the increased rate of deterioration presumably arose from other sources; increased leakage during imbibition (sec. 7.3) is an obvious factor that could exaggerate poor performance in low-vigor seeds.

It has been suggested that susceptibility to mechanical injury may be significant in determining species differences with respect to seed longevity (see sec. 3.5). MacKay et al. (1970) noted that in freshly harvested rye seeds (a notoriously short-lived species), damage to the radicle tip was fairly frequent; about 20 to 30% of all seeds tended to show some evidence of injury. In long-term storage a necrotic area was found to spread from the site of the initial lesion, eventually encompassing much of the axis. This type of injury was much less frequent in wheat and barley, and very rare in oats (a long-lived species). Although such effects cannot entirely explain species differences with respect to longevity, as ordinarily only a minority of the seeds in a batch are affected, they no doubt contribute to the overall pattern of decline.

3.4.2 Maturity, size, and specific gravity

Seed maturity has a pronounced influence on longevity. Immature material generally loses viability rather rapidly; such effects have been well documented in studies on parsnip (*Pastinaca sativa*) (Joseph, 1929), clovers (*Trifolium* spp.) (Brett, 1952), tobacco (*Nicotiana tabacum*) (Chirkovskii, 1960), winter rape (*Brassica napus*) (Schneider and Wiązecka, 1977), cabbage, carrot (Eguchi et al., 1958), and several species of the Solanaceae (Eguchi et al., 1958; Suzuki, 1969) and Gramineae (McAlister, 1943; Bass, 1965; Lawrence, 1967). In addition, premature harvesting of gymnosperm cones frequently produces seeds with inferior storage characteristics (Vincent and Freudl, 1931; Yanagisawa, 1965; Heit, 1967b), although Barnett and McLemore (1970) noted that this trend was not always apparent, as immature pine seed may sometimes ripen during the early stages of storage, so that germinability appears to increase during this period. Generally, though, harvesting of unripe material is not recommended. The only report to run seriously counter to this general experience is that of Ellis and Roberts (1981b), who studied the longevity of barley at two different levels of maturity, stored at 30–40°C with moisture contents of 14% or higher. In these experiments, no influence of seed maturity on aging could be detected.[5]

5. In an earlier contribution, Roberts and Ellis (1977) reported that seed of the same barley

For some species, smaller seeds within a seed lot may be shorter lived, Koval'chuk (1973a, 1973b), for example, examined three varieties of winter wheat and one of barley over six to seven years of storage. Seeds were sieved by size into three classes and viability and speed of germination were assessed. Smaller seeds were found to deteriorate more rapidly. Such effects could presumably be associated with lack of maturity; it is generally true, though, that undersized seeds in healthy seed lots often display less vigor and germinability even when they give no visual indication of immaturity (Delouche, 1980). Vaughan and Moore (1970) reported that small peanut seeds lost viability faster than larger seeds, although initial germinability of the smaller seeds was also somewhat lower; a similar phenomenon in sunflower was noted by Tewari and Gupta (1981) and in various forage legumes by Gáspár et al. (1981). Particularly extensive data collected by Ostromecki (1977) from several lots of cabbage seed held under two different storage regimes also clearly indicate a tendency of smaller seeds to lose viability more rapidly.

There have been several indications that smaller soybean seeds may have better storage characteristics than larger ones, especially under humid tropical conditions (Verma and Gupta, 1975; Singh, 1976; Wien and Kueneman, 1981), although Carvalho et al. (1980) found no marked effect of size on resistance to an accelerated aging regime (43°C, 100% relative humidity). Minor and Paschal (1982) have speculated that the superior storability of smaller soybean seeds may be attributable to the fact that such material is less susceptible to mechanical injury during processing. Lesions of this type have been documented to limit longevity in other large-seeded legumes (Toole and Toole, 1960); Singh and Setia (1974) reported that soybean seeds that had ruptured or wrinkled coats showed an earlier decline in viability than those with perfect coats. In seed lots of the wild soybean (*Glycine soja*) smaller seeds also tend to be longer lived (Hoshino et al., 1960), but in this case a linkage with hard-seededness (see sec. 3.4.3) could be established. Smaller seeds of wheat have also been reported to preserve viability better in long-term storage (Efeikin, 1961). In maize, however, no significant effect of size on aging resistance (at 42°C, 100% humidity) was noted by Costa and Carvalho (1983), and Usberti (1982) found a minimal relationship between size and aging for peanuts in open storage in southern Brazil.

The specific gravity of seeds has been used as a means of quality selection for centuries. It is frequently true that individual seeds of low specific gravity within a partially deteriorated seed lot are found to be nonviable (Clark, 1904). Vaughan and Delouche (1968), who confirmed the validity of this

---

cultivar lost viability more quickly when it was immature. This assertion was apparently erroneous (Roberts, personal communication).

assertion with various seed lots of clover (*Trifolium* spp.), intimated that such differences in specific gravity may arise from respiratory depletion in the absence of a decrease in seed size. Data obtained by Górecki (1982) from peas, however, suggest that larger and denser seeds normally deteriorate more slowly, and there are indications that similar considerations may apply to yellow lupine and broad beans (Górecki and Jagielski, 1982).

A less well-defined influence on longevity may be the nutrient status of the mother plant during seed maturation. In a study of carrot, lettuce, and sweet pepper, Harrington (1960) suggested that seeds that had matured under extreme conditions of potassium or calcium deficiency lost viability faster in storage than those from plants that received complete nutrients. The data presented to support this contention were rather erratic, and further corroboration of the effects is needed. Likholat and Lyubarskaya (1983) have reported that the nitrogen status of the maternal plant plays a similarly important role. When the nitrogen supply was varied, seeds high (20–21%) or low (12–14%) in protein were produced. In accelerated aging treatments (40°C, saturating humidity) low-protein grains were far more prone to lose viability.

3.4.3 Hard seeds

Hard-seededness is particularly significant in agronomically important legumes and in a few commercially significant species from other plant families (e.g., okra, in the Malvaceae). Hard-seededness generally helps to extend longevity, although for planting purposes it is frequently considered an undesirable characteristic. The percentage of hard seeds within a seed lot changes gradually according to the storage conditions. Generally, drier conditions tend to encourage the trait, whereas high relative humidity and low temperatures promote softening (Justice and Bass, 1978); in open storage in temperate climates, softening usually requires several years.

It is well documented that some hard seeds can remain unimbibed in the presence of water for long periods. An interval of 20 years has been recorded for *Astragalus siniacus* (Kondō, 1936), 23 years for *Abutilon theophrasti* (Anonymous, 1933), and 25 years for *Trifolium hybridum* (Munn, 1948). Even longer periods (up to 40 years) have been recorded for *T. pratense* (Nobbe, 1919; Chmelar, 1946; see also sec. 6.2). Many studies on legumes have indicated that the hard seeds within a particular seed lot retain viability longer in storage than their softer companions (e.g., Harrington, 1916; Wahlen, 1929; Crocioni, 1934; Lute, 1940; Crosier and Patrick, 1952; Carpenter, 1969; Flood, 1978; Gō et al., 1979; see Fig. 17). Caution may be warranted, however; studies by Vaughan and Delouche (1960) on clovers and by Filimonov (1952, 1958) on clovers and other legumes indicate that rapidly

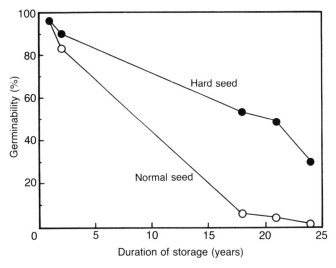

Figure 17. Longevity of hard and soft seed of *Lotus corniculatus* in open storage. Data of Crocioni (1934).

swelling seeds are usually inferior in quality in unaged as well as in aged seed lots.

Scarification of alfalfa seed before storage leads to a decrease in longevity (Graber, 1922; Battle, 1948), although this finding is presumably in part a consequence of mechanical injury. Scarification of *Pelargonium* seeds leads to similar decreases in longevity (Bachthaler, 1983). Within species, those lines or varieties that have a high incidence of hard-seededness tend to have superior storage characteristics, especially in humid environments. This has been well established in cotton (Christiansen et al., 1960; Patil and Andrews, 1985). Other studies have indicated a similar superiority for some lines of soybean with a tendency for hard-seededness stored under warm and humid conditions (Potts et al., 1978; Minor and Paschal, 1982).

3.5 Species differences in storability

Several summaries have listed the anticipated longevity of various commercially significant seeds under "normal" storage conditions. Harrington (1972), for example, has amassed numerous records and provided useful indications of storage life for cultivated species. Heydecker (1974) and Justice and Bass (1978) divided seeds into three broad groups according to relative storability, and Ullman (1949) provided a similar tabulation. Helpful summaries of many years of storage experience with crop seed collections in Leningrad have appeared (Khoroshailov, 1973; Khoroshailov and Zhukova,

1973, 1978), and useful contributions of a similar nature have been made by Stone (1930/33), James et al. (1964), Hafenrichter et al. (1965), Zelenchuk and Gelemei (1965), MacKay and Tonkin (1967), MacKay et al. (1970), and Szabó and Virányi (1971). Several extensive listings of flower seed longevity have also appeared (e.g., Anonymous, 1932; Goss, 1937; Nádvorník, 1949; Pidotti, 1952; Nesterenko, 1960; Bass, 1980b). Holmes and Buszewicz (1958) have collated much useful information on the storability of tree seeds, and other significant assessments of tree seed longevity may be found in the treatments provided by Schönborn (1964), Rohmeder (1972), Wang (1974), and the United States Forest Service (1974). Some of the more important surveys of seed longevity are summarized in an appendix to this volume.[6]

The general applicability of the life spans defined in compendia such as those cited above is rather limited, largely because absolute longevity data are apt to vary considerably from one storage station to another. Fortunately, the comparative or relative storability of species is less dependent on locality and some general trends can be defined. It is almost axiomatic, for example, that oats will remain viable much longer than rye under the same set of conditions. Thus oats may drop to 50% germinability in less than 10 years of open storage at one station (Bussard, 1935) and more than 20 years at another (Robertson et al., 1943), but at both localities rye deteriorates more than twice as rapidly as the oat seed. One recent attempt to collate some of the best comparative data on seed longevity is presented in the Appendix (Table A.9).

It has often been suggested that some relationship exists between the chemical composition of a seed and its storability. Commercial oilseeds, in particular, have developed a poor reputation. The relationship between some overall components of seed composition (protein, lipid, and starch) and storability (as defined by the P50 values given in Table A.9) is illustrated in Figure 18. There is no apparent correlation between protein content and storability (Fig. 18a), although there is a slight tendency for lipid-rich seeds to be relatively short-lived (Fig. 18b). Because of the preponderance of lipid in such species, their starch content is often relatively meager; consequently, some association between low starch levels and poor storability is evident in Figure 18c. A few high-lipid seeds store relatively well. Castor bean and tomato, for example, have oil-rich seeds (about 48% and 25%, respectively), yet both have good storage characteristics. It is important to note, however, that in some species a total compositional analysis may be misleading. In

---

6. Although not referred to in detail in this section, several surveys of seeds stored in less favorable climates, where excessive warmth and humidity tend to curtail longevity, have been published; e.g., Kondō, 1926; Sonavne, 1934; Rodrigo and Tecson, 1940; Akamine, 1943; Garrard, 1955; Abu-Shakra et al., 1969; Agrawal, 1980.

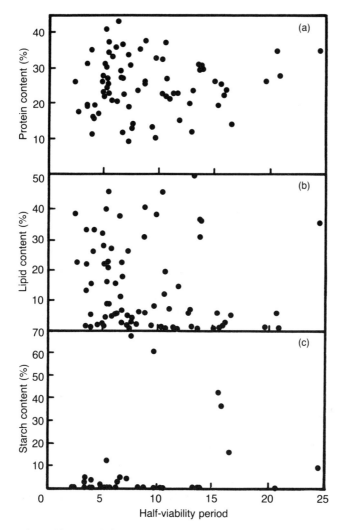

Figure 18. The relationship between seed composition and storability. Each point represents a separate species. Values for the half-viability period were obtained from Table A.9 in the Appendix. Data on protein and lipid content were taken from the tabulations of Spector (1956) and Gvozdeva (1971). Information on starch content is from Gvozdeva (1971) and from Mayer and Poljakoff-Mayber (1982). Additional compositional data are from lists provided by Earle and Jones (1962), Jones and Earle (1966), and Barclay and Earle (1974).

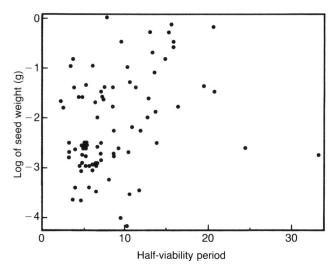

Figure 19. The relationship between seed weight and storability. Each point represents a separate species. Values for the half-viability period are from Table A.9 in the Appendix. Data on seed weight are from tabulations provided by the Association of Official Seed Analysts (1978) and from information published by Earle and Jones (1962), Jones and Earle (1966), and Barclay and Earle (1974).

cereal grains, in particular, lipid is highly concentrated in the embryonic region. Whole wheat kernels, for example, are about 3% lipid, but the embryo (germ) fraction contains on the order of 25–31% (Hargin and Morrison, 1980).[7]

An investigation of the relationship between seed size and storability reveals no very obvious correlation at the species level (Fig. 19), although smaller species tend not to be long-lived. In a survey of more than one thousand species (see sec. 4.1), Schjelderup-Ebbe (1935) reported that seeds that remained viable after several decades of dry storage all had diameters within the range of 1 to 7 mm, whereas very small seeds failed to survive. Why larger seeds should enjoy an advantage is not altogether evident, although Corner (1976) has noted that bigger seeds tend to possess elaborate and more massively developed coats, which may help to preserve viability (see sec. 3.4.3).

---

7. In his discussion of seed aging, Wiesner (1913) asserted that in addition to oil-rich seeds, species with pronounced tannin levels store poorly. This suggestion has apparently received minimal investigation, although the involvement of tannins in aging of lentil seeds has recently been proposed by Nozzolillo and De Bezada (1984).

3.6 Genetic effects on storability

3.6.1 Varietal differences in susceptibility to aging

There are clear indications, even below the species level, that seed longevity is in part genetically determined. Numerous accounts have suggested that some varieties store better than others under similar conditions. This has been held to be true, for example, in soybeans (Burgess, 1938; Mándy and Szabó, 1970a), *Phaseolus vulgaris* (Toole and Toole, 1953a; Mándy and Szabó, 1970a; Thomas, 1980; Roos, 1984), maize (Neal and Davis, 1956), rice (Jones, 1926; Dore, 1955; Madhava Rao et al., 1973; Radhakrishnan et al., 1976), vetches (*Vicia* spp.) (Mándy and Szabó, 1970b), melons (Bass, 1973b), proso and foxtail millets (*Panicum miliaceum* and *Setaria italica* (Mándy and Szabó, 1973), jute (Jain and Saha, 1971), alfalfa (Wilton et al., 1978), and wheat (Agrawal, 1979; Čuriová, 1984). Unfortunately, although some of these studies probably point to genuine genetic differences with respect to longevity, others no doubt are confounded by variability in growing or harvesting conditions; supposed differences in storability at the varietal level are easily confused with variation in stock quality (Beattie and Boswell, 1939). However, the investigations of varietal differences undertaken by James et al. (1967) on tomato, bean (*Phaseolus vulgaris*), pea, watermelon, cucumber, and maize are based on large samples and rest on statistically firmer foundations. This work indicates that significant, though rather small, differences in storability can be established between cultivars of each of the species examined. Similar variability of response within species evidently also extends to conditions of accelerated aging. According to Bourland and Ibrahim (1982), significant cultivar-by-aging interactions were noted in germination tests performed on cotton seed that had previously been exposed to 45°C and saturating humidity over several days, conditions that apparently favored fungal growth. Moreno-Martínez et al. (1978) also convincingly established differences between maize lines stored at 17% moisture content and 25°C for periods up to 120 days; in this case, enhanced longevity was clearly associable with resistance to invasion by pathogenic fungi of the genus *Aspergillus*.

As seeds of different varieties may be anatomically and compositionally dissimilar, it is perhaps unsurprising that susceptibility to aging should also vary. The effect of hard-seededness, in part genetically determined, has been discussed in section 3.4.3; but many other characteristics may exert an influence. The longevity of maize seed in particular has been reported to be highly dependent on its structural consistency. In a wide-ranging survey extending over nearly 20 years of open storage, relatively hard flint and dent varieties were found to remain viable longer than starchy and sweet forms (Yarchuk, 1966; Yarchuk and Leizerson, 1972). In closed storage, with seed moisture levels fairly constant, the differences between the various types of maize were much less evident. These observations suggest that "soft"-seeded maize varieties in open storage are probably more responsive to fluctuations in atmospheric humidity, and in consequence lose viability more rapidly (Yarchuk, 1974). This condition may extend to other species. Sahadevan and Narasinga Rao (1947), for example, suggested that rapidly deteriorating rice varieties in India were recognizable because they readily absorbed water.

The hypothesis that seed deterioration may be promoted by continual variations in atmospheric humidity, and hence seed hydration, is an old one; Blackman (1909) long ago suggested that "the ceaseless slight molecular changes involved in this process [might] slowly disorganize the viable protoplasm and in time cause the death of the seed." As Roberts (1972a) has pointed out, however, any deleterious effects of environmental fluctuations per se on seed viability have yet to be demonstrated. It should be noted that the relationship between longevity and seed moisture content (sec. 3.2.1) is such that a seed lot that experiences regular fluctuations in moisture content about a certain mean value should deteriorate more rapidly than another seed sample that oscillates with smaller amplitude but equal periodicity about the same hydration mean. In terms of the relationship between seed moisture content and storability, the "effective" level of seed hydration integrated over the whole storage period increases with the amplitude of the fluctuation, even though the mean temperature remains unchanged. There is therefore some theoretical reason to suspect that a limited advantage may accrue to seeds that are shielded from moisture changes or that respond only sluggishly to alterations in environmental humidity. This effect, though, is clearly unrelated to Blackman's concern over the cellular stresses imposed by unremitting fluctuations in the level of hydration.[8]

It has often been suggested that species other than maize may be similarly heterogeneous with respect to aging characteristics. The list is very diverse: Khoroshailov and Zhukova (1971), for example, intimated that varieties of *Linum usitatissimum* developed principally for oil (linseed) lost viability in storage faster than those cultivated primarily for fiber (flax). Within the taxonomically ill-defined *Delphinium* complex, Barton (1932) found that seeds of perennial plants lost viability more rapidly than those of annuals. Atabekova and Ermakova (1973) claimed that alkaloid-containing strains of *Lupinus angustifolius* were harder-seeded and longer-lived than "sweet" forms, although the data used to support this contention were rather variable. According to Gvozdeva and Zhukova (1971), chickpeas with black seed

8. A variant of this idea appears in a hypothesis developed by Aksenov et al. (1969) to explain the relatively pronounced longevity of starchy seeds. They propose that an embryo that is closely adjacent to a large body of starch (e.g., in cereal endosperm) will be buffered from rapid fluctuations in atmospheric humidity, and hence be longer-lived.

coats store better than lighter colored varieties, and Van der Maesen (1984) has also noted that pale-seeded "Kabuli" chickpeas are shorter lived than "Desi" types with thicker, harder coats. In addition, Starzinger et al. (1982) found that black-seeded soybeans were more resistant to storage under conditions of high humidity than pale varieties, a trait that was associated with a depression of fungal growth. Roos (1984) similarly reported that white-seeded lines of snapbean (*Phaseolus vulgaris*) deteriorated faster than colored ones under unfavorable storage conditions. Indonesian soybean varieties are also reported to store better than American ones, especially under tropical conditions (Wien and Kueneman, 1981; Russom, 1983), and this advantage is apparently not wholly ascribable to the color of the seed coat; indigenous soybean varieties evidently enjoy a similar superiority in India (Gupta, 1976). Adamova (1964) suggested that wrinkled peas lose germinability faster than smooth, and Khoroshailov and Zhukova (1973) found red wheat to be longer-lived than white wheat, and winter wheat longer-lived than spring wheat. Data of Van der Mey et al. (1982) also indicate that winter wheat stores better than spring wheat over periods of 15–20 years at 5°C. Others, however, have found no association between color and longevity in this species (MacKay and Tonkin, 1967), and Čuriová and Vlasák (1984) have indicated that varieties of spring wheat are far more stable in accelerated aging treatments (45°C, 95% relative humidity) than winter wheat lines. In several of the studies mentioned here, sampling sizes were evidently quite small. Further work is required to confirm that the trends reported are generally valid.

3.6.2 Genetic studies of susceptibility to aging

The genetic basis of susceptibility to aging has been best characterized in maize (*Zea mays*), although contributions to the problem have been relatively sporadic. This species has been reported to have several genes that can produce the *luteus* or yellow-seedling character independently. As Weiss and Wentz (1937) demonstrated, seeds homozygous for either the *luteus 2* or *luteus 4* genes suffered enhanced susceptibility to aging, and most died within two years under open storage. Seed material homozygous for other *luteus* genes displayed unimpaired longevity. The physiological basis for this difference remains obscure.[9] In unrelated work on maize, Lindstrom (1942, 1943) isolated a number of long- and short-lived inbreds. Reciprocal crosses indicated that the long-lived character appeared to be dominant, although a noncytoplasmic maternal plant influence could also be identified. Haber (1950) similarly investigated long- and short-lived lines of maize. Evidence for the

---

9. In pearl millet (*Pennisetum glaucum*), chlorophyll-deficient homozygotes have been reported to possess poor seed longevity in some studies (Burton and Powell, 1965), but not in others (Roos et al., 1978).

dominant character of the long-lived trait was obtained from an analysis of the storage characteristics of single-cross hybrids. More recently, Rao and Fleming (1979) performed studies on maize in which the same nuclear genotype was incorporated in genetically different cytoplasmic types. Seeds were stored for two years under normal conditions of room temperature and humidity in Georgia and then analyzed for germinability in the field. The results indicated a marked influence of cytoplasmic factors with respect to seed storability. The genetic basis of longevity in maize has also been addressed by Scott (1981), who used artificial procedures (42°C and saturating humidity) to select for seeds with pronounced resistance to aging. After three selection cycles, sensitivity to the accelerated aging procedure was significantly reduced. Whether such rapid procedures of recurrent selection will promote longevity under normal conditions of storage remains to be determined, but there is evidently good reason to believe that in maize, at least, genetic improvement for storability is feasible.

Genetic effects on seed aging have been explored to a limited extent in other species. In work on kidney beans (*Phaseolus vulgaris*), Nakayama and Saito (1980) used diallel-cross-analysis of homozygous lines to identify parental genotypes capable of retaining high germinability over seven to eight years of storage. In three separate analyses, one varietal genotype clearly transmitted superior storability properties to its progeny, and the genes that conferred pronounced longevity were of dominant character. From a study of seeds produced by crossing *Oenothera odorata* and *O. berteriana*, Schwemmle (1940) concluded that deterioration in storage was more marked in material that possessed one particular genetic complex. (In the specialized context of *Oenothera* genetics, a "complex" is a group of linked genes that is transmitted intact through an indefinite number of generations.) The physiological basis for this effect remains unclear. The heritability of seed longevity at the species level has also been defined in okra (*Abelmoschus* spp.) by Kuwada (1980), who studied an artificially created amphidiploid (2n = 192) produced from *A. esculentus* (2n = 124) and *A. manihot* (2n = 68). The new species displayed seed longevity characteristics similar to those of its *A. esculentus* parent and considerably superior to those of the other species involved. Autopolyploidy, on the other hand, is reported to decrease seed longevity in beets (*Beta vulgaris*). According to Perdok (1970), diploid seed is more resistant to aging than triploid, and triploid more so than tetraploid. Murín (1972), in contrast, claimed that increased ploidy in *Allium* was associated with lowered sensitivity to aging, although this conclusion was based on very limited sampling.

Recent work on the genetics of the storability of soybean seed is of considerable significance, as efficient use of this species in the humid tropics is greatly constrained by losses of vigor and viability (see sec. 3.2.2). Using genotypes selected for their storage characteristics (both superior and inferior), Kueneman (1983) was able to demonstrate by means of reciprocal crosses that the maternal plant exerted considerable influence on the longevity of F1 seeds. Kueneman assessed storability by using an accelerated aging regime (40°C, 75% relative humidity) that was designed to minimize fungal effects. The maternal influence identified presumably acts through characteristics of the seed coat, as this tissue is maternally derived. In subsequent analyses, Kueneman was able to distinguish a minor influence of the seed's own genotype on longevity, and suggested that any effect of cytoplasmic genetic factors on soybean storability was probably quite small. This work with soybeans, like that with maize, suggests that longevity, especially in the tropics, will be further improved by appropriate selection procedures.

Chapter 4

Surveys of the Longevity of Seeds Stored Dry in Collections

Systematic surveys of longevity based on collections of dry seeds have generally been of two types. The more useful—but usually less extensive—form of study is prospective, in which seeds of various species are stored under uniform conditions and tested at regular intervals to ascertain the limit of viability in each case. Results from several studies of this kind are summarized in the Appendix. For the most part, prospective studies have been limited to species of agronomic or horticultural significance. Generally, though, the broadest surveys of seed longevity have been retrospective: seeds that have been stored for different lengths of time in a particular collection are tested on a single occasion to determine whether any viability remains. Even if samples of the same species at different ages are available, variability in conditions of maturation, harvesting, and other factors frequently precludes any meaningful assessment of the decline in viability. Often in retrospective studies the material has not been dedicated to this particular aim from the outset of storage and it may have been exposed to unusual and possibly deleterious treatments. In herbaria especially, seed viability may have been compromised by contact with commonly applied insecticides, such as gaseous carbon disulfide or alcoholic mercuric chloride. Influences of this kind may well have a significant effect on the conclusions to be drawn from retrospective surveys. As will become apparent from the findings reviewed in this chapter, there is good reason to believe that long-lived seeds are frequently hard seeds. Nevertheless, when data related to herbarium specimens are assessed, it is prudent to bear in mind that hard seeds are probably better insulated from the potentially debilitating influences of applied chemicals

(Becquerel, 1907). Another persistent problem in surveys of longevity, especially those that consider wild species, is the possibility that seeds considered nonviable because of age were in fact merely dormant. In consequence, even the best surveys have generated rather imprecise data.[1]

4.1 Systematic surveys

Alphonse de Candolle (1846) has generally been credited with producing the first systematic survey of seed longevity. Germination tests were performed on 368 species of seed that had been sent to Geneva from the Botanical Garden in Florence 14 years previously, during which time the seeds had been stored "in a darkened room, secure from dampness and extreme temperature variations." Seventeen species from five families were found to germinate, including the Malvaceae (5 of 10 species), Leguminosae (9 of 45 species), Labiatae (1 of 30 species), Chenopodiaceae (1 of 8 species), and Balsaminaceae (the single species tested). The sample sizes that Candolle used for his assays were very small, and Ewart (1908) has argued that proper scarification would have increased the number of species that germinated. Nevertheless, Candolle's findings—which indicate unusually pronounced seed longevity among some members of the Malvaceae and Leguminosae—have been broadly confirmed by later workers. The nature of the species that survived storage also led Candolle to conclude that hard seeds tended to retain their viability longer.[2]

In 1840—six years before the publication of Candolle's pioneering work—an ad hoc committee of the British Association for the Advancement of Science was formed to investigate seed longevity. Seeds of known age were obtained from reputable sources and over the next 16 years germination tests were repeatedly performed on the samples to determine their longevity under conditions of open storage. The final report of the committee (Baxter, 1857) lists the longevities of 222 species determined according to this carefully planned scheme. The relatively pronounced life spans found among the Leguminosae and Malvaceae are again evident in the summary of data given in Figure 20. On the strength of a very limited number of observations, the Tiliaceae and Sterculiaceae can also be seen to possess species with considerable longevity. Both families are closely linked to the Malvaceae, all three being members of the suborder Malvineae.

---

1. A variety of taxonomic systems have been employed in papers surveying seed longevity. All data discussed in this chapter have been silently corrected to conform to the 12th edition of Engler's *Syllabus* (Melchior and Werdermann, 1954; Melchior, 1964).

2. According to Barton (1965), the following families are among those that are particularly noted for the incidence of hard-seeded species: Leguminosae, Malvaceae, Geraniaceae, Cannaceae, Chenopodiaceae, Convolvulaceae, and Solanaceae.

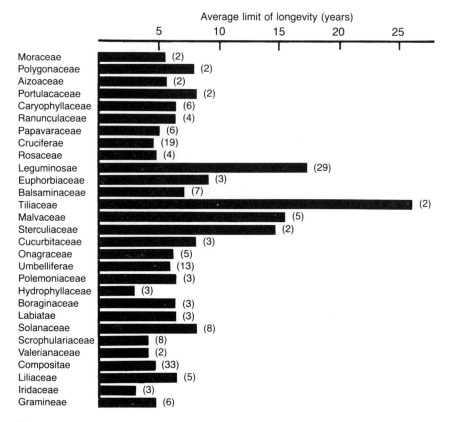

Figure 20. Results of a survey of seed longevity conducted by the British Association for the Advancement of Science (Baxter, 1857). This survey was designed to determine the extreme limit of longevity for species in open storage in the United Kingdom. Only those families are shown for which data on at least two species are available. Figures in parentheses represent the number of species investigated.

Becquerel (1907) investigated the germinability of 501 species of seed, with ages between 25 and 135 years, stored in the Museum of Natural History in Paris. Only 4 of the 45 families investigated in this retrospective study yielded germinable seeds: Leguminosae (16 of 85 species), Nymphaeaceae (3 of 5 species), Malvaceae (1 of 15 species), and Labiatae (1 of 17 species).[3]

---

3. There are minor differences in the paper between the tabulated data and Becquerel's own summary of it. The figures given here are taken from the tables. An earlier report of the same work (Becquerel, 1906) which summarized germination experiments on "about 550 species" displays similar internal inconsistency.

The three viable species classified under the Nymphaeaceae in Engler's system all belonged to the genus *Nelumbo*. Strictly, this number should be reduced to two species if taxonomic synonymy is taken into account. There is also a recent tendency to recognize this genus as

In further experiments, Becquerel removed the seed coats from some of the old viable seeds in order to investigate their permeability properties. He was unable to detect any permeation of air through the coats during a period of six months. Further, all the viable seeds failed to swell in water without scarification. He concluded that a characteristic of long-lived seeds was their effective isolation from the surrounding environment. Becquerel retested the germinability of a number of long-lived species 28 years later (Becquerel, 1934). Five species, all legumes, with ages ranging from 81 to 115 years, were still viable.[4] The oldest germinable seeds were of *Cassia multijuga* (Leguminosae), collected 158 years previously and analyzed for the first time in 1934.[5]

The bulky monograph on seed aging published by Ewart (1908) has had a remarkably prolonged impact on the field. In a single tabulation ranging over 176 pages, he attempted to summarize both his own extensive observations on seed longevity and sundry earlier records.[6] Unfortunately, the entries in Ewart's massive tabulation were of highly variable quality. Further, the 4,338 individual records were minimally organized; the sole concession to order was an alphabetization of the results by genus and species. Ewart's records came from three main sources. Of most interest are the germination tests that he performed on more than 600 species of seeds that had been identically stored over 50 years. He notes that

> shortly after my arrival in Melbourne, in a locked cupboard... two large packages of named seeds were found, including over 600 different sorts. The packets were... accompanied by a list... marked 'Seeds of Kew'. On further investigation it was found that when the University of Melbourne was founded, Baron von Mueller sent these seeds to Prof. McCoy for the University garden, but as the garden was not ready the seeds were placed on one side and replaced later by fresh sendings. The original 1856 sets of

---

entirely separate from the Nymphaeaceae, forming a monogeneric family (the Nelumbonaceae) in a distinct order, the Nelumbonales (Snigirevskaya, 1964).

4. There are further minor discrepancies between the data in this paper and that of Becquerel (1907).

5. Becquerel's tabulated data include a column for "probable longevity" as well as one for "determined longevity." Thus the seeds of *Mimosa glomerata* had a "determined longevity" of 81 years and a "probable longevity" of 221 years. The latter figure was estimated on the basis of the probability that the seeds, then 81 years old, would drop an additional 10 percentage points in germinability for every 28 years of further storage. In view of the great uncertainty surrounding Becquerel's estimate for the rate of viability loss and also the potentially misleading nature of the term "probable longevity," these appraisals, though widely adopted by later reviewers, are probably better forgotten.

6. Despite Ewart's concern to make his tabulation "as comprehensive as possible," he made no reference to the copious longevity records assembled by the British Association Committee 50 years earlier. It seems very unlikely that Ewart would have made a deliberate omission; presumably he was unaware of the committee's existence.

seeds became subject to Prof. McCoy's remarkable powers of collecting and storing material, and remained unopened and untouched in a dry, airy, dark cupboard... until the University's 50-year Jubilee in 1906.

The findings from these 50-year-old seeds are analyzed in greater detail below.

Ewart also conducted a series of tests on "about 200 sets of 10-year-old seeds" obtained from various sources and approximately 2,000 tests on material from "the National Herbarium, either from dated specimens or from stored dated seed which had been used in the past for exchange purposes."

The third component of Ewart's work consists of a motley collection of 1,372 records derived from earlier workers. Many of these records refer to seed buried in the soil or immersed in saltwater rather than stored dry. Several lack important details. It is difficult, for example, to see the usefulness of an entry for *Senecio sylvaticus* which gives the age of the sample as "indef.[inite]" and percentage of germination as "some."

In reviewing his data, Ewart arbitrarily divided seeds into one of three classes "according to their duration of life under optimal conditions." The longevity of microbiotic seeds did not exceed 3 years, mesobiotic seeds lived 3 to 15 years, and macrobiotic seeds "may last from 15 to over 100 years." This mode of classification has proved influential, although critics have repeatedly pointed out its most obvious failing: the storage conditions investigated by Ewart were far from optimal. Specifically, under more favorable conditions a microbiotic seed may show itself as mesobiotic, or even macrobiotic. Such criticisms are obviously valid; nevertheless, a system such as Ewart's, despite its undue simplicity, does serve as a broad relative measure of seed longevity under conditions of open storage in a temperate climate.

Ewart identified 180 macrobiotic species, most of them among his herbarium specimens. Almost all of the macrobiotic species were accounted for by three families: the Leguminosae (137 species), Malvaceae (15 species), and Myrtaceae (14 species). To these should be added the Nymphaeaceae (3 species), Labiatae (3 species), Iridaceae (2 species), Euphorbiaceae, Polygonaceae, Geraniaceae, Sterculiaceae, and Tiliaceae (one species each). The prominence of the Leguminosae in this list is to some extent a simple reflection of the large number of legumes that were tested. A more balanced comparison can be obtained by an examination of data for McCoy's "Seeds of Kew" collection—seeds stored for 50 years in Melbourne under identical conditions. Fortunately, the original data set can be recovered almost intact from Ewart's tabulation. In his listing there are 675 records, corresponding to 607 species, that refer to 50-year-old seed.[7] Inspection of this unique data

---

7. According to Ewart, McCoy's collection consisted of "over 600 different sorts." Some of the records for 51-year-old seed may refer to material from the same collection, but these records have not been included in the analysis that follows.

set shows that only 14 of the 607 species retained any viability after 50 years. Thirteen of these survivors were legumes—out of a total of 117 species of legumes tested. One of the 13 malvaceous species investigated was also still viable. A more detailed analysis of the legume species reveals a further trend that Ewart did not comment on. Of the 117 species of legumes tested, 26 are of Australian distribution (determined by reference to *Index Kewensis*). Ten of the 13 long-lived legume species were derived from this geographical group, the 10 species being divided among 9 genera: *Acacia, Albizia, Gompholobium, Goodia, Hardenbergia, Hovea, Indigofera, Kennedia,* and *Oxylobium*. In marked contrast, of the 91 species of non-Australian legumes tested, only 3—from the genera *Medicago* and *Melilotus*—retained viability. An obvious inference is that the occurrence of great longevity among the Australian legumes may indicate some special adaptive significance.[8] The two oldest viable species that Ewart recorded, *Goodia lotifolia* and *Hovea linearis* (105 years at the time of germination), are of Australian distribution, but this may reflect no more than Ewart's considerable reliance on Australian collections. As systematic studies of seed longevity after Ewart's time have paid rather scant attention to Australian species, no additional information is currently available. In considering the systematic implications of his survey, Ewart concluded that the incidence of longevity was especially marked among the Leguminosae because of a tendency within the group to hard seed coats and to relatively stable starchy seed reserves.

A systematic survey of seed longevity published by Schjelderup-Ebbe (1935) is the only one that even remotely rivals Ewart's in magnitude. His tabulated data incorporate findings from two quite distinct seed collections preserved in Oslo, Norway. The various categories of data are readily identifiable in his paper, and are probably best considered separately. The smaller group (the Blytt collection) consisted of 133 species in corked glass tubes. The age of the tested seeds ranged from 30 to 40 years, although the great majority were 37 to 40 years old. Among legume seeds, 5 of 13 species germinated; only 2 of 120 nonlegume species still remained viable. More interesting are the data for 1,039 species of the Moe collection, the seeds of which, preserved in paper bags, ranged in age from 40 to 117 years. Of the 167 legume species tested, 28 (17%) were still viable. Of 46 species of the Malvaceae, 12 (26%) germinated, and of 14 species of the Convolvulaceae, 3 (21%) were viable. This analysis by species is necessarily somewhat approximate, as some samples were identified only to the generic level and

---

8. Floyd (1966) states that seeds of several of these genera are well suited to environments that are subject to burning at irregular intervals (see sec. 5.4). A more prosaic explanation for Ewart's finding could be that his description of the seeds was incomplete. It seems quite likely that fresh locally grown Australian species may have been added to the "seeds of Kew" following a lengthy transshipment of the latter across the equator.

others were probably duplicated as a result of synonymy. Among the remaining 812 species studied, only 3 samples gave any germination.

Several longevity surveys of more moderate scope have appeared. Dent (1942) investigated the viability of 45 species after 20 to 62 years of storage at Dehra Dun, in Uttar Pradesh, India. Nine species of the Leguminosae (of 32 tested) and 2 (of 3) species of the Malvaceae were germinated. More recently, two longevity surveys resulting from analyses of Danish seed collections have been published. Jensen (1971) tested the germinability of 148 species from 39 families collected 43 to 66 years previously and stored dry in paper packets. Five species still retained limited viability, including two members each of the Leguminosae and Compositae and one member of the Geraniaceae. Subsequently, Buchwald and Jensen (1974) examined a 60-year-old collection of 148 species of seed stored in corked glass tubes. Five species germinated fairly feebly: two from the Leguminosae and one each from the Malvaceae, Chenopodiaceae, and Cruciferae. A relatively small-scale longevity survey, mostly of mountain plants, has also been published by McDonough (1974), who examined 36 species of 14 families after 41 to 44 years of open storage in Utah. Four families yielded viable material: one species each of the Compositae and Polemoniaceae, both species of the Labiatae tested, and 3 of 12 members of the Gramineae. The Leguminosae and Malvaceae were not surveyed.

Studies by Vainagii (1971, 1973) on 78 wild species of seed collected in the Ukrainian Carpathians and stored under laboratory conditions merit attention because of their unusual prospective character. Viability was assessed, usually at monthly intervals, for rather less than four years. The stability of the legume species (*Trifolium* spp., *Lotus corniculatus, Vicia cracca*, and *Anthyllis vulneraria*) was particularly marked. The superior longevity of legume seeds was also evident in a survey conducted by Hull (1973) on material stored in the western United States for periods of 14 to 41 years (mostly about 20 years). Of 39 species of the Gramineae, 18 (46%) still showed some viability, compared to all 10 legumes tested. Similar trends were also identified by Meshcheryakov (1957), who investigated 104 species of legume seeds in the collections of the Timiryazev Academy in Moscow. Most of the specimens, which included both wild and cultivated seeds, ranged in age from 20 to 84 years. Of 43 species found to have retained viability, all but one—a sample of alfalfa—showed some degree of hard-seededness. Pronounced longevity among legumes was also noted by Bakir et al. (1970), who analyzed multiple seed accessions of 46 range and forage species in a Turkish collection, some acquired 32 years previously but most aged 14 years or less. Of 22 legume species, 6 were at least 90% viable after 14 years storage, but this was true of only one of the 23 species of the Gramineae examined: *Sorghum sudanense*.[9] Only one survey has failed to reveal marked seed longevity among members of the Leguminosae: De Vries (1891), who tested 82 species of seeds after almost 17 years of storage at Leiden in the Netherlands, found that only 2 species displayed any viability, and none of the 11 legumes tested was germinable. This finding, though, clearly runs counter to the norm.

Turner (1933) designated the Leguminosae as a "macrobiotic family" because of the relatively high incidence of long-lived seeds within the group. Unfortunately, he extended the term to other families, such as the Compositae and Gramineae, for which the evidence is far less assured. In all, on very poorly developed grounds, he recognized 30 "macrobiotic families." Others have since followed Turner's example. Quick (1961), for instance, developed a shorter and rather more defensible list of families possessing long-lived seeds. Unfortunately, our knowledge of the systematics of seed longevity is so scanty that such attempts are more likely to deceive than to enlighten. A graphic example is provided by Sprague's appendix to the paper written by Turner (1933). Sprague attempted to test the hypothesis that species with macrobiotic seeds are preferentially represented in desert floras, a commonly held opinion (e.g., Rowntree, 1930; Went, 1957), reasoning that an area of highly infrequent rainfall may well lead to a development of this particular habit. Sprague analyzed the flora of the Atacama Desert in Chile and was rewarded by finding that 65% of the 414 angiosperm species belonged to one of Turner's "macrobiotic families." Turner's groupings are so broadly defined, however, that other floras of altogether different character would have given a similar result. If Sprague had examined the nearly 1,600 species currently recognized as forming the native flora of the British Isles (Clapham et al., 1962), for example, he would have found that about 64% belong to one of Turner's "macrobiotic families." Clearly, this concept is so debased as to be practically worthless.

The findings reviewed here justify the assertion that the Leguminosae and the Malvaceae include a scattering of species whose seeds show a marked tendency to longevity in the dry state. Some other families, such as the Myrtaceae, Labiatae, and Convolvulaceae, also include species that have this trait, but many more data are required before broad generalizations can be made. In a stimulating review of this question, Hanelt (1977) has pointed out that ecological factors tend to outweigh taxonomic considerations in determining seed longevity. The Sterculiaceae, for example, contain both

---

9. Seeds of *Sorghum bicolor* also have exceptional longevity. Thornton (1963) tested a seed lot of this species annually for 37 years, and found at the end of that time that its germinability was still around 90%. This behavior, however, is exceptional for members of the Gramineae.

the recalcitrant and short-lived seeds of humid tropical habitats and the long-lived seeds of extreme xerophytes. Hanelt has argued that ecological adaptation has a lesser influence on the incidence of longevity in the Leguminosae and Malvaceae. According to this view, hard-seededness—and hence pronounced longevity—is a primitive characteristic in these groups, of rather less adaptive value than in other taxa.

4.2 Records of extreme longevity

In theory, the longevity of seeds from herbaria or museums is determinable with considerable precision. Viable old seeds recovered from natural deposits or archeological sites, on the other hand, can be assigned only inexact dates, even with the assistance of radiocarbon measurements (see chap. 6). Unfortunately, there has been considerable confusion as to the identity and age of the oldest known seed germinated from a dated collection, despite the objectivity and precision that are supposed to prevail in such studies. In the two most recent reviews of this topic, Justice and Bass (1978) choose *Nelumbo nucifera* (237 years) as having the longest recorded life span, whereas Bewley and Black (1982) prefer to recognize *Cassia multijuga* (158 years).

As we shall see in chapter 6, there is convincing evidence of the extreme longevity of *N. nucifera* seeds (more strictly, dry fruits) recovered from the Pulantien peat deposit in China. For many years it has also been evident that seeds of this species are very long-lived in collections. Unfortunately, some doubt has arisen over longevity records for the oldest museum samples of *Nelumbo* because in two critical cases full details of successful germination experiments were never published, and our information about these tests is derived entirely from scattered secondary sources. The collation that follows is necessarily tentative: the strands of the story have become so hopelessly tangled through time that it is doubtful that any entirely successful resolution can now be achieved.[10]

Between about 1843 and 1855 Robert Brown performed a famous series of germination experiments on samples of *Nelumbo* seeds (both *N. nucifera* and *N. luteus*) from the Sloane collection of the British Museum, which had been assembled about 100 to 150 years earlier. Brown failed to publish details of this work, but he did preserve his germinated samples. Subsequently, Candolle (1895) and Ohga (1926a, 1927c) provided brief accounts of Brown's experiments, and Exell (1931) published a photograph of some of his germinated seedlings. All of these later authorities asserted that Brown's specimens were "about" or "more than" 150 years old, a figure that seems

---

10. I am particularly indebted to A. O. Chater of the British Museum for providing information on the *Nelumbo* specimens still preserved in the Sloane collection.

to have been more of an estimate than a precise dating. Brown's material was obtained from Sir Hans Sloane's collection of "Vegetables and Vegetable Substances," assembled between 1701 and 1749 (Dandy, 1958). As accession numbers run chronologically, reasonably precise dates can be assigned to many of the samples in this collection. According to Ohga (1926a), Brown used three specimens: no. 506, *Nelumbo nucifera*(?), about 1701–1705; no. 8110, *N. nucifera*, 1718 or earlier; and no. 8517, *N. luteus*, about 1727. The dates these samples were acquired have been provided by Britten (in Dandy, 1958) and Chater (personal communication).[11] The only precise information about Brown's experiments apparently comes from Britten, who noted that a seed of no. 8517 (*N. luteus*) germinated in 1848, yielding a minimum age of about 121 years.

In 1942 a single *Nelumbo* seed from the Sloane collection was successfully germinated by Ramsbottom. A rather opaque account of this event appeared at secondhand in *Nature* (Anonymous, 1942a). Bewley and Black (1982) chose to discount this observation because they were not satisfied that the seed was from the same old stock that Brown had used. This criticism seems unwarranted, however, because another source (Anonymous, 1942b) clearly indicates that Ramsbottom's sample came from an old receptacle previously illustrated by Exell (1931) in reference to Brown's work, and described by Exell as *N. nucifera*. This unnumbered receptacle is still preserved in the Sloane collection, although the year of its accession cannot now be ascertained with precision (Chater, personal communication). Conservatively, even if we assume the latest possible acquisition date for the receptacle (1749), the seed that germinated in 1942 would have been 193 years old.[12] Unfortunately, even the specific identity of this famous old seed seems to be in question. According to an editorial note by Polunin in Barton (1961), Exell later considered it possible that the sample was of *N. luteus* rather than *N. nucifera*. Certainly the two species of *Nelumbo* are highly vicarious and the fruits are easily confused.

---

11. Specimen no. 506 is described in the Sloane catalog as "Traffle Malaice. Herm. 2342." Paul Hermann, who died in 1695, specialized in the collection of East Asian plants. No. 8110 is entered as "Faba Aegyptiaca wt. the seeds in it. From P.," and was acquired from the collection of James Petiver, who died in 1718. The description for specimen no. 8517 runs "Lotus Aegyptiaca from Carolina." This latter accession appears in a photograph published by Whitehead (1981). (Information from Dandy, 1958, and Chater, personal communication.)

12. Justice and Bass (1978) supply an age of 237 years for this seed, thereby implying 1705 as its year of acquisition. This same date (1705) was also given by Godwin and Willis (1964a, 1964b), but with no obvious justification. The report in *Nature* (Anonymous, 1942a) states that Brown's seeds were "not less than 150 years old" (although the evidence of the Sloane catalog suggests that they may well have been younger) and that 87 years had passed since Brown's last experiments in 1855. The figure of 237 years apparently resulted from the addition of these two numbers.

Table 2
Germination of seeds stored in the dry state for 100 years or more. "Open" storage conditions include loosely closed containers that were not hermetically sealed. Two observations by Gérardin (1810) are absent from this list. His claim that seeds of *Phaseolus vulgaris* obtained from Tournefort's herbarium were still viable after more than a century is rather unlikely and has generally received short shrift (e.g., Becquerel, 1907). Gérardin's other assertion that seeds of *Mimosa pudica* remained germinable after 100 years of hermetic storage seems plausible, although the details he supplied are sparse.

Species	Family	Minimum age (years)	Germination	Storage location	Reference
Open storage					
Nelumbo	Nymphaeaceae (Nelumbonaceae)	193	1/1	London, U.K.	Anonymous (1942b)
Cassia multijuga	Leguminosae	158	2/2	Paris, France	Becquerel (1934)
Albizia julibrissin	Leguminosae	147	?/?	London, U.K.	Anonymous (1942b)
Nelumbo luteus	Nymphaeaceae (Nelumbonaceae)	121	?/?	London, U.K.	Dandy (1958)
Cassia bicapsularis	Leguminosae	115	4/10	Paris, France	Becquerel (1934)
Goodia lotifolia	Leguminosae	105	2/26	Melbourne, Australia	Ewart (1908)
Hovea linearis	Leguminosae	105	2/12	Melbourne, Australia	Ewart (1908)
Lotus uliginosus	Leguminosae	100	4/600	Kew, U.K.	Youngman (1951)
Trifolium pratense	Leguminosae	100	4/600	Kew, U.K.	Youngman (1951)
Closed storage					
Hordeum vulgare	Gramineae	124	15/75	Nuremberg, F.R.G.	Aufhammer and Simon (1957)
Avena sativa	Gramineae	124	8/82	Nuremberg, F.R.G.	Aufhammer and Simon (1957)

In addition to the famous seeds of the Sloane collection, other old museum samples of *Nelumbo* have proved to be viable. Becquerel (1907) germinated a 56-year-old sample of *Nelumbo luteus*, and Poisson (1903) demonstrated viability in a sample of the same species that was at least 54 years old. He also germinated a 42-year-old sample from India, presumably *N. nucifera*. In 1925, however, Ohga (1926a) was unable to obtain any germination from 11 fruits of the Sloane material he tested or from four seeds derived from a collection that was then only 87 years old. Dent (1942) found 62-year-old seeds to be nonviable, and Rees (1911) also obtained negative results with material that was then only 13 years old. Obviously some but by no means all *Nelumbo* fruits are destined for an extended career.

Table 2 provides a summary list of seeds that have been shown to germinate after a century or more of dry storage. The high incidence of leguminous seeds in this compilation is very evident. All of the records for seeds in open storage were established by herbarium or museum specimens, the only unusual entry among this group being *Albizia julibrissin*, which represents a spontaneous germination following fire and water damage to the British Museum collection in 1940. The results for barley and oats in sealed storage reported by Aufhammer and Simon (1957) demand to be considered apart from the other records in the table. According to their account, four undamaged glass vials were recovered from the foundation stone of a demolished theater in Nuremberg. Samples of barley, oats, wheat (various species), and lentils had been sealed in the four vials, each container having a volume no greater than 65 ml and holding about 20 g of seed. The exceptional preservation of viability in the barley and oats over 124 years seems to be due partly to the low water content of the seeds (7.3% and 8.0%, respectively); but darkness, the exclusion of air, and the generally cool environment also presumably made significant contributions. Seeds of the Gramineae are not generally noted for extreme longevity in the dry state.

At least one long-term storage experiment in progress will probably yield evidence of even greater longevity in seeds than any so far recorded. In 1949 Went initiated an experiment in which 98 species and varieties of the Californian flora were sealed into glass vials under vacuum (Went and Munz, 1949). Data collected over 20 years indicate that the seeds are extremely well protected (Went, 1969). Sufficient samples remain for regular testing throughout the next millennium.

Chapter 5

The Longevity of Seeds in the Soil

The longevity of seeds in the soil is of considerable importance to both agronomists and ecologists. Many buried weed seeds can remain viable for years, posing serious problems of control for the farmer. To the ecologist, on the other hand, the persistence of seed banks in the soil is a major component of the phenomenon of plant succession and plays a substantial role in the evolution of plant communities (Grime, 1979; Cook, 1980; H. A. Roberts, 1981). This chapter considers evidence for the survival of seeds in soils over periods of up to a century. Rare instances of quite extreme longevity in buried seeds are discussed separately in chapter 6.

Our knowledge of the way in which seeds persist in the soil is derived from two principal forms of study. One type involves an analysis of buried seeds recovered from undisturbed ground. If the previous vegetational history of the site is known, some estimate of the longevity of certain species may be attempted. The second form of approach, superficially far more objective, involves the organized burial of seeds for later retrieval and evaluation, mostly by means of a germination test. It is unfortunate that although the most interesting experiments have been performed on weed seeds from arable land, germination data in such cases are particularly likely to mislead, because the many and varied dormancy mechanisms associated with weedy species preclude any simple corresponence between germinability and viability. Indeed, after reviewing the copious literature on seed viability and dormancy, Kolk (1962) came to the disturbing conclusion that the great majority of studies on weed seed longevity were fundamentally deficient in this respect; a negative result in a germination test is of restricted usefulness when the longevity of such species is being investigated. This limitation inevitably applies to many of the data considered below.

5.1 Evidence from studies of seed banks

Not infrequently a plant will appear spontaneously in an unexpected locality. Such events have sometimes prompted the conclusion that the seed has lain dormant in the soil from an earlier time when a different vegetation prevailed. The older botanical literature is pervaded by such suggestions; in the last century particularly, large engineering projects stimulated numerous reports of the successful unearthing of supposedly long-buried seeds. Unusual plants that sprang from excavations for railroads (Penney, 1842), canals (Goiran, 1893), roads (Pugsley, 1928), and bridges (Aigret, 1909) have all excited comment. In many reports of this kind, enthusiasm often outran caution. Few commentators, for example, remarked on the large amounts of fodder that were presumably distributed over such sites in times when horses were widely used as a power source. Observations of newly dug graves (Melsheimer, 1876) and excavations under pavements (Ernst, 1876) and buildings (Arthur, 1882) also led to estimates of considerable seed longevity in the soil. Dredging of estuaries (White, 1882/84) and harbors (Salter, 1857) and even the salvaging of shipwrecks (Barrington, 1905) were all claimed to have resurrected dormant seeds. Doubtless in some such cases the interpretation is accurate; but for the most part reports of this nature defy objective assessment. It seems probable that many such instances are merely testaments to the powers of seed dispersal with which some species are peculiarly endowed.

Sometimes the manifestation of new growth at a disturbed site is so marked that there seems little doubt that plants have arisen from seeds buried in the soil. Such would seem to be the case for poppies and other agricultural weeds that appeared in abundant flushes on denuded battlefields during World War I (Hill, 1917; Shenstone, 1923). The generally accepted explanation of this phenomenon is that churning of the soil by the impact of shells, as well as the activities of trenching and grave digging, all led to an upward displacement of buried seeds from former wheatfields. Thiselton-Dyer (1889) similarly noted that poppies appeared in profusion following the Battle of Waterloo in 1815.

Unfortunately, a fairly sudden flush of plants is in itself no guarantee that they originated from buried seed. A classic instance is provided by the aftermath of the Great Fire of London in 1666, when Ray (1670) noted that London rocket (*Sisymbrium irio*) appeared profusely within the rubble of destroyed houses. Both Home (1757) and Candolle (1832) interpreted this phenomenon as evidence for the sudden germination of long-dormant seeds, formerly hidden within or under the walls. With modern hindsight it seems clear that the *Sisymbrium* was expanding in rather spectacular fashion to fill a newly opened ecological niche. After the burning of London in 1940, the closely related species *S. orientale* spread in comparable manner (Gilmour and Walters, 1973).

It has long been noted that destruction of forests may lead to the growth of an unusual flora from what was once the forest floor; one inference frequently made is that the new plants derive from long-lived seeds that have persisted in the soil from before the establishment of the stand (Dureau de la Malle, 1825; Fliche, 1905). Obviously, casual observations of this sort are insufficiently rigorous to support the suggestion, as rapid dispersal of seeds may provide an equally tenable explanation. The first careful studies of the seed flora under forests of known age were initiated by Peter (1893, 1894), who found that soils from forests of relatively recent date (20–46 years) that had grown up over formerly cultivated land still retained many seeds of agricultural weeds. From long-established stands of trees, however, only seeds typical of a forest environment were normally found. Summarizing the findings of numerous studies, he concluded that some seeds were able to survive in the soil for periods considerably in excess of half a century. Peter's pioneering contributions are as remarkable for their discernment as for the invective that was heaped upon them by his contemporaries. Cieslar (1904), for example, found Peter's suggestion of prolonged longevity to be "risky" and "not at all necessary," and Ewart (1907) derided Peter's reasoning as "childish in the extreme." Enough studies of a similar kind have since appeared to vindicate Peter's position completely.

Snell (1912) investigated soil samples from forests of known age growing on formerly arable land. The number of viable weed seeds recovered was smaller in stands 30 to 50 years old than in younger plantings, and soils from still older forests were devoid of such species. Oosting and Humphreys (1940) examined the soil flora at several sites in North Carolina where old fields were undergoing successional change to pine stands. Seeds of weed species generally decreased as the succession progressed, but a longevity of several decades was indicated in some cases. Broadly similar conclusions were reached by Livingston and Allesio (1968) when they investigated dated pine stands in Massachusetts; likewise, Guyot (1960) in a study of abandoned fields in northern France encountered evidence for the persistence of weed seed over 20 to 50 years. Brenchley (1918) found viable seed of agricultural weeds under 58-year-old ungrazed pasture at Rothamsted in England, and Chippendale and Milton (1934) encountered a similar phenomenon when they examined seeds under a 68-year-old pasture in Wales. Fivaz (1931), in an unusual study of the root balls of wind-fallen trees, concluded that *Ribes* seeds could remain viable for periods of up to 70 years, and Ødum (1969) adduced similar evidence for seeds of *Calluna vulgaris* preserved under 120-year-old beeches. In Japan, Kasahara et al. (1965) excavated a former rice paddy and concluded that viable seeds of several species, including a rush (*Juncus effusus*) and a sedge (*Cyperus difformis*), had survived for more than 50 years. More recently, Ødum (1978) has produced evidence from numerous sites in Denmark suggesting the persistence of ruderal species in the soil over several decades, and Hill and Stevens (1981) have described the survival of moorland seeds after about 30 to 50 years under forestry plantations in Wales and Scotland. In view of some uncertainties in dating, and especially because of the possibility of contamination, individual observations of this sort will always be open to dispute. Taken together, however, the mass of data indicate quite convincingly that seeds of many species may remain viable in the soil for half a century or more.

One important consequence of the extended longevity of some buried seeds relates to the question of species extinction. In several well-documented instances species that were considered lost to regional or national floras have subsequently reappeared, seemingly as a result of the germination of seeds preserved in the soil. Such is suggested to have been the case for *Cyperus fuscus* in the Nordhausen district of Germany (Wenzel, 1922) and for *Viola persicifolia* at Wicken Fen in England (Rowell et al., 1982). Walters (1974) presented particularly strong evidence for the reappearance of *Senecio paludosus* from buried seed following trenching operations in Cambridgeshire. The species had previously been considered extinct in Britain. Apart from such dramatic isolated recursions, seed longevity often has a key role to play in the survival of small populations of narrowly endemic plant species. Baskin and Baskin (1978) have emphasized that the seed bank in such cases increases the size and genetic diversity of the effective breeding population, thereby assisting in its preservation.

5.2 Buried seed experiments

The most venerable of all buried seed experiments was initiated by Beal in the fall of 1879 in East Lansing, Michigan (Beal, 1879). Fifty seeds of each of twenty species were mixed in moist sand and packed in stoppered glass pint bottles (0.47 l). Beal's selection of species included 19 common weeds and one example of crop seed, white clover (*Trifolium repens*). Twenty similarly prepared bottles were buried at a depth of about 46 cm, mouth downward, with the long axis inclined at about 30° to 45° from the vertical. For a century Beal and his successors have exhumed single bottles at regular intervals and examined the germinability of the seeds within. Over the first 40 years germination tests were made every five years, but thereafter bottles were recovered only once a decade (Beal, 1885, 1889, 1894, 1899, 1905,

1910; Darlington, 1915, 1922, 1931, 1941, 1951; Darlington and Steinbauer, 1961; Kivilaan and Bandurski, 1973, 1981). After 100 years of the experiment, six bottles currently remain.

The Beal experiment suffers from several limitations. Critics have noted that the mode of burial was unrealistic. In addition, at least one bottle in the series was found to be cracked (Darlington, 1941), so that water might percolate through the sample and perhaps unduly influence the result. Some experimental material appears to have been misidentified. Darlington (1922) noted the unanticipated presence of an extra species of *Amaranthus*; an additional and similarly unexpected species of *Verbascum* has also been germinated from later samples (Darlington, 1931; Kivilaan and Bandurski, 1981). More significant, Beal's sample sizes were fairly small and the difficulties associated with breaking of dormancy appear to have proved insurmountable on occasion. Beal himself was keenly aware of this problem: "The results have been far from satisfactory. . . . I have never felt certain that I had induced all the sound seeds to germinate" (Beal, 1910). Later workers have discussed this point at length on the basis of modern knowledge of the dormancy characteristics of some of the species concerned (Baskin and Baskin, 1977). Despite these cautions, Beal's experiment has proved invaluable in demonstrating conclusively that at least seven species are capable of retaining viability in the soil for half a century (Fig. 21). At the 100-year stage, three species were still viable: one seed of *Malva pusilla*, one of *Verbascum thapsus*, and 21 of *V. blattaria* were germinated (Kivilaan and Bandurski, 1981). As over 40% of the seeds of the latter sample were still producing seedlings, its demise still seems quite far off.

A more ambitious burial experiment, though of rather limited duration, was initiated by Duvel (1905). In 1902 he buried 107 species in pots at three depths (20, 60, and 107 cm) in Arlington, Virginia, and at intervals over the next 39 years the samples were exhumed and tested for germinability (Goss, 1924; Toole and Brown, 1946). About a quarter of the 107 species were common crop or pasture species; the remainder were weeds or wild species. As with Beal's series, it seems likely that lack of proper treatments to overcome dormancy may have led to underestimates of viability, especially in the earlier stages of the experiment. The decay of viability through time for the 107 species shown in Figure 22 may also be rather atypical of most seeds because of the high proportion of weedy species investigated. A summary of the findings by family is shown in Figure 23. Most of the relatively small number of crop seeds buried by Duvel failed to survive one or two years, presumably because they lacked effective dormancy mechanisms. Marked exceptions were tobacco, celery, and red clover (*Trifolium pratense*), all of which retained some viability until the termination of the experiment in 1941. The two most tenacious survivors were the weedy species *Phytolacca*

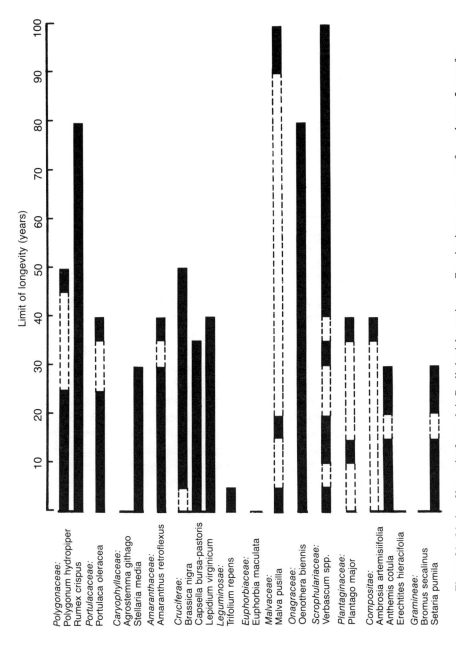

Figure 21. Limits of longevity for seeds in Beal's burial experiment. Germination tests were performed every 5 years for the first 40 years and every 10 years thereafter. Dashed lines indicate test periods when no positive germinations were noted. "*Verbascum* spp." indicates *V. thapsus* and/or *V. blattaria*.

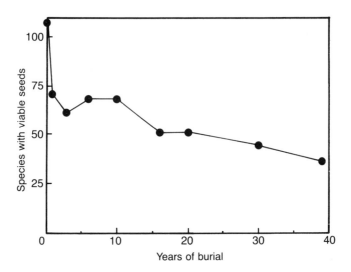

Figure 22. Results of Duvel's buried seed experiment. Plotted from data provided by Toole and Brown (1946).

americana, 81–90% of which still germinated after 39 years, and *Solanum nigrum*, which had maintained 79–83% germinability.

A third large-scale buried seed experiment was started in Denmark in 1934 (Dorph-Petersen, 1934/35). Seeds of 29 species, about one-third of which were crops and the remainder weeds, were buried in pots at a depth of 20 cm. Samples have usually been tested at yearly or five-yearly intervals. Results for the first 26 years of the experiment have already appeared (Kjaer, 1940, 1948; Madsen, 1962) and data for a further 23 years have been obtained, although they have not yet been published (H. A. Jensen, personal communication). As with Duvel's experiment, weed seeds generally far outlived crop seeds. After 25 or 26 years of the experiment, however, only four species still gave germination above 10%: lambs-quarters (*Chenopodium album*), field poppy (*Papaver rhoeas*), charlock (*Brassica kaber*), and *Vicia hirsuta*. As with the two earlier buried seed experiments, year-to-year variation in the Danish series was frequently considerable. Variability within species between stations is also often highly significant in buried seed experiments. Lewis (1958, 1973, in press) investigated the viability of 39 species of crop and weed seeds in two soil types: a mineral soil of fairly neutral pH and a highly organic, acidic peat soil. Most species deteriorated more rapidly in the peat. As in Duvel's experiment, crop seeds were particularly quick to lose viability. Generally, those species that survived four years of burial were found to persist through the full 30-year period of the experiment. Similar conclusions can be drawn from Goss's (1939) experiment initiated in California in 1932, when 12 species of locally significant weeds

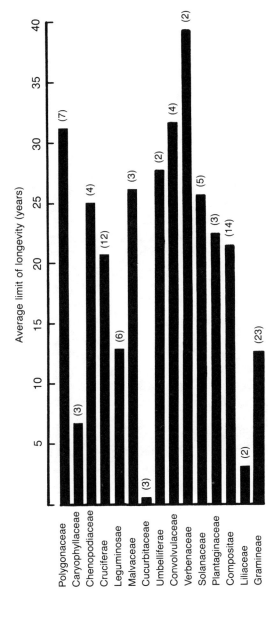

Figure 23. Results of Duvel's buried seed experiment, by family. Only those families are shown for which data on at least two species are available. Figures in parentheses represent the number of species used. The experiment was terminated after 39 years, when 36 species were still viable. Data of Toole and Brown (1946).

were buried. Most lost viability within the first five to ten years, although more than 50% of the hard-seeded bindweed (*Convolvulus arvensis*) still germinated in 1958 (Bruch, 1961).

When one reviews the body of data related to the longevity of seeds in soil, some trends are readily apparent. Most crop seeds fail to survive for more than a year or two, presumably because they are incapable of remaining dormant. In the case of wheat, for example, Kjaer (1940) found clear evidence that the seeds had germinated at too great a depth for successful emergence and had decayed in the soil. On the other hand, crop seeds capable of some dormancy may survive for several decades; volunteering of individuals from a previous year can even be a significant problem for seed producers concerned with varietal contamination, particularly for those who grow grasses and clovers (Rampton and Ching, 1966, 1970).

Weeds of cultivated fields can also be divided broadly into two groups: annuals that return large numbers of persistent seeds to the ground and so-called cultivated weeds that are sown as impurities in poorly cleaned crop seed (Kjaer, 1940; Salisbury, 1964). Seeds of the second group are generally much shorter-lived in the soil and evidently do not have appropriate dormancy mechanisms. Thus Salzman (1954) noted in a buried seed experiment in Switzerland that cultivated weeds like the corn cockle (*Agrostemma githago*), rough dogtail (*Cynosurus echinatus*), and chess (*Bromus secalinus*) all had very restricted longevity in the soil, whereas such tenacious weeds as the field poppy (*Papaver rhoeas*) and pennycress (*Thlaspi arvense*) were highly persistent over the 11-year course of the experiment. Similar trends are evident in data from a Canadian buried seed experiment conducted by Chepil (1946). Weedy species that lack appropriate dormancy mechanisms, such as Russian thistle (*Salsola kali*), showed very poor persistence in the soil; in upper horizons the seeds sprouted readily, whereas at lower depths they rotted before they could emerge.

The kinetics of seed viability loss in the soil have been explored by H. A. Roberts and his coworkers (Roberts and Feast, 1973a, 1973b). For many weedy species, the number of viable seeds tends to decrease in an exponential manner (Fig. 24), so that

$$S = S_o e^{-gt}$$

in which S represents the number of viable seeds in the soil at time t, S_o is the number of seeds in the original population, and g is a constant that varies with both species and burial environment (E. H. Roberts, 1972d). In Figure 24, the slope of the regression line yields a value equal to $-g$. Under conditions that promote the rate of viability loss, values of g increase. Distinct differences in g are evident between species exposed to the same environment (Cook, 1980).

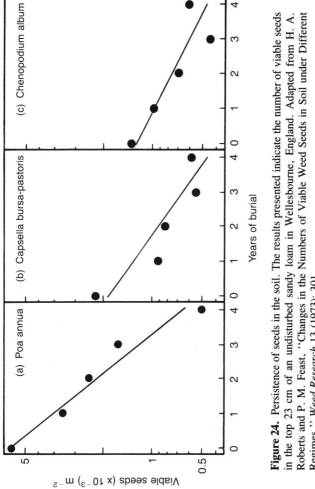

Figure 24. Persistence of seeds in the soil. The results presented indicate the number of viable seeds in the top 23 cm of an undisturbed sandy loam in Wellesbourne, England. Adapted from H. A. Roberts and P. M. Feast, "Changes in the Numbers of Viable Weed Seeds in Soil under Different Regimes," *Weed Research* 13 (1973): 301.

5.3 Factors influencing persistence of seeds in the soil

Seeds that survive in the soil for long periods fall fairly naturally into two groups. One is comprised of hard seeds that fail to imbibe until the resistance of the coat is overcome. Familiar examples of this type include hard-seeded legumes and some species of the Malvaceae. The other group consists of seeds that survive in a fully or partially hydrated state under conditions of restrained metabolism. The two survival strategies are not mutually exclusive within species; in some cases, after the impermeability of the coat is removed, dormancy of the imbibed seed may be maintained by some other mechanism. This is apparently the situation for such common weeds as *Rumex crispus* and *Chenopodium album* (Lewis, 1973). It seems unlikely, however, that the degree of hard-seededness in these weeds would provide an enduring barrier to imbibition; Kolk (1962) concluded that other dormancy mechanisms were probably of greater significance for the survival of these species. It is important to note that the difficulties involved in categorizing the seeds of weedy species are often compounded by a high incidence of somatic polymorphism (Harper, 1977). A single plant may produce structurally or physiologically quite diverse seeds, each type endowed with a distinctive degree of persistence in a given soil environment (Robocker et al., 1969). Marked genetic variability between geographically distinct populations of a single species can also influence longevity in the soil (Zorner et al., 1984).

As section 3.4.3 indicated, in such families as the Leguminosae the degree of hard-seededness is often highly variable within a seed lot. Within such a sample, relatively permeable seeds tend to germinate spontaneously in the soil, whereas the harder seeds undergo a period of softening before imbibition and germination occur (Lewis, 1958). Details of the softening process are poorly understood (Rolston, 1978), but in the well-studied case of seeds of snowberry (*Symphoricarpos albus*), the soil fungi evidently play an important role (Pfeiffer, 1934). Such environmental factors as warming and freezing may also be significant; slow temperature oscillations, entailing cycles of expansion and contraction, are thought to be especially important in softening subterranean clover (*Trifolium subterraneum*) in the soil (Taylor, 1984). Whatever mechanisms are at work, they are evidently effective in distributing germinations through time. Thus Winkler (1891), who observed the emergence of seeds of musk mallow (*Malva moschata*) from soil over 12 years, found new seedlings appearing annually as coat integrity was overcome. Extensive investigations of this phenomenon were reported by Kling (1931); Kling's data for *Vicia hirsuta* and bindweed (*Convolvulus arvensis*) are shown in Figure 25. In these species, softened seeds apparently either germinated rapidly or decomposed in the soil; as time progressed, only the most resistant seeds remained. Similar conclusions emerge from a 20-year study of mesquite

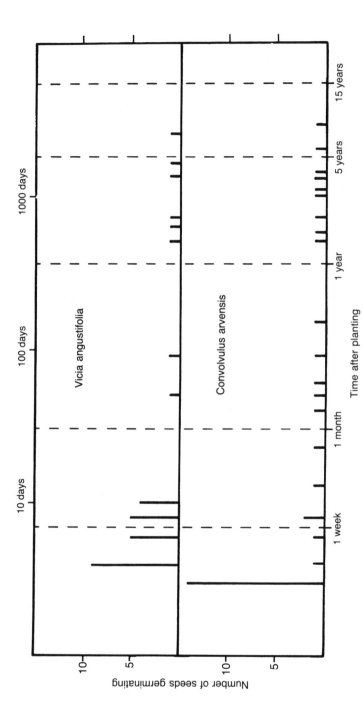

Figure 25. Germination of hard-seeded weeds. Fifty seeds of each species were planted in moist sand in a germinator. The germination of *Vicia angustifolia* was complete after 7 years, although 4 hard seeds remained in the sand for a further 14–16 years before decomposing (Kling, 1942/43). Data of Kling (1931).

(*Prosopis juliflora*), a troublesome leguminous weed of rangeland in the western United States (Martin, 1970). A revealing example of this phenomenon was also provided by Bier (1925a, 1925b), who investigated seeds of yellow lupine (*Lupinus luteus*) recovered after 55 years in the soil under a pine plantation. These seeds, which yielded a very high percentage of germination following scarification, were much smaller and harder than freshly grown material.

Seeds that survive in the soil for long periods in an imbibed state do so by remaining dormant, and their dormancy may be due principally to either external (exogenous) or internal (endogenous) factors. In the former case, germination is inhibited by the absence of one or more of the conditions essential for its initiation. A state of endogenous dormancy, on the other hand, is characteristic of many newly matured seeds, which will not germinate even when all the appropriate conditions are supplied. Schafer and Chilcote (1969) have proposed a simple equation to describe the status of a population of buried seeds at a particular point in its history. The sum total of seeds (S) can be partitioned into persistent seeds in a state of enforced dormancy (P_{ex}), persistent seeds with endogenous dormancy (P_{end}), seeds that have germinated in situ (D_g), and seeds that are nonviable (D_n). The status of the buried seed population may then be described as:

$$S = P_{ex} + P_{end} + D_g + D_n$$

With further splitting of the four classes, Roberts (1972d) generated an expression with eight terms to the right of the equality sign, although most workers have used the simpler version given here. Recent data published by Zorner et al. (1984) exemplify the successful use of the relationship to document the fate of a population of buried wild oat seeds (*Avena fatua*). The conclusion to be drawn from this and other studies (Roberts, 1972d) is that germination following loss of dormancy (i.e., the component D_g) is usually the major factor leading to depletion of weed seeds from soils.[1]

Some form of metabolic restraint seems mandatory for prolonged survival of seeds in a moist substrate, as respiratory rates characteristic of nondormant seeds would otherwise lead to a relatively rapid depletion in storage reserves. Respiration is often markedly depressed in imbibed dormant seeds (Sherman, 1921; Powell et al., 1983). There has been much speculation regarding the factors that impose dormancy on buried seeds. Darkness is an important control in many cases, but temperature and aeration also appear to have significant functions in some species (Wesson and Wareing, 1969a). A high

---

1. To add complexity to this deceptively simple analysis, there is good evidence that the dormancy characteristics of buried seeds fluctuate in a cyclical fashion according to season (Karssen, 1982).

level of carbon dioxide in soils generated by the activity of microorganisms may encourage dormancy (Salisbury, 1964; Wareing, 1966), as may lowered oxygen concentration. Bibbey's (1948) work on pennycress (*Thlaspi arvense*) and charlock (*Brassica kaber*) showed germination to be inhibited when oxygen fell below 16% or carbon dioxide increased above 5%. Similarly, Brown and Porter (1942) found that imbibed seeds of leafy spurge (*Euphorbia esula*), horse nettle (*Solanum carolinense*), and hoary cress (*Cardaria draba*) all became increasingly dormant as oxygen levels fell below about 10%. According to Kolk (1962), this may be generally true of other weed species, but some doubts remain as to whether such highly unbalanced atmospheres occur sufficiently widely in soils for them to be a significant source of control (Karssen, 1980/81).

Besides influencing dormancy (both P_{ex} and P_{end}), the peculiar characteristics of the soil environment may also affect seed longevity (especially D_n) in other ways. Many ruderal species, for example, possess lipid-rich seeds, and it has been suggested that low levels of oxygen in the soil help limit deleterious oxidations of the lipid reserves (Gola, 1906; Korsmo, 1930). The soil microflora must also exert a profound influence on seed preservation. In a study of wild oats (*Avena fatua*) over three years of burial, Kiewnick (1964) indicated that initial surface sterilization of the seed and steam sterilization of the soil both promoted longevity. In the absence of any control, mortality was much higher.

The depth at which seeds are buried has been noted to influence survival to a small but apparently significant extent in some experiments. Mechanical disturbance of the top layer of soil often leads to a depletion of seeds through loss of dormancy and germination (Karssen, 1982); but there are indications that even in undisturbed plots, increasing depth of burial exerts a stabilizing influence. In a six-year investigation of buried weed seeds, for example, Dorph-Petersen (1911) found germinability better preserved at a depth of 20 or 30 cm than at 8 cm. Further, according to Toole and Brown (1946), of the total number of seedlings to emerge in tests during the first 30 years of the Duvel buried seed experiment (see above), 27% came from seeds buried at a depth of 20 cm, 34% from 60 cm, and 38% from the lowest horizon, at 107 cm. Not all workers have noted a beneficial effect of deeper burial, however. In a 20-year experiment in a mineral soil in Wales, Lewis (1958, 1973) found little consistent difference in survival at 13, 26, and 39 cm, and Egley and Chandler (1983) likewise have found no marked association between depth and longevity during the course of a weed seed burial experiment in Mississippi. The factors that may assist in the preservation of viability at the greater depths are uncertain; nevertheless, greater stability of such influences is presumably assured at lower horizons. Deeper burial certainly seems to lead to enhanced dormancy in some species. Petersen and Lund (1944), for example, investigated 12 species of weeds (including *Chenopodium album, Brassica kaber*, and others) buried at depths of 0, 2, 4, 6, and 8 cm. At the lowest depth for all species, and at 4 and 6 cm for some, the seeds became dormant and did not germinate. Wesson and Wareing (1969b) similarly found that *Plantago lanceolata* and *Spergula arvensis* seeds became dormant when they were planted a few centimeters below the soil surface, apparently as a consequence of limited aeration. Taylorson (1970) also found depth and dormancy to be associated in the preservation of viability in weed seeds.

One seldom-considered factor influencing the persistence of seeds in burial experiments is the previous storage history of the material employed. Koch (1968) placed seeds of *Brassica kaber* and *Alopecurus myosuroides* in moist soil after various periods of dry storage up to five years (range of viabilities: 87–100%). Levels of seed mortality (D_n) were monitored over the next four months; high viability was maintained by samples of fresh seeds, but older material succumbed more readily. In *A. myosuroides*, in particular, four-year-old seeds were dead within three months in the soil, whereas fresh material still retained 78% viability.

Some comment is in order regarding comparisons in the literature between seed longevity in the dry state and in the soil. Reviewers have repeatedly suggested that some species survive longer in the soil, despite hydration levels that normally would be considered unacceptably high for conventional storage (e.g., Crocker, 1938; Wareing, 1966; Villiers, 1973). By extension, this is usually taken as evidence for the operation of some form of repair process within the imbibed seed in the soil. The evidence from controlled experiments for the existence of such mechanisms in seeds is strong, and further consideration is given to these data in section 7.11. A comparison of viability in the storeroom and in the soil, however, provides insufficient evidence for the assertion that repair processes are specifically responsible for a marked extension of longevity underground, as the two environments are so manifestly different in temperature, aeration, and microflora as well as moisture. If these uncertainties are laid aside, data available for some highly adapted weed species suggest that the germinability found in seeds recovered from the soil may indeed sometimes surpass that of seeds kept in dry storage (e.g., Lewis, 1973); but in other studies the trend has not always been very evident (e.g., Goss, 1924; Muenscher, 1935), and it may be wiser to refrain from any detailed comparison of the longevity of dry seeds with the longevity of seeds in the soil. It is sufficient to emphasize that many wild species have the capacity to persist for extended periods in the buried state.

5.4 Seed longevity in the soil: taxonomy and ecology

In an extensive review of the data from buried seed experiments, Koch (1969) noted that taxonomic affiliations are often evident. This is especially clear at the generic level: *Chenopodium album* and *C. hybridum* and *Rumex crispus* and *R. obtusifolius* provide obvious examples of pairs of long-lived species in the same genus. At the family level, Koch recognized that prolonged soil persistence was often associated with members of the Polygonaceae, Chenopodiaceae, Leguminosae, Solanaceae, and, to a lesser extent, Cruciferae. This conclusion must inevitably be tentative, as so few species have been surveyed for the trait. It is noticeable that of these five families, only the Leguminosae are widely recognized for marked longevity in dry storage (chap. 4).

Long-lived seeds can be considered to be dispersed through time as well as space, an obvious ecological adaptation. Seeds that persist in the soil in the imbibed state tend to germinate after the soil is disturbed or in response to a particular type of climatic fluctuation (Karssen, 1982). On the other hand, several reports have suggested that the germination of some long-lived hard seeds in nature may be specifically triggered by fire. Ewart (1908) and Cunningham and Cremer (1965), for example, noted the ability of seeds of *Acacia* to recolonize land devastated by bush fires in Australia, and *Rhus ovata* in the Californian chapparal apparently has similar characteristics (Stone and Juhren, 1951). Lying quiescent in the soil for long periods, and presumably scarified by the passing fire, seeds of this type germinate and rapidly reestablish the parental species. Observations of a similar nature have been made by Allers (1922) for long-lived seeds of yellow lupine (*Lupinus luteus*), by Dahlgren (1923) for *Geranium bohemicum*, and by Gratkowski (1973) for *Ceanothus sanguineus*.

Fire-related dispersal strategies have also evolved in some coniferous species, although seeds are retained on the tree rather than in the soil (Baldwin, 1942; Lotan, 1974). In such pines as *Pinus serotina, P. attenuata*, and *P. contorta*, fire prompts the cone to release seeds. Although precise studies are lacking, there are indications that viable seeds may be extracted from cones that are several decades old (Coker, 1909; Blumer, 1910), and it has frequently been assumed that the closed structure of the cone in some way enhances seed longevity. Crossley (1955), however, observed that seeds of *Pinus contorta* were unusually long-lived in open storage and suggested that the seeds' inherent longevity could be closely linked to a strategy of dispersal through time. Mirov (1946) also noted that seeds of closed-cone pine species were particularly stable in storage.

Seed longevity is also intimately linked to ecological features in other types of tree (Hanelt, 1977). At one extreme are the recalcitrant or very short-lived seeds of trees of primary rain forests; seedlings of such canopy formers are adapted to growth in dense shade and germination begins almost immediately upon dispersal. Occasional catastrophic destruction of the forest canopy, however, leads to the development of a secondary succession, composed of species with much longer-lived seeds. Generally the seeds of primary tree species are recalcitrant, whereas those of secondary species are orthodox (Whitmore, 1983). Hardwood trees in temperate zones also produce fairly brief-lived seeds, but this characteristic is not so strongly developed as it is in the tropics; in particular, seasonal inclemencies frequently necessitate the development of seeds capable of overwintering by means of complex dormancy mechanisms. There are notable exceptions: willows (*Salix* spp.) and poplars (*Populus* spp.) tend to flower and seed early in the growing season, and their propagules, which germinate almost immediately, are difficult to store (Zasada and Densmore, 1977). Another well-known example occurs within the genus *Acer* in North America. The silver maple (*A. saccharinum*) matures seed between April and June; these seeds tend to germinate with minimal delay, and their storage is problematic. The sugar maple (*A. saccharum*), in contrast, produces seeds in the summer months (July–September), and these seeds overwinter and germinate the following spring; sugar maple seeds can be stored under artificial conditions for several years (United States Forest Service, 1974).

The very hard long-lived seeds of the aquatic *Nelumbo* species (chap. 6) may have developed in response to the periodic drying of lakes or ponds. Toyoda (1974) has suggested that scarification—and eventual germination—may result from the cracking of the dark pericarp in hot sunlight. Extremely arid climates also seem to favor plants with long-lived seeds (Went, 1957). Faced with highly intermittent rainfall, such species spend most of their life cycle in the seed state, passing rarely and rapidly through their active phase only when conditions permit.

Chapter 6

Ancient Seeds

Reports of extreme longevity in seeds have never failed to excite curiosity, but unfortunately this area of seed research has become considerably burdened with overstatement and misconception. Before the advent of the radiocarbon method, datings rested on stratigraphic or cultural considerations and were rather imprecise at best. Even radiocarbon measurements can be deceptive unless the seeds themselves are examined. Dating of material or artifacts associated with seeds is not usually definitive, as confusion may result from previous disturbance of the site by animals, people, or other factors. A truly conclusive proof of viability in ancient material, therefore, involves germination and radiocarbon dating of the same seed (Godwin, 1968). Very few studies have fulfilled this criterion.

In investigations of ancient seeds, whether the material is living or not, some of the most useful information is normally lacking; rarely are the conditions of storage known with any degree of precision. The chronological age of a sample may be determinable with considerable accuracy, but this value alone is a poor measure of deterioration. Generally the best preserved seeds have been found either in very dry climates, in very cold ones, or in environments deprived of oxygen. Thus in the Egyptian desert or the arid American Southwest, ancient biological material often retains considerable structural integrity. Archeological sites in these two areas have provided abundant remains of finely preserved seeds, although germinability has always been long lost. This type of material is considered in section 6.1. Less well known but equally impressive are the samples recovered from frozen tombs in the Altai Mountains of Siberia (Rudenko, 1953), although here, too, no reports of viable seeds have yet been forthcoming. Evidence pertaining to germinable seeds obtained from permafrost is considered at some length in section 6.2. A third environment that leads to excellent preservation of ancient seeds, that of the anaerobic peat bog, is also discussed in section 6.2.

6.1 Seeds recovered from dry archeological sites

Seeds from archeological sites have provided a wealth of information concerning the lifestyles of ancient peoples and the evolutionary development of present-day crops. Studies have also described the anatomy and chemistry of very old material, giving added insight into the effect of long periods of time on seed structure and organization. Inevitably, as ancient crop seeds are never found to be viable, most of the obvious differences between very old and fresh seed are due to postmortem effects. At best such comparisons should serve to highlight those features of seed structure that are relatively stable and those that are comparatively labile.

A serious problem in interpreting the deterioration of very old seeds from archeological sites lies in the possibility that some were pretreated in deleterious ways before being deposited. In particular, great confusion arises in discussions of carbonized remains. Except in unusual circumstances, old seeds recovered from the soil of archeological sites are normally of a charred or carbonized appearance. Archeologists seem generally to be of the opinion that uncarbonized seeds are unlikely to be preserved in moist soil for more than a few hundred years at most. The further assumption is usually made that charring of seeds has been brought about by some human agency: overheating in grain piles, excessive parching of unthreshed grain before fires, cooking accidents, funerary practices, or simply the destruction of dwellings by fire (Helbaeck, 1963; Coles, 1973; Walker, 1973; Renfrew et al., 1976; Keepax, 1977).

Many plant scientists, on the other hand, have assumed that carbonization is a direct result of aging (Carruthers, 1895; Percival, 1936; Mercer, 1948; Lupton, 1953; Justice and Bass, 1978). According to this view, the deterioration of seeds in storage may be considered a form of very slow combustion, resulting finally in a charcoal replica of the original structure. Ancient seeds do often show a pronounced darkening (see below), and the conclusion has not infrequently been drawn that carbonization is an extreme case of the same phenomenon. There is some theoretical justification for this position. The oxidation (or combustion) of biological materials involves a decrease in free energy, so that the reaction should progress inevitably to equilibrium given sufficient time. It is also true, however, that the process usually occurs at an almost imperceptible rate at normal temperatures. Thus calorimetric analysis of Egyptian emmer (estimated age 3,700 years) yielded a value that was closely similar to that of modern wheat (Snyder, 1904). Further, in dry maize seed, 1,500 years of aging is insufficient to oxidize even the labile lipid dienes (Priestley et al., 1981), and unsaturated fatty acids have been detected in vastly older fossil seeds of Eocene and Oligocene age (Niklas et al., 1982). If these examples are representative, the complete carbonization of seeds by spontaneous oxidation is likely to be a very extended process indeed. Direct studies of old carbonized grains by electron spin resonance confirm that their state was induced by exposure to high temperatures (Hillman et al., 1983); similar conclusions arise from anatomical analyses of the surface of charred wheat (Tillez and Ciferri, 1954). Täckholm and Täckholm (1941) described a case in Egypt in which both uncarbonized and wholly carbonized seeds were found lying together in contact with uncarbonized wheat straw, and Renfrew (1973) similarly states that charred and uncharred grains are sometimes found together in the same storage pit. In sum, there is ample evidence to support the view that carbonized seeds in archeological remains were charred before or during deposition, but there is little to indicate that pronounced carbonization is ever a direct consequence of prolonged storage unless it is induced by overheating of grain piles. The distinction is important, as some germination studies and chemical assays of very old seeds have clearly been performed on charred material. Findings of this type probably contribute little to our knowledge of how seeds store over long periods of time.

6.1.1 Seeds of ancient Egypt

The facts related to the supposed germination of seeds from ancient Egypt—so-called mummy seeds—can be easily summarized. Many respectable scientists of earlier times, particularly in the first half of the nineteenth century, gave some credence to reports suggesting that ancient material could still germinate. Professional opinion in our own time is quite unanimous that such claims are entirely erroneous. Regrettably, the earlier view has become so deeply entrenched in popular thinking that it can probably never be successfully eradicated. One indication of this polarization of views is the automatic disinterest displayed by reputable scientists when they hear any new report of the germination of ancient Egyptian seeds. It is ironic, perhaps, that many professional accounts of this topic have been plagued by misconception and inaccuracy.

The seeds that have been recovered from ancient Egyptian sites are very heterogeneous; samples of wheat, barley, and pea have probably provoked most attention. The only two species of wheat that are significant in this historical context are emmer (*Triticum dicoccon*), cultivated in Egypt from the 5th millennium B.C., and durum (*T. durum*), which arose in the first millennium B.C. (Peterson, 1965). In addition to taxonomic considerations, there is a vast range of temporal variation, from the samples of remote predynastic times (fifth millennium B.C.) studied by Hallam (1973) to the grains that Percival (1921) assigned to the Ptolemeic period (first century B.C.). The best-preserved seeds are largely unaltered in outward aspect, except for a relatively slight color change. Well-conserved cereal grains are characteristically reported to have a reddish or brownish tint (Kunth, 1826; Gain, 1900; Whymper, 1913; Percival, 1921; Anonymous, 1935; Åberg, 1950). Not infrequently, though, ancient Egyptian seeds have been exposed to the influence of embalming fluids (Duchartre, 1867; Bernard, 1878; Whymper, 1913), which probably rendered them nonviable at an early stage. Still others are partially charred or completely carbonized (Duchartre, 1867; Bernard, 1878; Carruthers, 1895; Täckholm and Täckholm, 1941; Barton-Wright et al., 1944). Seeds altered in this fashion could never have been considered germinable at any stage in their storage career.

Reports of a number of negative germination assays performed on old Egyptian seeds have appeared in the literature, but it is not always evident whether the seeds succumbed to the effects of long-term storage or to some other cause (Unger, 1859; King-Parks, 1885; Anonymous, 1894; Snyder, 1904; Petrie, 1914; Wittmack, 1922; Turner, 1933; Caton-Thompson and Gardner, 1934; Anonymous, 1931, 1934; Täckholm and Täckholm, 1941). Similar concerns apply to samples of ancient Indian wheat and barley, tested for germinability by Luthra (1936). Since these seeds were clearly carbonized, the test was largely meaningless. Our present knowledge of seed longevity suggests quite unequivocally that healthy seeds of any of the species considered in these reports are most unlikely to have survived in open storage for longer than a few decades (see the Appendix). Museum and herbarium samples of cereal grains dating back no more than a few centuries, for example, have consistently failed to germinate (Wittmack, 1888; D'Africa, 1940; Simon, 1958; Bogdán et al., 1963). Ultimately it is evidence of this sort that provides the only basis on which we can refute the observations of those earlier workers who reported the germination of ancient seeds.

Four noteworthy accounts in reputable nineteenth-century publications sought to document viability in wheat from ancient Egypt. The most consequential of these accounts was probably that of the Count von Sternberg, dating from the 1830s, as the mythology of mummy-seed viability seems to stem initially from his observations.[1] The impact of Sternberg's report was considerable because of his undoubted prestige within the scientific com-

---

1. In 1830 the journal *Flora* carried a report of a viable onion bulb supposedly recovered from an Egyptian mummy (Anonymous, 1830). A purely anecdotal account of the germination of Inca maize collected from Peruvian tombs was published even earlier (Anonymous, 1824).

munity.[2] Commentaries on his experiments with ancient Egyptian wheat have almost always relied on a brief published abstract (Anonymous, 1835), although Sternberg's own account is fortunately still extant (Sternberg, 1836). In initial attempts to induce germination, he noted that "the grains decomposed after a few hours and were transformed into a milky fluid." Undaunted, "I consulted with my gardener, and we hit upon the idea of dipping them in oil, coating them with dust and planting them in holes about 2 inches deep in a pot containing earth . . . ; on the twelfth day the green sprouting tips of two grains appeared." The mature plants resembled Talavera wheat, a well-known Spanish form of *Triticum aestivum*. That some, at least, of Sternberg's sample was genuinely ancient is suggested by the fact that the seeds disintegrated when wet, an experience noted by others (e.g., Wittmack, 1873; Pierret, 1875). On reflection, the unbiased observer is forced to conclude that the gardener was probably the villain of this particular piece. A recollection of Wittmack (1922) suggests that some of Sternberg's contemporaries shared the same opinion: "Dr. Alexander Braun, who was present if I am not mistaken [at the meeting in Stuttgart at which Sternberg presented his findings], told us as students in 1865, that the gardener had wished to please the Count. When he saw that the mummy seed had not emerged, he planted a living grain in the soil and this accounts for its similarity to Talavera wheat."[3] Although alternative suggestions have been put forward to explain the mystery (e.g., Whymper, 1913), Braun's explanation rings sufficiently true to the original account.

A second incident of similarly unusual character occurred in England less than a decade later. Tupper (1840) reported on germination experiments performed with wheat and barley procured from a collection of Egyptian antiquities belonging to Thomas Pettigrew, and derived ultimately from excavations made by Sir Gardner Wilkinson.[4] According to a report of a lecture given by Pettigrew (Anonymous, 1840), Wilkinson found the seeds in a vase that was "hermetically sealed." Despite the unfavorable appearance of the grains, described as brown and shriveled, Tupper observed the germination of a single wheat seed. After careful tending, it gave rise to considerable progeny. Tupper's success was widely publicized and gained the influential

---

2. Kaspar Maria Graf von Sternberg (1761–1838), an associate of Humboldt and Goethe, was a founder of the National Museum in Prague. He is still considered as perhaps the preeminent paleobotanist of his time.

3. Braun's published comments on the matter (Braun, 1878) are couched in less explicit terms.

4. Sir Gardner Wilkinson (1797–1875) was a renowned explorer and distinguished Egyptologist. Thomas Pettigrew (1791–1865), surgeon and antiquarian, was a noted authority on mummification and suspended animation. Martin Farquhar Tupper (1810–1889) attained enormous popularity in the middle years of the nineteenth century as a prolific author of sententious doggerel.

approval of Lindley (Anonymous, 1843).[5] The eminent physicist Michael Faraday also displayed keen interest and presented a report of Tupper's work to the Royal Institution of London in 1842 (Tupper, 1886). The only possible explanation of Tupper's finding for the modern reader is that, by accident or design, fresh seed was added to the old sample. Coincidentally, perhaps, the wheat that germinated was characterized as a "semi-bearded Talavera" (Anonymous, 1843; Tupper, 1886), reminiscent of Sternberg's experience. Bewley and Black (1982) suggest that Tupper was the victim of a hoax, but a more innocent explanation is just possible. Many years after the event, Henslow (1860) noted that Wilkinson had also once provided Robert Brown with a sample of ancient seed for analysis. Among the genuine material, Brown encountered a few grains of maize, which Henslow suggested could only have been the result of accidental contamination. Tupper's sample may have become similarly adulterated with modern material during storage or transit.

A third example of the mummy-wheat phenomenon was afforded by the Marchese Cosimo Ridolfi in a communication to the Accademia dei Georgofili in Florence in 1853. The seeds in question had supposedly arrived in Ridolfi's hands through the agency of the British Museum. On reviewing the evidence in this case, Cifferi (1942) pointed to clear evidence of fraud, attributing guilt to the keepers of the museum in London. With hindsight, however, it seems probable that Cifferi's dogmatic opinion was directed by the temper of the times in which he wrote. An unprejudiced solution to this particular puzzle is likely to remain beyond reach, although the seeds that germinated were clearly not ancient.

The fourth and last example of any note was instigated by Guérin-Méneville (1857), who laid before the French Academy of Sciences remarkable yield figures for so-called Drouillard wheat. This variety was reputed to have originated from five disentombed Egyptian seeds germinated in 1849 by an associate of Drouillard de la Marre.[6] Details of the supposed germination are too meager to give any insights to the truth, although cupidity may have had a role to play. According to Guérin-Méneville, following the prodigious field trials on Drouillard's estates, this particular variety was sold to farmers at five or six times the normal price of the best seed corn. Thereafter, interest in the wheat seems to have declined even more precipitously than it arose (Becquerel, 1907). It is relevant to note here that a variety of *Triticum*

5. Bewley and Black (1982) have recently reprinted this article in facsimile, though without identifying John Lindley as its author. Lindley frequently failed to sign his contributions to his own journal. An extract of correspondence between Lindley and Tupper concerning the seeds was provided by Hudson (1949).

6. N.-M.-H. Drouillard de la Marre (1791–1856) was a prominent financier and landowner. His brief political career ended abruptly following prosecution on a charge of electoral corruption.

turgidum known as miracle or mummy wheat, characterized by compound ears, has been grown on a small scale around the Mediterranean for some centuries. Its origins are obscure, but it seems generally to have been regarded as a curiosity rather than a profitable crop (Candolle, 1883; Lenglen, 1949).

Within the scientific literature, the myth of viable mummy wheat was relatively short-lived. Unger (1846) cited Sternberg's work without question and Gärtner (1849) displayed similar credulity, but thereafter a succession of reviewers, including Duchartre (1867), Candolle (1883), de Vries (1899), and Griffon (1901), clearly documented the weaknesses of such claims. A pernicious variant of the wheat myth, the miraculous mummy pea, also enjoyed a considerable vogue. Potent criticisms of the pea fable were provided by Lindley (Anonymous, 1849) and Carruthers (1895), yet like the legendary wheat, this story seems to have gained a life of its own.[7]

6.1.2 Other ancient seeds recovered from dry archeological sites

Samples considered in this section have generally received attention of a less sensational nature than that which clings to old Egyptian seeds. Nevertheless, many pitfalls still abound for the unwary. In examining seed remains from Pompeii, for example, Wittmack (1903) found a few that were uncarbonized, including a remarkably modern peach stone whose authenticity he was forced to question. Similarly, Degen (1925) recounted a cautionary tale concerning recent seed of *Convolvulus arvensis* found to be contaminating an ancient sample of barley recovered from a Hungarian tumulus. With such precedents, skepticism becomes the first prerequisite for any assessment of extraordinary claims in this field. The nineteenth century proved to be the heyday for assertions of lingering viability in samples from Roman and pre-Roman Europe as well as from Egypt. A notorious New World canard concerning viable Inca maize was also championed in this same period by Kunzé (1881). Two celebrated instances of apparently inordinate seed longevity, one in France and one in England, provoked a good deal of curiosity in their own time (e.g., Darwin, 1855c) and still merit some mention.

The first case concerns a Gallo-Roman grave opened by the antiquarian Audierne in 1834 near Bergerac, France. Two quite independently written accounts of his findings have survived, both agreeing in almost all details (Des Moulins, 1835; Jouannet, 1835; see also Des Moulins, 1846). When the grave was opened, the skull of the occupant was found to be resting on a slab; directly beneath the skull a small hole, bored into the slab, contained

---

7. A relatively modern example (Anonymous, 1945) was subjected to a withering rebuttal by White (1946). Unfortunately, White's efforts at extermination were not entirely successful. The same story in somewhat borrowed garb recurs in Sturtevant (1951). A highly fasciated variety of *Pisum sativum* known as "mummy pea" has existed for some centuries (Compton, 1911), but it has no relevance to studies of seed longevity.

seeds, a funereal practice not apparently uncommon in that region. At least two species germinated in some quantity: *Heliotropium europaeum* and *Medicago lupulina*. *M. lupulina* has been known to retain viability over 50 years of storage in herbarium conditions in Melbourne (Ewart, 1908). Assuming that the temperature in the tomb was some degrees cooler, a longevity of several centuries is quite credible for this species, if not for the other. The details provided by Des Moulins (1835), however, suggest that the seeds were sprouting almost as soon as they were exhumed, and it seems quite possible that they were in fact modern seeds deposited in the tomb by an animal.

A rather similar discovery near the English town of Dorchester in the same year of 1834 was reported by Loudon ("Conductor," 1836).[8] During excavation of a tomb probably dating to pre-Roman times, a hardened mass of fecal matter was found preserved in situ among some human remains. Soon after excavation, part of the dried and matted conglomerate was turned over to Lindley, who found numerous raspberry seeds within it, many badly decayed, but with some superficially intact. From these seeds a few plants of wild raspberry were raised. Unlike the Count von Sternberg, Lindley took due precaution against the interference of overzealous gardeners, and his account is not easily dismissed. Lindley was an acerbic and discerning critic of those who sought to claim extreme longevity for seeds on trivial pretexts; his botanical qualifications are also beyond question. If he was duped by the antiquarian who provided the sample, then the hoax was indeed brilliantly contrived, both in the preparation of a specimen of authentic appearance and in the extensive affidavits of witnesses (Anonymous, 1855). However, it might be easier to believe that a few seeds of many hundred in the gut had retained viability over such an extended period if we had other good evidence of pronounced longevity for this species. Ewart (1908), for example, tested three different herbarium specimens (ages 18, 20, and 50 years) without finding any sign of viability.[9] In the absence of additional information, Lindley's case is likely to be judged not fully proven. Uneasy remarks by Hooker (1855) fairly summarize this and comparable incidents: "That some unknown and unsuspected cause of error may lurk in this case too, I am strongly inclined to suspect; and though I cannot point to any such, . . . I have been too often deceived myself, and have been sent after so many mares'-nests by honest but inexperienced observers, that I have learnt by experience to mistrust all paradoxes in science." Lindley's contemporaries focused their

8. Many extra details were contributed by Lindley almost twenty years later (Anonymous, 1855). See also Smith (1851) and Anonymous (1852).

9. In the same paper Ewart cites Peter (1893) as suggesting that raspberry seed survives in the soil for 100 years, but this supposition seems to result from a misreading of Peter's account.

doubts on the embarrassing possibility that he had accidentally contaminated the old seeds with fresh ones during public exhibition, a suspicion voiced by Henslow (Anonymous, 1860). Lindley evidently rejected this charge as completely specious.[10] The same accusation, but attributed to Hooker, appeared many years later in Petrie (1914). This latter account, however, is a confused conflation of two entirely separate incidents and is altogether untrustworthy. Further, Hooker was in all probability merely repeating family gossip: Henslow happens to have been his father-in-law.

The best-substantiated incident of longevity in a seed from an archeological site is that of *Canna compacta* (Cannaceae), found within a walnut shell rattle at a pre-Hispanic site in northwestern Argentina (Sivori et al., 1968). On the basis of cultural evidence and carbon dating of associated remains, an age of about 550 years was determined. Following criticism by Godwin (1968) that this measurement was too unspecific, the nutshell itself was eventually radiocarbon dated to A.D. 1362 ± 73 (Lerman and Sigliano, 1971). The *Canna* seed was therefore about 600 years old at the time of germination. The plant grown from it was somewhat anomalous in its development compared to modern material. Judging from photographs published by Sivori and Nakayama (1973), the emerging radical was normally georesponsive; however, not all rootlets of the developing seedling were positively geotropic and they tended to display precocious ramification. Further, the adult plant had an abundance of tillers (Nakayama and Sivori, 1968), and seeds produced by self-fertilization of this plant were also reported to show anomolous geosensitivity. The survival of the *Canna* seed over six centuries has been ascribed to the aridity of the site and the extreme hardness of the seed coat (Sivori and Nakayama, 1973).

Recently Spira and Wagner (1983) presented the results of tests carried out on seeds recovered from dry adobe walls of historic buildings in California and northern Mexico. The seeds had been collected from the adobe about 50 years before germination was attempted and had been stored in vials at room temperature in the interim. One seed of *Medicago polymorpha* (Leguminosae) germinated at an estimated age of 200 years. Several other seeds that did not germinate were reported to possess some measure of vitality on the basis of their reaction to tetrazolium.[11]

---

10. As an editorial insertion within Henslow's critical remarks (Anonymous, 1860) Lindley added, "This is quite a mistake."

11. A parallel can be drawn with the largely forgotten work of Lowe (1940), who analyzed the viability of seeds from the walls of a sod house in Kansas. Sod (or turf) was once used as a convenient construction material on the prairies, where wood and stone were scarce. Sod is less enduring than adobe; the seeds recovered and germinated by Lowe were no more than a few decades old.

6.1.3 The structure and chemistry of ancient dry seeds

Analyses of the chemical and anatomical properties of ancient uncarbonized crop seeds from dry archeological sites suggest that the level of structural organization may still be surprisingly good, although well below that needed for the retention of viability. Careful studies of the internal integrity of wheat and barley from ancient Egypt were first provided by Gain (1900), who examined sections of imbibed seeds of widely different ages (fifth to twenty-first dynasties, *c.* 2494–945 B.C.). The material he investigated was clearly in an excellent state of preservation, the only unusual external feature being a reddish-brown tint. On closer examination, however, the embryo was often found to be physically separated from the endosperm and tended to display a distinctly "resinous" coloration. Furthermore, within the embryonic radical, rows of cells were frequently separated by long cracks, suggesting a massive loss of structural integrity. The material was extremely fragile. Whymper (1913) published light micrographs of sections made from similarly well-preserved wheat (estimated date 1500 B.C.). The seeds investigated were described as fairly plump and dark rusty red (although smelling of embalming unguents), rather friable, and somewhat shriveled about the embryonic axis. In the endosperm, starch grains—but little else—were structurally well preserved. In seeds of this age, starch typically survives virtually unaltered. Its reactivity to iodine has been established for both wheat (Bonastre, 1828; Gain, 1900; Brahm and Buchwald, 1904; Miehe, 1923; Hallam, 1973) and pea (Unger, 1866), and its susceptibility to amylase is said to be unimpaired (Gain, 1900). In the study by Whymper (1913), individual starch granules lacked pitting, suggesting that enzymic digestion had been minimal.[12] Chemical analyses of the percentages of total fat, protein, starch, and soluble carbohydrates showed that they differed little from those of modern wheats. The ground meal was considerably more acidic, however, presumably because of hydrolysis of lipid to fatty acids. A similar increase in acidity was noted in prehistoric *Phaseolus* beans from Peru (De Palozzo and Jaffe, 1965). Barton-Wright et al. (1944) also reported pronounced acidity in Egyptian barley (*c.* 1350 B.C.) but this finding may be less meaningful, as the seeds were extensively carbonized. A sample of emmer wheat (*c.* 1400 B.C.) examined by Percival (1921) had recognizable aleurone and endosperm, but the embryo was almost completely disorganized. Similarly, he found that durum wheat dating to the first century B.C. displayed excellent preservation of starch in the endosperm. Cell walls, however, had generally disappeared, except in the aleurone layer, and the embryo was dark brown and shriveled, with little structure. Ohga (1951) reported examining a large quantity of grass

---

12. Starch has been reported to persist even in much older fossil seeds (Baxter, 1964).

seeds (either *Panicum* or *Echinochloa*) recovered from a pillow within a coffin belonging to a member of the Fujiwara family. The seeds, estimated at over 800 years old, were dark and without evident embryonic structure. Derbyshire et al. (1977), working on maize from the American Southwest (estimated to date from the first to the thirteenth century) found embryos to be discolored and deranged compared to the relatively intact endosperm and aleurone. Inca maize (*c.* 16th century) from Peru has been described in similar terms (Reiss and Stubel, 1887).

Deterioration of the embryo has also been reported in seeds of more recent age. In rye, a grain noted for its poor storability, samples dating to 1642 possessed badly disrupted embryos, according to Wittmack (1888). Kondō and Okamura (1938), reporting on the appearance of 100-year-old rice taken from a granary in Japan, likewise described the embryo as shrunken and dark, and Simon (1958) found wheat, rye, and barley to be similarly affected after more than 80 years in a seed collection. Gain (1901) investigated embryonic discoloration in a wide variety of seeds of the Gramineae, dating from modern times to 3,000 years earlier. Up to 120 years there was an increase in embryonic browning with age, the growing tips being particularly sensitive to modification. After about two centuries of storage the degree of browning remained fairly constant. Gain ascribed the browning to alterations in the lipids, as cereal embryos are relatively rich in oils. A more recent view is that embryonic discoloration is probably related to Maillard reactions between sugars and proteins (see sec. 7.5).

Ultrastructural studies of wheat embryos from Thebes (3000–2000 B.C.) and El Fayum (4441 ± 180 B.C.) were presented by Hallam (1973; Hallam and Osborne, 1974). Seed material used for these investigations, although brittle, was otherwise normal in appearance, without indications of carbonization (Hallam, personal communication). Cellular ultrastructure was remarkably well preserved; nuclei, with nucleoli and aggregated chromatin, and lipid bodies were especially evident. Membranes were considerably disrupted, however, suggesting total breakdown in cellular compartmentation, and the plasmalemma had completely retracted from the cell wall in places. Similar shrinkage of the cytoplasm was earlier noted by light microscopists (Brahm and Buchwald, 1904). Shewry et al. (1982) published a micrograph of Egyptian barley (1000 B.C.) in which they detected protein bodies squeezed between massively predominant starch grains. Derbyshire et al. (1977), in an examination of the aleurone layer in a maize sample of more recent date (first to thirteenth century) described endoplasmic reticulum, intact nuclei with dark-staining chromatin, and protein and lipid bodies.

Gluten, the mixture of storage proteins that enables leavening, has been found to be degraded in old Egyptian grain samples (Whymper, 1913; Percival, 1921). Recently more sophisticated studies have confirmed that protein structure has usually been markedly transformed in very old seed. Shewry et al. (1982) found that the amino acid composition of Egyptian barley (estimated dates: 1000 B.C., 1900 B.C., and 3000 B.C.) was still close to that of modern seeds. Arginine and lysine in all three samples and methionine and glycine in the two oldest were depleted but it is unclear whether the differences between the old and recent seeds are due to genotype or age. Despite the excellent preservation of amino acids, sodium dodecyl sulfate–polyacrylamide gel electrophoresis of the seed proteins gave rise to no distinct bands, suggesting a fairly complete modification as a result of the considerable age of the sample. Zeven et al. (1975), on the basis of starch gel electrophoresis, reported a similar degeneration of protein fractions from Egyptian emmer (300 B.C.) and Judaean barley (100 B.C.–A.D. 73). The hordein fraction from Egyptian barley (1000 B.C.), a heterogeneous mixture of storage proteins characterized on the basis of solvent extractability, gave no immunological reaction with antisera raised against modern hordein polypeptides (Shewry et al., 1982). De Palozzo and Jaffe (1965), however, working on Peruvian *Phaseolus* (dated about A.D. 1000) reported a very weak antigenic response of the storage protein fraction to corresponding antisera. Derbyshire et al. (1977) also found proteins to be rather well preserved in the endosperm of an old maize sample (first to thirteenth century). Storage proteins were resolved into subunits on sodium dodecyl sulfate–polyacrylamide gels. There was some similarity, although certainly not complete identity, with profiles derived from modern maize. The amino acid composition of the seeds was remarkably close to that of modern maize, but as in the study on barley conducted by Shewry et al. (1982), lysine and methionine were somewhat depleted.

The studies reviewed here give every indication that even in the best-preserved crop seeds from archeological sites, the conformational structure of most proteins has been lost over several centuries of storage. It is not surprising, therefore, that all attempts to demonstrate enzymic activity have failed (Brocq-Rousseu and Gain, 1908; Whymper, 1913; Miehe, 1923; Hallam, 1973; Derbyshire et al., 1977). Degradation usually extends to the lipid fraction. Priestley et al. (1981) investigated the lipids of maize seeds of various ages (the oldest was estimated to date from the first to the third century) obtained from sites in the American Southwest. Linoleic acid, although generally still present even in the most ancient samples, had undergone considerable atmospheric autoxidation. The overall level of lipid polyunsaturation was consequently far below that normally found in viable maize seed. To complete the picture of total physiological incapacity, Osborne et al. (1974) report that nucleic acid polymers are wholly lacking in ancient cereal grains.[13]

13. The culinary virtues of very old seeds have probably engendered almost as much

6.2 Ancient seeds recovered from moist or frozen ground

The difficulties inherent in determining the age of seeds taken from the soil have been noted in section 5.1. In the absence of a strict physical dating, any claim related to the unearthing of a viable ancient seed is liable to considerable suspicion. Extreme caution is certainly merited in such cases, as historically the imaginative license displayed by the hopeful discoverer has often led to quite startling claims.

In England it seems to have been something of a folk myth that application of fossiliferous limestone to fields led to the germination of weed seeds preserved within it (Mackenzie, 1841), and rather similar views even found their way into responsible journals. Kemp (1844), in a published letter to Darwin, recounted the viability of seeds deeply buried in glacial sands, and Lees (1851), in a breathtaking leap of the imagination, suggested that a living seed of the maritime plant *Glaucium flavum* had been recovered from the deposits of a long-dry Jurassic seashore now lying in the English Midlands. The same species also figured in a quite separate and much better-known incident. While collecting around cleared spoil heaps near ancient silver mines in Attica, Greece, Heldreich encountered specimens of *Glaucium* which he subsequently described as a new species, *G. serpieri* (Heldreich, 1873, 1876). On the basis of scant evidence, he suggested that the seeds had lain dormant for 1,500 to 2,000 years. Later opinion has not been kind to Heldreich's claim. Haussknecht (1892) relegated *G. serpieri* to the well-known *G. flavum*, and on floristic grounds Turrill (1933) found nothing to preclude the sudden appearance of modern plants of this species at the Attica station.

Many reports of a comparable nature occur in the older scientific literature but none withstand serious scrutiny. In modern times the most detailed and scrupulous studies have been those of Ødum (1965, 1974, 1978), who analyzed the viable seeds contained in old undisturbed soil horizons at various sites in Denmark and Sweden. Among his most provocative observations are the discoveries of viable seeds of *Chenopodium album* and *Spergula arvensis* taken from a depth of 120–146 cm (1,700 years?) at an Iron Age site and a single seed of *Verbascum thapsiforme* germinated from grave filling dated

---

misrepresentation as their supposed powers of germination. Analyses of Harris and Walster (1953) make it clear that even well-stored wheat loses its capacity to be made into bread over a few decades, primarily as a consequence of protein denaturation. Anecdotes abound nevertheless. Poncelet (1779) claimed, for example, that 132-year-old wheat was baked into bread on the orders of Louis XIV. (Another report by Loiseleur-Deslongchamps [1843] places its age at 155 years.) The king, whose taste in such matters was no doubt impeccable, is said to have judged the loaf to be excellent. More bizarre, perhaps, is an account of Tsountas and Mannatt (1897): extensive deposits of pea seeds were recovered during Schliemann's excavation of Troy and some attempt at pea soup was evidently undertaken. The palatability of this dubious concoction has passed unrecorded.

to the eleventh century and recovered from under a church floor. Other findings noted by Ødum indicate that occasional species may retain viability in the soil for a few centuries, and many others for decades. In view of the extent and meticulousness of his investigations, there can be little doubt that Ødum's conclusions are broadly correct; nevertheless, in the absence of direct radiocarbon dating, individual instances of seed longevity are open to challenge. All three of the species noted above, for example, have fairly small seeds (diameters less than 2 mm). In some of Ødum's studies up to 100 kg of soil was removed from open sites and screened for germinable seeds. Even with the greatest care, occasional contamination by modern seeds must be a distinct possibility. Additionally, as Ødum (1964) noted, in some cases seeds are probably subject to vertical transport through the soil. Darwin (1881), for example, commented on the ability of earthworms to carry seeds to lower levels, and there is also the possibility that recent material could be washed down burrows.

In relation to Ødum's claims for extraordinary longevity in *Verbascum* seeds, it is relevant to note that Garboe (1951) similarly adduced evidence for the successful excavation of viable seeds of *V. thapsiforme* from a monastery garden dating to the Middle Ages. Certainly it is evident from Beal's buried seed experiment (sec. 5.2) that seeds of *Verbascum* species may be remarkably well preserved even after a century in the soil. The exponential deterioration equation of section 5.2 describing loss of weed seeds from soil can be used to provide a very approximate estimate of persistence for two of the species investigated by Ødum. Given g values of 0.105 for *Chenopodium album* and 0.340 for *Spergula arvensis* (Cook, 1980), and assuming a fairly generous initial seed population (20,000 m^{-2}) and environmental conditions similar to those in the studies of Roberts and Feast (1973a), the following values are generated: For *Chenopodium album*, the viability of the original seeds will drop to 1 per m^2 after 94 years and 1 per 100 m^2 after 138 years. For *Spergula arvensis*, the values are 1 per m^2 after 29 years and 1 per 100 m^2 after 42 years. Increasing the initial seed population tenfold in such calculations extends persistence by only a few years in each case. Other extrapolations from short-term burial experiments suggest that some weed seeds may persist for longer periods, however. Burnside et al. (1981) used a rectangular hyperbolic function to fit nine-year burial data for various weed species. When the equation was solved for the number of years required for viability to reach 1%, smooth ground cherry (*Physalis subglabrata*) and fall panicum (*Panicum dichotomiflorum*) yielded nominal longevities in excess of 200 and 300 years respectively. Obviously, though, the reliability of such extrapolations is open to considerable doubt.

Uncarbonized seeds are not infrequently found in archeological horizons, even in humid climates. It has been usual to suggest that this type of material results from modern contamination (Keepax, 1977; Lopinot and Brussell, 1982); the opposing view, that such seeds may be genuinely ancient, has received less support (e.g., Kaplan and Maina, 1977). We may hope that the recent development of accelerator techniques for radiocarbon dating at the milligram level (Hedges and Gowlett, 1984) will help to remove doubts of this nature from future studies.[14]

The possibility that seeds may lie preserved in snow or ice for many centuries has long engendered considerable speculation. In the last century, Candolle (1870) suggested a search for seeds or other vegetable matter from interglacial times trapped under Alpine snows. More recently, Kjøller and Ødum (1971) screened old soil horizons in permafrost near Fairbanks, Alaska, but without finding evidence for especially long-lived seeds. On the other hand, Porsild et al. (1967) reported successful germination of arctic lupine seeds (*Lupinus arcticus*) recovered from frozen silt in the Canadian Yukon. The seeds were discovered by a mining operative 3–6 m below the surface in a former burrow. Associated with them were remains of the collared lemming, an animal no longer found in that region. By analogy to similar rodent remains elsewhere, Porsild and his colleagues suggested that the seeds were probably 10,000 years old. Unfortunately, no direct radiocarbon dating of the seeds (or even associated remains) was provided, and the assertion of the seeds' great longevity has in consequence been subjected to considerable criticism (Wester, 1973; Justice and Bass, 1978; Bewley and Black, 1982). It is true that as a hard-seeded legume, the arctic lupine would be a good candidate species for pronounced longevity. Further, seeds buried deeply in continuous permafrost would be in an excellent storage environment. In the American arctic, for example, seeds buried at 10 m would be exposed to a fairly unchanging temperature of around -8 to $-10°C$ (Price, 1972), although nearer the surface annual temperature oscillations would result in seasonally higher and lower values. Even for crop seeds, extrapolation from the equations of Ellis and Roberts (1980) suggests that at a constant temperature of $-10°C$ and at a low moisture content (about 5%), longevities on the order of many thousands of years are indeed conceivable. Unfortunately, in the absence of direct dating of the seeds themselves, the claim for the arctic lupine is unlikely to command widespread acceptance.

Section 4.2 noted that seeds (more strictly fruits) of *Nelumbo* have been germinated after 100–200 years of dry storage in museums. After considerable controversy, it now seems certain that seeds of *Nelumbo nucifera*, the

---

14. In addition to Ødum's claims, Heit (1967a) recorded an instance in which seeds of sumac (*Rhus* sp.) were recovered at a depth of about 1 m from a grave excavated by archeologists in North America. The seeds, which were about 80% viable, were estimated to have been buried "over 200 years," but further details are lacking. Hardness is a common trait among seeds of this genus.

Asiatic lotus, can also survive for some centuries in their natural habitat. The most significant example of this phenomenon was first described more than 60 years ago at a site in what was then Manchuria; but it is only recently that the antiquity of the seeds has been demonstrated beyond reasonable doubt. The presence of a peat deposit containing viable lotus seeds near the town of Pulantien (Pulandian, near modern Xinjin [Hsinchin], Liaoning province of China) was evidently first subjected to serious investigation around the turn of the present century. A few seeds were germinated by Japanese botanists and a description of the deposit was made by geologists of the South Manchuria Railway Company (Minami Manshū Tetsudo Kabushiki Kaisha, 1915). To the northeast of Pulantien a lake, about 1.5 km in diameter, had at some stage formed in a basin between hills. Growth of vegetation resulted in a thick layer of peat on the lake bed. Later the lake drained, leaving behind a peat deposit 0.3 to 0.7 m thick. Wind-blown dust (loess) from the Gobi Desert then covered the peat under an additional thickness of 0.5 to 1.5 m. Finally, a river draining across the basin had cut down through about 13–14 m of sediment, leaving the peat layer and its lotus seeds exposed high up on the river bluffs.

The site and the seeds were studied in great detail over a number of years by Ohga (1923, 1926b, 1926c, 1926d, 1927a, 1927b, 1935).[15] In attempting to ascertain the age of the peat deposit, he relied on several lines of argument. It was evident that the seeds were unlikely to have been of very recent origin, since *N. nucifera* no longer grew in that area. From documentary and monumental evidence, Ohga determined that the former lake had existed at least 160–250 years earlier. Furthermore, the circumference of poplar trees growing in the basin made an age in excess of 150 years seem mandatory. From the rate of erosion of the basin sediments, Ohga also estimated that it would have taken the river about 400 years to cut the bluffs then present. These pieces of evidence taken together pointed to an age of at least several centuries for the lotus seeds. This dating has proved highly contentious, although with hindsight Ohga's reasoning seems eminently sensible.

One extreme position on the Pulantien lotus was taken by Chaney (1951), who considered the peat layer to be almost unquestionably of Pleistocene age in view of the prolonged series of events that was needed to explain the present geology of the site. That this view was vastly overstated was revealed soon after when Libby (1951) performed a single radiocarbon test on a number of Ohga's seeds and arrived at a date of 1040 ± 210 B.P. (before present; by convention, "present" = A.D. 1950.) Subsequently, however, Godwin and Willis (1964a), using the same carbon sample, were completely unable

---

15. Ohga's earlier papers on the Pulantien lotus, with substantial additions, were reprinted collectively in Ohga (1927c).

to repeat Libby's dating. In their hands the specimen was indistinguishable in age from a modern carbon sample (100 ± 60 B.P.). Godwin and Willis (1964b) also put forward alternative geological arguments to explain the relatively rapid formation of the features noted by Ohga at the Pulantien site, and according to this assessment the lotus seeds might be of rather recent origin. In his review of the data then available, Wester (1973) disputed their conclusion, as it was clearly at variance with the historical and other evidence that Ohga had assimilated. In Wester's opinion, lotus seeds had probably accumulated in the former lake over many centuries before it drained about 400–500 years ago. Because of the scarcity of Ohga's collected material in recent years, additional data to help resolve these differences have not been immediately forthcoming. New determinations of age were eventually published by Priestley and Posthumus (1982), who obtained four seeds from a second collection made at the site in the early 1950s. They dated two seeds individually, one living and one dead, obtaining ages of 430 ± 100 and 340 ± 80 years, respectively. These determinations are broadly consistent with Wester's scenario; Libby's (1951) dating would then be valid for some of the older seeds in the deposit, whereas Priestley and Posthumus (1982) presumably examined some of the most recent material.[16] Subsequently, Shen-Miller et al. (1983) have reported a radiocarbon age of 705 ± 165 years for a single viable seed from this site.

Following his careful examination of the Pulantien deposit, Ohga (1935) reported the recovery of ancient lotus seeds from other Chinese sites, in Liaoning and Hopeh provinces. In Japan, because of the religious associations of this species, its seeds have been allowed to accumulate over many centuries in ponds and lakes belonging to temples. Ohga (1936) described the recovery and successful germination of apparently old lotus seeds from shrines in Kagoshima and Chiba prefectures and an additional site in Saitama prefecture, although their dates are again uncertain. Subsequently he reported on the germination of a single lotus seed from a cache contained within an unglazed jar from Namegawa (Chiba prefecture), estimated to be 1,200 years old (Ohga, 1951). A further deposit of viable lotus seeds at a depth of about 6 m in peat at Kemigawa (Chiba prefecture) was associated with the remains of a boat whose wood was radiocarbon dated to 3,052 ± 200 and 3,277 ± 360 B.P. in two separate determinations (Libby, 1955). Ohga (1953) suggested that the seeds might be somewhat younger than the boat, in the order of 1,500 to 2,000 years. The extreme age of the Kemigawa lotus has been disputed on the grounds that the seeds themselves were not dated (Godwin

---

16. According to Shen-Miller et al. (1983), in 1975 the Institute of Archeology in Beijing performed a radiocarbon dating on a sample of Pulantien lotus seeds of unknown viability and recorded an age of 915 ± 80 years.

and Willis, 1964b) and also because the published age antedates the probable introduction of the species to Japan (Wester, 1973). More recently, Toyoda (1974) has described the spontaneous germination of old *N. nucifera* seeds unearthed from deep mud at Gyoda (Saitama prefecture), which produced plants of distinctly different morphology from that of the modern lotus. Unfortunately, radiocarbon dating of the fruits themselves has been somewhat equivocal, although associated material has yielded an age of 1,390 ± 65 years (Toyoda, 1980).

We are fortunate in having available a number of anatomical and physiological studies performed on lotus seeds recovered from the peat at Pulantien, which give some insight into their pronounced longevity. All *Nelumbo* seeds, including those from Pulantien, are surrounded by a massively thickened pericarp that appears to be completely impermeable to water. Both *N. nucifera* and *N. luteus* fruits can be immersed in water for well over a year without swelling (Jones, 1928; Barton, 1965). Germination in these species is usually initiated by rigorous scarification (24 h in concentrated sulfuric acid or removal of the end of the coat by sawing), although in the natural condition gradual fissuring of the pericarp may trigger imbibition (Toyoda, 1974). In germination experiments with the Pulantien material extending over several years, Ohga (1930) found that a small percentage of seeds were capable of immediate germination without artificial treatment. The pericarp of the Pulantien lotus is characterized by the absence of an epidermis, presumably stripped away during its long sojourn in the peat (Ohga, 1926a), but otherwise it is as well protected as modern material.

Although the peat layer at Pulantien was drained when Ohga observed it, it presumably provided a highly reducing environment over a long period. Such conditions are excellent for the conservation of biological material (Erdtman, 1954). Helbaek (1958), for instance, described numerous seeds with a high level of structural integrity within the stomach of Grauballe Man (third to fifth century), whose corpse was preserved in a Danish peat bog. In addition to the anaerobic properties of the peat at Pulantien, it seems likely that the thick pericarp of the *Nelumbo* seed may also have served to restrict the access of air. Ohga (1926a) considered it likely that the pericarp was as impervious to oxygen as it was to water, and Toyoda (1958), working on Japanese lotus fruits, came to a similar conclusion.

According to Ohga (1923), the water contents in the embryo and endosperm were about 10% and 11% respectively, and following scarification he was unable to detect any evidence of respiration at such low levels of hydration (Ohga, 1926b).[17] About 0.15 to 0.2 ml of gas are normally trapped within

---

17. The suggestion made by Went (1974) that the seeds had remained viable "while fully saturated with water" is incorrect.

intact lotus fruits. According to Ohga (1926b), the composition of this gas in Pulantien fruits was about 18.3% oxygen, 0.7% carbon dioxide, and 80.9% nitrogen. He obtained very similar results with modern fruits, as did Toyoda (1958) with old and new lotus fruits from Japan. The most evident feature of these analyses is the buildup of carbon dioxide to over 20 times its concentration in normal air, although it is unclear whether this finding is due to true respiratory activity or some form of degassing. Whatever its origin, the carbon dioxide was evidently no more prevalent in ancient seeds than in modern ones. It is also apparent that the Pulantien lotus seeds maintained their viability despite prolonged exposure to oxygen.

The biochemical composition of the Pulantien material has not been reported in detail, but according to Blasdale (1899), *N. nucifera* fruits are characterized by a high starch content (45–57%), about 18–20% protein, and very little lipid (2–3%). Priestley and Posthumus (1982), who examined the fatty acids present in Pulantien and modern seeds, found high levels of polyunsaturation in both. Linoleic acid (36–49% of the total) was predominant in the old seeds and small amounts of linolenic acid (3–5%) were also present, indicating that they had undergone little oxidative degradation over several centuries. Ultrastructural studies of the lotus seeds have indicated a high degree of structural preservation (Hallam and Osborne, 1974; Zhukova and Yakovlev, 1976).

In his examination of many thousands of Pulantien lotus seeds, Ohga found very few that were non-viable. Further, he noted that the old seeds were generally more vigorous than recently harvested material (Ohga, 1926c). Catalase activity and respiratory rate in imbibed material were also higher in the old seeds. In an attempt to project the future longevity of the Pulantien material, he exposed seeds to various heating regimes (70–120°C) and noted the time required for loss of viability. Extrapolating his results to a temperature of 10°C (close to the annual mean in Manchuria), he predicted an additional 2,475 years of viability for the seeds (Ohga, 1927b). Although the precision of so radical an extrapolation may be doubted, the experimental data that Ohga collected at 70°C can be readily compared with the behavior of modern crop seeds. At this temperature the Pulantien lotus seeds completely lost viability in about 13.45 days (Ohga, 1927b). Their level of hydration was reportedly close to 10%. When the predictive equations of Ellis and Roberts (1980, 1981a) discussed in section 3.2 are employed, fresh crop seeds are found to compare very unfavorably with the Pulantien lotus seeds under the same conditions of elevated temperature and at a comparable level of hydration. Assuming 99% initial germinability, the time taken for viability to drop below 1% would be 2.4 days for cowpea, 1.5 days for chickpea, about 17 hours for barley, 10 hours for soybean, and 3 hours for onion. Clearly, under such extreme storage conditions, the fresh seeds of these species are considerably less stable than the ancient lotus seeds. Presumably the advantage enjoyed by the lotus fruit over the other species at 70°C would also be significant under less rigorous storage conditions.

Chapter 7

Morphological, Structural, and Biochemical Changes Associated with Seed Aging

The nature of the aging process in a desiccated system differs significantly from the forms of deterioration encountered in physiologically active organisms. A relatively dry seed, with a hydration level corresponding to sorption zones I or II (sec. 2.1), lacks a well-developed metabolism and is evidently incapable of extensive self-repair functions (sec. 7.11.1). Seed viability and vigor are dependent on the integrity of cellular macromolecules and the orderly compartmentalization of the cell; but in a seed stored in the dry state, degradation of these structural and functional elements will progress irreversibly as a result of adventitious chemical reactions or other forms of denaturation. Aging of a desiccated organism, therefore, involves an inexorable trend to disorder. Defense mechanisms innate to the seed—structural and chemical features that are characteristic of a particular species—may serve to limit the rate of this decay (Bartosz, 1981), but in the absence of imbibition, metabolism, and repair, the level of disorganization must steadily increase.

At some point, degeneration of a dry seed will have progressed so far that repair processes normally activated upon hydration are incapable of effective cellular restoration, and death will ensue. Hinton (1968) has usefully characterized the nature of the deteriorative process in desiccated systems: "All mechanical or other injuries sustained by an organism in a state of cryptobiosis are strictly cumulative.... [In a physiologically active condition] an organism can sustain a series of injuries over a span of time that, had they occurred simultaneously, would have caused its immediate death. In a state of cryptobiosis only the amount of damage is of concern and not the rate at which it is inflicted."

A wide range of studies have described potentially debilitating changes in seeds that can arise as a consequence of aging, but it often remains difficult to assess their significance. Investigators of seed deterioration have repeatedly looked for physiological or biochemical deficiencies at the subcellular level in attempts to associate these changes with loss of vigor and viability; often such accounts reveal an implicit assumption that the alteration observed is closely linked to the mechanism of seed deterioration, and even that it may represent the primary lesion responsible for aging. Much of the literature to be reviewed in this chapter attests to the popularity of such an outlook. Nevertheless, arguments based solely on association of phenomena are often inconclusive when one attempts to unravel a complicated web of cellular interactions; with most studies of this nature we are left no wiser as to whether the effect observed is a cause or a correlate of seed aging.

Rosen (1978) has persuasively argued that attitudes toward the study of aging have been influenced by our experiences with deteriorating mechanical or electronic systems. In such cases, the loss of total system performance is always associated with the failure of some specific localized component or subsystem. A consideration of the manifold systems and interactions present within a seed militates against the acceptance of such a simple hypothesis to explain aging; it is more probable that several important cellular systems become significantly degraded with age, so that inefficiencies in any one function amplify imperfections in others. Moreover, if different storage regimes favor particular types of cellular debilitation (sec. 3.2.3), the concept of a primary or principal lesion responsible for all types of seed deterioration becomes even less defensible.

Difficulties of interpretation also exist when observations are made at higher levels of seed organization. The vitality of the embryo, for instance, is obviously of basic concern when one assesses structural or biochemical changes induced by aging, as a seed that lacks a viable embryonic axis will not germinate; but the degeneration of other tissues is often of undeniable significance. One proposed vigor test, for example, considers the physiological status of the aleurone layer in cereals (Kietreiber, 1975; Perry, 1981). Other vigor assays measure cellular deterioration in terms of total seed leakage during imbibition (Perry, 1981; Association of Official Seed Analysts, Seed Vigor Test Committee, 1983), a parameter that is considerably influenced by the condition of nonembryonic tissues. Enhanced release of exudates is particularly important in the field, as it stimulates soil pathogens. From a practical viewpoint, therefore, both embryonic and nonembryonic tissues appear to be of importance in the evaluation of the level of deterioration.

Transplantation experiments with cereal grains—in which young or old embryos are implanted into endosperm tissue of different age—have been attempted on several occasions. Early use of the technique by Carruthers

(1911) confirmed the importance of embryonic viability for seed germination, and subsequent workers have attempted to use this protocol to unravel the relative significance of embryonic and nonembryonic tissues in loss of vigor.[1] Kikuchi (1954) found some indication that when young wheat embryos were transplanted into deteriorated endosperms, they grew less vigorously than control implants. Floris (1966, 1970) reached similar conclusions with durum wheat, although the condition of the embryo was evidently of greater significance than that of the endosperm in determining performance of the whole seed. In rice, aged endosperms were found to retard the growth rate of vigorous embryos considerably (Mandal and Basu, 1981). Slesaravichyus (1971), working with rye, on the other hand, found the level of endosperm deterioration to be of no influence, the condition of the axis alone determining embryonic performance; similar conclusions were reached by Harrison and Perry (1976) when they investigated barley. Using a different mode of analysis, Aspinall and Paleg (1971) concluded that the vigor of isolated wheat embryos could not be restored simply by nutrients, but other evidence (sec. 7.10.1) led them to conclude that deterioration of the aleurone was nevertheless partly responsible for a depressed rate of germination. On balance, embryo manipulation experiments suggest that aging of the nonembryonic tissues of cereals may exert a minor but often significant influence on the rate of germination.

Topographical viability testing by means of tetrazolium salts and other agents (sec. 7.6.2) has generally confirmed that function is lost heterogeneously throughout the seed (e.g., Germ, 1956; Banerjee, 1978; Purkar and Negi, 1982). In a particularly extensive examination of 41 species aging under a variety of storage conditions (5–25°C, 20–83% relative humidity), Bulat (1963) was able to demonstrate that each species was characterized by a particular mode of deterioration. In many cases, necrosis was first detected in the embryo, particularly around the radicle tip; but in onion (*Allium* sp.), for example, the nonembryonic tissues died while the embryo remained fully viable. Patterns of necrosis also tended to display family affiliations. In a group such as the Leguminosae, the embryonic axis was generally found to deteriorate at a faster rate than the cotyledons, the radicle tip being the most sensitive area.

In chemical or biochemical studies of aging seeds, some precision is gained by separately analyzing particular structures—the axis, endosperm, and co-

---

1. The debilitating influence of the aged endosperm has been viewed from two quite separate perspectives. Some workers have proposed that deteriorated endosperm tissue may contribute mutagens to the axis, threatening its integrity (sec. 7.8.3). Others have considered that a failing supply of nutrients from the endosperm may depress embryonic growth. At present there is little evidence to support the former contention (Corsi and Avanzi, 1969; Slesaravichyus, 1971).

tyledons, for instance. Each of these structures is in itself distinctly heterogeneous, however; thus the evidence of topographical viability testing clearly suggests that various parts of the embryonic axis deteriorate at different rates. Similar conclusions arise from autoradiographic analyses of ^{3}H-thymidine and ^{3}H-uridine incorporation into different areas of aged embryonic tissue (Innocenti and Floris, 1979), and Innocenti et al. (1983) have reported that relatively vigorous parts of the embryo from nonviable durum wheat seeds can be excised and cultured to form plantlets *in vitro*. Practical difficulties inevitably dictate that much of this heterogeneity of response among tissues with respect to aging is lost in conventional biochemical investigations, and deteriorative changes are usually integrated over a diverse array of cell types within a particular seed structure. An additional complication arises because compositional and enzymic features are sometimes found to vary considerably between peripheral cell layers and areas deeper within the seed structure (Barber, 1972). Complexities of this nature constitute yet another limitation to conventional modes of analysis.

A final caveat to be added to this list of cautions concerns the imprecision that can arise in studies of a diverse seed population. Seeds in storage do not die simultaneously; rather, individual seeds within a sample age at different rates. Consequently, a partially deteriorated seed lot may span a spectrum of debility, including seeds of high and low vigor as well as those that are no longer germinable. The prevalent assumption in studies of seed aging is that population effects mimic alterations at the level of the individual seed. In many cases the supposition is no doubt valid, but it is important to recognize that it provides one more constraint on the interpretation of chemical and biochemical studies. Postmortem effects are not of primary relevance to our understanding of seed aging, yet their contribution is undeniable whenever seed lots that are less than fully germinable are analyzed.

7.1 Morphological changes

The features of the testa (or the pericarp of dried fruits) are among the most visually evident characteristics of stored seed, and in many species the coat darkens or colors with age. Although it is unlikely that these alterations are closely linked to diminished performance, they have long been used as a simple index of seed quality. Before the days of organized seed testing, for example, when the price of clover seed was often established by its color, Girardin (1848) recorded an ingenious abuse whereby old, dark seeds were bleached with sulfuric acid to make them more marketable.

Legume seeds are particularly prone to display changes in storage: apart from the development of brown, red, yellow, or purple coloration in clovers (e.g., Jenssen, 1879; West and Harris, 1963; Vaughan and Delouche, 1968), darkening occurs in *Vicia faba* (Garner and Sanders, 1935; Lowig, 1970), *Phaseolus vulgaris* (Hopkins et al., 1947; Barton, 1966a; Grange et al., 1973), *P. coccineus* (Lowig, 1970), peanuts (Marzke et al., 1976), and soybeans (Friedlander and Navarro, 1972; Saio et al., 1980), among others. Filter (1932) provided a list of 33 species of legumes that he found to darken with age, and a further list of 12 species that did not. Nonleguminous species may also exhibit browning: seeds of *Lepidium sativum* (Lowig, 1970) and lettuce (Harrington, 1973) are among many that suffer such effects.

In some cases, discoloration of the seed coat evidently results from oxidative reactions (e.g., Marzke et al., 1976), which are promoted by elevated temperature and humidity (Hopkins et al., 1947; Hughes and Sansted, 1975). Color change of legumes in particular, however, can be quite heterogeneous within a seed lot. There are indications that those seeds that maintain their original color in storage also tend to preserve greater vigor (Filter, 1932; West and Harris, 1963); but seed darkening can also be artificially induced by ultraviolet or cool-white fluorescent lights without any marked reduction in seed quality (Hughes and Sandsted, 1975).

In addition to discoloration, prolonged storage can give rise to other visual changes in the seed coat. Most are apparently trivial: Mercer (1948), for example, noted that seeds of *Poa* spp. tend to lose their soft hairs as a consequence of aging. Czyżenski (1958) observed an aggregation of material in the oil ducts of the pericarp of old carrot seeds (*Daucus carota*), which he ascribed to the polymerization of terpenoid materials, and Merkenschlager (1924) reported that the coats of old *Oenothera* seeds became impregnated with oil, causing them to darken; but in all of these cases the changes observed were presumably consequences or correlates of the aging process and of limited significance in themselves.

Although the seed coat is primarily responsible for the seed's changing appearance during storage, discoloration of internal tissues may be an additional factor in those species that have a relatively translucent covering. Certainly color changes are by no means restricted to the seed coat: in unfavorable storage conditions, lettuce seed can develop red necrotic lesions in the cotyledons (Bass, 1970), and in cotton the endosperm may turn green (Anthony and Tarr, 1952). Lentils characteristically turn yellow in the course of prolonged storage (Nozzolillo and De Bazada, 1984), and browning of the embryonic axis in cereals exposed to excessive humidity (the so-called sick grain phenomenon) has been well documented (Christensen and Kaufmann, 1969; see also sec. 7.5). Less evident are changes that can be detected only under ultraviolet light. Using cut or crushed seeds, Niethammer (1929a) reported a wide variety of age-related changes: peas and flax, for example, gave characteristic fluorescences that diminished with time of storage. Lupines, on the other hand, failed to fluoresce when fresh but brightened with age. Although fluorescence tests have been advocated as a means to measure seed deterioration in a broad range of species (e.g., Karyakin et al., 1956; Firsova, 1959; Zheng and Yan, 1964), they have not generally found great favor. In some cases, at least, *Aspergillus* infections can give rise to seed fluorescence (Ashworth and McMeans, 1966).

7.2 Ultrastructural changes

A number of ultrastructural investigations of aged seeds have appeared in recent years and progress in this area has received attention from several reviewers (Roberts, 1972b, 1973b; Koostra, 1973; Villiers, 1980; Roberts and Ellis, 1982). The unusual difficulties of adequately fixing and observing fine structure in seeds have been emphasized in section 2.3; but when deteriorated seeds are subjected to investigation, the complexities of interpretation are still further compounded. Although some of the analyses reviewed in this section have been made on "unimbibed" seeds fixed in aqueous glutaraldehyde solution, many workers have relied heavily on micrographs obtained from tissue fixed after some minutes or hours of imbibition. When a particular abnormality in cellular structure is encountered in such cases, it is difficult to assess whether it represents a cause or a consequence of cellular debility. Nevertheless, ultrastructural studies have proved useful to the extent that they have helped localize aberrant functions within deteriorated cells. Greatest emphasis has been laid on membranes, either through observations of the fine structure of the membranes themselves or from a consideration of cellular compartmentation, a feature that is contingent on proper membrane function.

One of the most commonly observed aberrations reported from unimbibed aged seeds is a coalescence of lipid bodies. This phenomenon has been noted in embryos of wheat seeds (Anderson et al., 1970) and pea (Harman and Granett, 1972), and in the endosperm of old nongerminable seeds of *Pinus sylvestris* (Simola, 1974) and *P. pinea* (Fernández García de Castro and Martínez-Honduvilla, 1984). Further reports suggest that it occurs in the embryos (Villiers, 1972; Smith, 1978), but not in the cotyledons (Smith, 1983), of deteriorated lettuce seeds. Lipid body coalescence has also been observed in the embryonic root tips of aged seed of both *Protea compacta* (Van Staden et al., 1975) and *P. neriifolia* (Van Staden, 1978; Van Staden et al., 1981); in this genus there was a marked tendency for nuclei to appear necrotic, with condensed patches of chromatin, and the plasmalemma was retracted from the cell wall. Withdrawal of the plasmalemma has similarly been reported in nonviable rye seed root primordia, aged over several years at 15% hydration (Hallam, 1972; Hallam et al., 1973), in *Pinus pinea* seeds after prolonged storage (Fernández García de Castro and Martínez-Honduvilla, 1984), and in wheat (Anderson et al., 1970) and pea (Harman and Granett, 1972) subjected to accelerated aging. In the rye seed there was also evidence of an unnatural distension of the outer mitochondrial membrane and irregularities in nuclear chromatin. Overall, ultrastructural studies of this kind suggest quite convincingly that some degree of membrane disruption may be evident in deteriorated seeds, even from the earliest stages of imbibition.

Differences in fine structure between vigorous and aged material become more obvious as imbibition progresses. In healthy tissues the first hours of rehydration are marked by structural alterations that accompany physiological development. In aged seeds, on the other hand, development progresses more slowly, and in severely deteriorated material, necrosis and autolysis are evident. Particularly extensive documentation of the wide range of ultrastructural changes involved has been assembled by Berjak and Villiers (1970, 1972a, 1972b, 1972c) in their studies of maize seed subjected to accelerated aging at 14% hydration and 40°C.

The first 24 hours following the onset of imbibition in unaged maize root cap cells are marked by internal organizational changes, cell division being postponed to the second day of development. After 12 hours, mitochondria have a circular outline with limited invaginations of the inner membrane to form cristae. Dictyosomes are also present at this time, presumably having formed during the first hours of imbibition (see sec. 2.3). Ribosomes aggregate into polysomes within six hours from the onset of hydration (Berjak and Villiers, 1970). After 20 days of accelerated aging, when germinability dropped below 20%, maize embryos were divided into three arbitrary classes, depending on their degree of cellular degeneration. In the least deteriorated embryos, which were presumably still viable, ultrastructural irregularities were observable during the early stages of imbibition, although there was some evidence that these deficiencies were remedied within 48 hours (Berjak and Villiers, 1972a). The catalog of abnormalities touches on many areas of the cell: nuclei were lobed, mitochondria and plastids displayed distorted profiles, dictyosomes were less frequent and of distorted appearance, and polysomes were slower to form. Embryos of the second class were more deteriorated, and probably no longer germinable. At this stage, structural irregularities in nuclei, mitochrondria, and plastids were more evident, with little indication of restoration or repair within 48 hours from the onset of imbibition. Polysomes were apparent by 24 hours, but dictyosomes were absent (Berjak and Villiers, 1972b). In the third class of embryos, those showing the most extreme evidence of deterioration, polysomes and dictyosomes failed to develop and membrane-bound structures appeared swollen and disorganized, presumably indicative of incipient decompartmentalization (Berjak and Villiers, 1972c). Histochemical evidence suggested that the later stages of deterioration were accompanied by a release of acid phosphatase into the cytosol.[2] In a recent study of maize seed stored for up to 12 years (Berjak et al., in press), a substantially similar sequence of deteriorative events has been established.

Ultrastructural studies of other species confirm that cellular organelles of aged seeds show considerable irregularity and distortion during imbibition. Electron micrographs of deteriorated embryonic tissues of lettuce (Villiers, 1972), rye (Hallam et al., 1973), rice (Vishnyakova et al., 1976, 1979), and peanut (Fu et al., 1983) all provide evidence of disturbance or degradation, which most observers interpret as autolytic. Old nonviable gymnosperm seeds show similar dissolution of cellular structure over several days of hydration (Simola, 1974, 1976); once this stage of disorganization is attained, there can presumably be little question of repair or recovery.

7.3 Cell membranes and permeability

Deficiencies in membrane integrity that can be visualized at the ultrastructural level (sec. 7.2) also become manifest through a breakdown in normal cellular permeability properties. The inability of the plasmalemma to contain solutes within badly deteriorated cells entails two immediate consequences: the cell is unable to respond osmotically, failing to maintain proper turgor, and a substantial efflux of seed metabolites is likely to stimulate potentially pathogenic organisms in the microflora within and about the seed (Presley, 1958; Matthews and Bradnock, 1968; Short and Lacy, 1976; Norton and Harman, 1985).

The failing integrity of cells in aged seeds following imbibition was long ago visualized under the light microscope by Niethammer (1929b, 1942). Using seeds of wheat, tomato, and other species that had been hydrated for several hours, she was able to demonstrate that the ability of cells to plasmolyze was closely linked to the age and viability of the material. Doroshenko (1937), apparently working independently, reached similar conclusions from radicles of soybean and other legumes. A recent and elegant demonstration of the inability of cells to maintain turgor in deteriorated seed material was made by Parrish et al. (1982), who investigated the elastic modulus of cotyledons from soybeans subjected to various periods of accelerated aging (41°C, saturated humidity). Changes were evident in seeds aged for only two days, before viability and vigor had dropped significantly. In all of these studies, though, it is unclear whether the plasma membranes of abnormal cells are completely ruptured or simply lacking in selective permeability.

Alterations in membranes may become evident in less direct ways. For

---

2. An extensive tabulated summary of this work has been published by Roberts (1972b).

many years researchers attempted to measure viability and vigor by placing individual imbibed seeds directly between two electrodes, completing a circuit, and observing the effects of an applied voltage. The excitation responses detected, which may depend in part on action potentials generated by cells with functioning membranes, have not been fully elucidated, but differences between young and old tissue are certainly evident (Waller, 1901; Johnson, 1907; Fraser, 1916).[3]

An alternative approach to the study of cellular integrity involves observation of the influx of reagents that would normally be excluded from healthy seed tissue. Nelyubov (1925), a pioneer in the use of staining techniques in seed testing, identified deteriorated areas in old and otherwise deficient seeds in this way, after coloration with indigo carmine and acid violet, and Niethammer (1929c) employed methylene blue to similar effect. Among many other stains that have since found favor are acid fuchsin, which remains colorless in contact with non-living cells (Effman and Specht, 1967), and Evans blue, which readily penetrates and stains dead cells (Duke and Kakefuda, 1981; Schoettle and Leopold, 1984). Further staining procedures have been listed by Overaa (1984). Other workers have used X-ray techniques to analyze the uptake of barium salts in deteriorated seeds (Simak, 1957; Simak et al., 1957; Ryynanen, 1980). A significant advantage of such vital staining procedures is that the information obtained is both topographical and semiquantitative.

A variety of leakage tests have been devised to measure solute release from seeds during the first hours of imbibition. One of the earliest techniques was that of Lesage (1911, 1922), who used dilute KOH to color seed exudates. Leachates from viable seeds gave only slight coloration, whereas older material generated yellow solutions. Kugler (1952), working with seeds of crucifers, recommended measurement of the release of fluorescent substances, whereas others investigating aging in sorghum (Perl et al., 1978) and soybean (Schoettle and Leopold, 1984) have assayed leachates that absorb ultraviolet. Takayanagi advocated the use of urinalysis papers to measure glucose exudation from deteriorated seeds (Takayanagi and Murakami, 1968, 1969; Takayanagi, 1975), and Perl et al. (1978) monitored the release of ninhydrin-reactive materials and inorganic phosphate. By far the most widespread protocol, however, has been the conductimetric analysis of leached electrolytes (Hibbard and Miller, 1929). This technique has been increasingly recognized as providing a rapid estimate of seed vigor, and custom-designed

---

3. A method of viability testing in which imbibed seeds are exposed to a low constant voltage between two electrodes has also been described (Levengood et al., 1975). In this case, only the conductivity of the seed is in question.

commercial equipment for seed leakage testing has become available (Bondie et al., 1979; Steere et al., 1981).

Many studies indicate that the release of solutes during the first hours of imbibition can be broadly correlated with aging (e.g., Ching and Schoolcraft, 1968; Ching, 1972; Pehap, 1972; Martínez-Honduvilla and Santos-Ruiz, 1975; Short and Lacy, 1976; Ovcharov and Sedenko, 1976; Powell and Matthews, 1977; Ray and Gupta, 1979; Loomis and Smith, 1980; McDonald and Wilson, 1980; Ghosh et al., 1981; Kaur and Srivastava, 1982; Nagy and Nagy, 1982; Grzesiuk and Tłuczkiewicz, 1982; Sreeramulu, 1983a; see Fig. 26). In some cases, enhanced leakage from aged material is detectable during the earliest stages of rehydration (Parrish and Leopold, 1978; Nikolova and B"chvarov, 1982; Dupart and Le Deunff, 1983), prompting the suggestion that deficiencies in membranes are present in the dry condition, and probably do not develop secondarily as a consequence of physiological dysfunction in the newly metabolizing seed.

The relationship between imbibitional leakage and aging cannot be assumed for all species or all types of aging, however. Pesis and Ng (1983), for example, found that for seeds of melon subjected to accelerated aging (45°C and 100% relative humidity), declines in germinabilty preceded signs of an increase in electrolyte efflux. Abdul-Baki and Anderson (1970) also reported that leakage of sugars from barley seeds subjected to accelerated aging (40°C and 100% relative humidity) was unrelated to the early stages of deterioration, although the same could not be said of seeds that had lost quality during prolonged storage under laboratory conditions. Although Halloin (1975) noted an increase in leakage when cotton seeds were subjected to 50°C and saturating humidity, it was poorly correlated with loss of viability in seeds maintained at 20% moisture content and 35–50°C, and Coolbear et al. (1984) found that tomato seeds subjected to aging at 45°C and an unusually high moisture content (70%) leaked more sugars and amino acids (possibly because of reserve hydrolysis), but electrolyte release did not increase. For reasons that are not entirely clear, Halder and Gupta (1980, 1981) reported that conductivity correlated positively with viability in sunflower seeds stored at 28°C and various moisture contents.

Despite its evident usefulness for the assessment of seed quality in most cases, doubts have sometimes been expressed concerning the physiological relevance of the leakage phenomenon. First, seed tissues commonly crack under the stresses of hydration, especially when the material has previously been maintained relatively dry (see also sec. 3.2.1). This kind of cellular rupture may confound alterations attributable to aging. On the basis of germination tests and leakage data, for example, Srivastava and Gill (1975) maintained that soybeans stored below 10% hydration tended to deteriorate at an increased rate, but it seems unlikely that aging per se was the principal

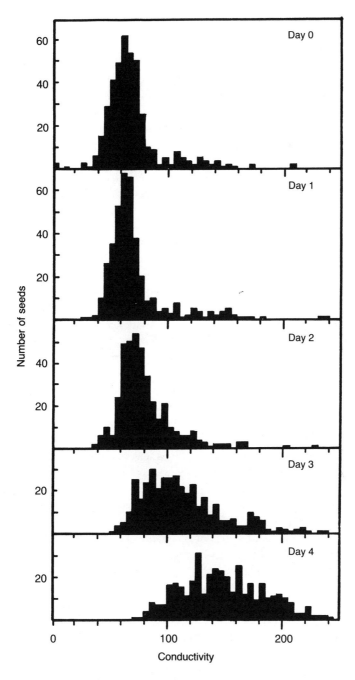

Figure 26. Leakage of electrolytes during imbibition of soybeans previously subjected to accelerated aging. To induce aging, seeds were subjected to 41°C and saturating humidity for up to 4 days, a treatment that was sufficient to reduce viability to below 10%. The electrolytes released were quantified by means of an ASA-610 analyzer, each seed being separately assessed. Conductivity is expressed in arbitrary units. Adapted from M. B. McDonald and D. O. Wilson, "ASA-610 Ability to Detect Changes in Soybean Seed Quality," *Journal of Seed Technology* 5 (1980): 62.

factor in the efflux of solutes in this case. The release of electrolytes from soybeans during imbibition increases markedly if their level of hydration is below 10% at the outset (Parrish and Leopold, 1977; Tao, 1978). A second complication is that metabolic events within the seed probably have a marked influence on the nature and extent of the solutes exuded into the imbibing medium. Takayanagi (1977) found that although glucose was usually exuded from old rape seed in considerable quantities, a preliminary heat treatment of the seed sufficient to inactivate invertase led to the release of sucrose instead. Abdul-Baki and Anderson (1970) also raised questions concerning measurements of carbohydrate efflux from barley, claiming that in partially deteriorated seeds with a relatively active metabolism, sugars released in the earliest stages of imbibition were later reincorporated and metabolized.

An important and somewhat contentious issue is the exact physiological condition of the seeds and cells that generate the exudate. In most conductimetric vigor assays, leakage from a population of seeds is measured, and the relative contribution of viable and nonviable individuals is therefore uncertain. Perl et al. (1978) found little evidence that declining vigor in sorghum seeds was related to increased leakage; in their view, the release of solutes is instead controlled primarily by the number of dead seeds in a sample. Thus in a deteriorated seed lot comprising both viable and nonviable individuals, the germinable seeds will tend to be of low quality and leakage of the total population will be relatively high. According to Perl and his associates, a negative correlation between vigor and leakage has often been established in this fashion, but most studies fall short of demonstrating that partially deteriorated seeds are themselves leaky. Nevertheless, some investigators (e.g., Parrish and Leopold, 1978) have noted an increase in leakage rates before a decline in germinability.

Similar problems of interpretation arise if leakage from individual seeds is considered. Experience gained with plasmolytic and vital staining techniques has indicated that patches of nonviable cells tend to coexist alongside healthier tissues. An obvious inference is that enhanced seed leakage with age may be due to an increased proportion of nonviable cells throughout the tissue—although a gradual increase in the permeability of living cells certainly cannot be excluded. Thus on the basis of tetrazolium tests (see sec. 7.6.2), Powell and Matthews (1977) suggested that increased leakage can occur from partially deteriorated peas that lack necrotic areas. The implication is that living cells contribute significantly to the leakage pattern. However, these complex issues concerning the cellular basis of seed exudation are still imperfectly resolved. It is possible that variability between species or aging treatments may be in part responsible for the different views that have developed; but at present the only certainty lies in the rather lame conclusion that further studies in this area are needed.

One consequence of membrane failure in aged seeds is that imbibitional characteristics may change. Parrish and Leopold (1978), for example, examined soybean seeds that had been previously subjected to accelerated aging at 41°C and 100% relative humidity. Water uptake was far greater in fresh seed than in the aged material during the first few hours, and an inverse relationship with solute loss was demonstrated. Deficiencies in membranes, however, are unlikely to be the only factor to control swelling. After some hours of hydration, further accumulation of water in many species is apparently contingent on an increase in osmotic potential, presumably as a result of reserve hydrolysis. The inability of aged rye embryos to swell to the same degree as vigorous ones has been ascribed to a failure of this "active uptake" of water (Hallam et al., 1973; Sen and Osborne, 1977). Studies on pine (*Pinus sylvestris*) and spruce (*Picea abies*) have also suggested that aged seeds imbibe less readily than fresher material (Vincent, 1929; Göksin, 1942); age, however, does not always alter the capacity to imbibe. In a study of radish and cauliflower seeds, Vasil'eva and Lazukov (1970) found that old seed of low viability swelled to the same extent as fresh material, or even slightly more. Aged nonviable seeds of bambara groundnut (*Voandzeia subterranea*), on the other hand, have been reported to absorb water much faster than fresh material during the first 12 to 14 hours of imbibition, although the reasons are unclear (Sreeramulu, 1983a, 1983b).

The biochemical basis of membrane lesions in aged seeds has been subjected to considerable scrutiny, as these deficiencies not only threaten cellular integrity directly but also may account for declining membrane-associated metabolic activities, such as respiration (sec. 7.7). Alterations in lipids have been most commonly invoked to explain deterioration of membrane function.

7.4 Lipids

7.4.1 Changes in lipid content

Lipids may constitute as little as 2% of the dry weight of seeds of some species, whereas in others they may exceed 50%. Most seeds contain a relatively small amount of phospholipids (about 1–2% of dry weight is common), and these polar (or amphipathic) lipids form the basic bilayer structure of cellular membranes. Amounts of storage lipids vary greatly among species, however, and even among different parts of the same seed. These reserves are composed primarily of apolar (or neutral) lipids, triacylglycerols being most frequently encountered. The apolar character of the storage lipids favors their sequestration in oil bodies, well away from the polar environment of the cytosol. There is a widespread perception, substantiated to some extent by the evidence considered in section 3.5, that lipid-rich seeds tend to have limited longevity. It is not surprising, therefore, that hypotheses of seed aging based on lipid degradation have enjoyed a considerable vogue.

A decrease in total lipids has sometimes been noted in oily seeds during prolonged storage. Rape, sunflower (Ivanov and Berdichevskii, 1933), onion (Czyżewski, 1963), cotton (Abdelmagid and Osman, 1975), and bambara groundnut (Sreeramulu, 1983b), for example, have been reported to show such effects. Storage lipids presumably decrease as a consequence of slow metabolism by the seeds (or the seed microflora) under relatively humid storage conditions. The extent of this metabolic depletion is unlikely to threaten viability, though; in soybeans stored at 13% hydration for six months at 35°C, for example, about 15% of the oil fraction is lost (Nakayama et al., 1981). Use of the deterioration equations of section 3.2.1 suggests that viability would have disappeared much more rapidly—within about two months—under these conditions.

Loss of membrane lipid components is probably of much greater significance than a decrease in storage reserves. During storage at high humidity, phospholipid levels have been reported to decline 50% or more in seeds of cucumber (Koostra and Harrington, 1969), peanut (Pearce and Abdel Samad, 1980), and pea (Powell and Matthews, 1981b). Less severe declines were noted in soybeans by Priestley and Leopold (1979), in peas by Yang and Yu (1982) and Powell and Harman (in press), and in sunflower by Halder et al. (1983). At 13% hydration and 35°C, Nakayama et al. (1981) found that 45% of lipid phosphorus was lost in soybeans over six months. Phospholipid content has also been found to decline in embryonic and nonembryonic parts of tomato seeds aged at 45°C and 70% moisture content (Francis and Coolbear, 1984).

There are fewer reports of a decrease in phospholipids in seeds undergoing aging at low levels of hydration (sorption zones I and II as defined in sec. 2.1), although Matsuda and Hirayama (1973) noted a small decline in dry-stored rice, and Pearce and Abdel Samad (1980) made a similar claim for peanuts held at 5°C and 30% relative humidity; in soybeans stored at 10–11% hydration, though, Paulsen et al. (1981) found that more than 50% of lipid phosphorus was lost as germinability declined. Koostra and Harrington (1969) maintained that in cucumber seeds the responses of phospholipids to accelerated aging and to deterioration in long-term dry storage differed quite markedly; phospholipid levels dropped precipitously when the seeds were exposed to high humidity, whereas under dry conditions, the lipid levels were stable for many years, despite declining seed germinability. Similar conclusions arise from data recently presented by Petruzzelli and Taranto (1984): wheat embryos tend to lose phospholipid when they are stored for short periods at high humidity, whereas this constituent is more stable when humidity is low. A report by Chapman and Robertson (1977) stands in some contrast to this general consensus. Phospholipid levels were reported to double in soybeans exposed to 35°C and 85% relative humidity for 6 days, and increased approximately threefold over 27 days. No germination data were given. Powell and Harman (in press) also demonstrated an increase in phospholipid content in pea seeds exposed to 45°C at 20% moisture content; under these conditions, however, seed performance increased slightly, possibly because of repair functions (sec. 7.11.1).

Two suggestions have been offered to explain decreased phospholipid levels in aged seeds: the lipids may have been subjected to peroxidation (a proposition that is considered further in sec. 7.4.3) or else they may have been degraded by lipolytic enzymes, particularly the phospholipases. Phospholipase D—an enzyme that cleaves the polar head group from phospholipids to leave phosphatidic acid—is present in dry seeds (Quarles and Dawson, 1969), and minor accumulations of phosphatidic acid in soybeans exposed to unfavorable storage conditions have been reported (Priestley and Leopold, 1979; Nakayama et al., 1981). Some studies have also found lysophospholipids to increase (Matsuda and Hirayama, 1973; Nakayama et al., 1981). In animal and microbial systems, the hydrolytic removal of a single fatty acid from a phospholipid to form a lysophospholipid is normally attributed to an enzyme of the phospholipase A type, but the existence of phospholipase A in higher plants is questionable, and such functions may be assumed instead by an unspecific lipid acyl hydrolase (Galliard, 1980). The amounts of phosphatidic acid and lysophospholipid accumulating in seeds provide rather unsatisfactory measures of lipid degradation, however, as presumably these products are themselves subject to further breakdown.

The source of the lipolytic activities responsible for initiating the degradation of phospholipids during aging is uncertain, as most observations have been made at relatively high levels of hydration, equivalent to sorption zone III. Under such conditions, enzymic activity endogenous to the seed may obviously be significant, but fungal effects cannot be entirely discounted. In an extreme case, Daftary and Pomeranz (1965) demonstrated that wheat seeds stored at 18–22% hydration at 49°C suffered a massive loss of phospholipids that was evidently associated with molding. It may therefore be premature to suggest that phospholipids are necessarily degraded by the seed's own enzymes during exposure to elevated humidity. Moreover, it remains unclear whether all cells lose phospholipids equally during accelerated aging, or if such loss is restricted to a few necrotic areas. Imbibition of old, nonviable maize seed, for example, leads to a massive loss of lipid phosphorus (Sedenko and Ovcharov, 1969), presumably as a result of a complete breakdown in cellular compartmentation. If, on the other hand, phospholipids are degraded within viable cells as a result of aging, it is unlikely that they could be immediately replaced during hydration. In seeds of both high and low vigor, membrane lipid synthesis occurs at low levels during the first 12 hours of imbibition, although it increases greatly thereafter (McDonnell et al., 1982). Declining phospholipid levels, apparently attributable to phospholipase activity, have been previously reported during drying of yeast (Harrison and Trevelyan, 1963); in that system, loss of lipid entails a considerable increase in leakage during rehydration.

In addition to phospholipids, one other class of membrane lipid component has received recent study with respect to seed aging. Although sterols and sterol derivatives are present in relatively small amounts in seeds, their influence on membrane function may be disproportionately great (Mudd, 1980). According to Bhattacharyya and Gupta (1983), who worked with sunflower and chickpea, free sterols and steryl glycosides increased as seeds were aged (40°C and saturating humidity), whereas steryl esters declined. The significance of these observations for membrane performance remains unclear, however.

7.4.2 Fat acidity

The hydrolysis of the ester linkages between fatty acyl chains and glycerol in seed triacylglycerols liberates fatty acids, a reaction sponsored by lipases. One consequence of the accumulation of "free" fatty acids within seeds is that the pH of aqueous or alcoholic seed extracts is lowered. Fat acidity, measured in this way, has been recognized for many years as an index of seed quality; over a century ago, for example, Cugini (1880) proposed a protocol for testing the acidity of alcoholic extracts from oil seeds as a rapid substitute for a germination assay. Measurements of fat acidity have been most frequently used to monitor the deterioration of cereals (which have a relatively lipid-rich germ) and oilseeds (e.g., Zeleny and Coleman, 1938; Täufel and Pohloudek-Fabini, 1954), although oily gymnosperm seeds show similar changes (Vincent, 1929). In addition to providing a measure of deterioration in storage, assays of fat acidity can yield useful information concerning heat or frost damage (Baker et al., 1957). As with conductivity testing (sec. 7.3), there can be little doubt of the practical utility of assays for fat acidity, but questions can be raised concerning the source and physiological relevance of the effects observed.

All observers agree that elevation of temperature and moisture content in stored seeds leads to an increased level of fat acidity (Fig. 27), but there is some dispute as to whether the hydrolysis arises from the action of seed or fungal lipases. In a comparison of sterile and inoculated wheat, for example, Hummel et al. (1954) concluded that almost all of the rise in acidity in storage was due to mold. In uninoculated seeds, fat acidity increased only if levels of hydration were in excess of 24%. Further studies on the same species found that surface sterilization depressed both the formation of mold

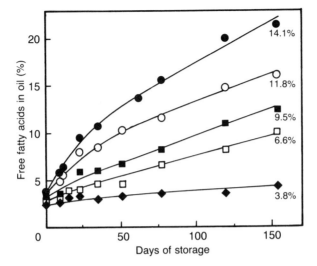

Figure 27. Development of free fatty acids in brown rice samples of different moisture contents stored at 25°C. Adapted from I. R. Hunter et al., "Development of Free Fatty Acids during Storage of Brown (Husked) Rice," *Cereal Chemistry* 28 (1951): 235.

and the production of free fatty acids in seeds stored at 13–14% hydration (Sorger-Domenigg et al., 1955). Similar conclusions arise from work on cotton: fat acidity rises steadily under conditions of commercial storage (Robertson and Campbell, 1933), even though lipase activity is negligible in the ungerminated seed of this species (Olcott and Fontaine, 1941). Increased fat acidity is not always a concomitant of fungal invasion, however: Dorworth and Christensen (1968) reported no change of acid value in some samples of aging soybeans, despite loss of viability and complete invasion by storage fungi.

In addition to the considerable evidence that fungal lipases can make significant inroads into the lipids of stored seeds, there are indications that increased fat acidity may in some cases be generated primarily by seed enzymes. A few studies have suggested that fat acidity increases even when seeds are stored at sorption levels corresponding to zone II (sec. 2.1), and therefore below the level of hydration at which fungal effects are usually significant (sec. 1.2). Inferences of this sort can be drawn from the data of Hunter et al. (1951), who worked with rice (Fig. 27), and of Agrawal and Siddiqui (1973), who used soybean. A modicum of endogenous lipase activity can be extracted from unimbibed seeds of both species (Matsuda and Hirayama, 1975; Lin et al., 1982). Other studies suggest that seed lipases are capable of activity in model systems at very low levels of water sorption (sec. 2.2), although this finding in no way implies that the enzymes are functional within seeds under these conditions. Furthermore, lipase activity is barely detectable or absent in extracts from ungerminated seeds of most species (St. Angelo and Ory, 1983).[4]

Elevated levels of free fatty acids, which are toxic to most cells, are not found in healthy seed tissue. It has been suggested that the germinability of cotton seeds, for example, will be negligible if free fatty acid content is in excess of 3% of the total oil fraction (Hoffpaiur et al., 1950). Free fatty acids have particularly deleterious effects on membranes, probably because of their ability to act as detergents. Isolated plant mitochondria show swelling and uncoupling of oxidative phosphorylation in the presence of fatty acids (Baddeley and Hanson, 1967; Earnshaw et al., 1970), and chloroplasts suffer inhibition of the Hill reaction (Krogmann and Jagendorf, 1959). In addition, soluble enzymes may be sufficiently denatured to cause loss of their activity (Tortora et al., 1978).

Most fatty acids within seeds have relatively long chains (C14 to C22), but short-chain acids occur to a minor extent and may exert considerable physiological influence. Several reports have indicated that short-chain fatty acids (generally C5 to C11) inhibit the metabolism and germination of seeds (Berrie et al., 1975; Buller et al., 1976; Metzger and Sebesta, 1982), but on present evidence, it seems improbable that they have a significant role to play in seed aging; on the basis of a very limited survey by Carugno et al. (1961), Wittmer (1961b) suggested that the aging of tobacco seed (both *Nicotiana tabacum* and *N. rustica*) was accompanied by losses, rather than increases, of short-chain fatty acids (C3, C4, C5, and C9).[5]

7.4.3 Lipid peroxidation

Many polyunsaturated fatty acids found in seeds are highly susceptible to peroxidative degradation. As a result, not only is the lipid itself destroyed, but a complex series of reactions generates a variety of potentially toxic products. In stored material, peroxidation may arise either as an atmospheric autoxidation or through the agency of lipoxygenase (or lipoxidase), an enzyme present in many unimbibed seeds. Although it is still questionable whether such changes are necessarily linked to aging, hypotheses of seed deterioration based on oxidative degradation of lipids have been current for

---

4. Castor bean is unusual among cultivated species in displaying high levels of lipase activity in the unimbibed seed (Huang and Moreau, 1978). It also happens to be abnormally long-lived for an oilseed.

5. As a saponification step was included in these analyses, it is possible that the short-chain fatty acids existed in esterified form *in vivo*.

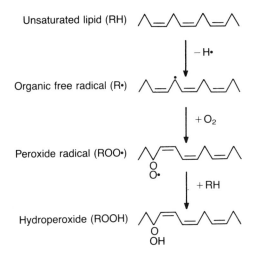

Figure 28. Peroxidation of a polyunsaturated fatty acid. A reaction scheme is illustrated for a fatty acid such as 9,12,15-octadecatrienoic (linolenic). The fragment shown here would extend from carbons 7 through 18.

many years (e.g., Van Tiegham, 1884; Mirov, 1944) and show no signs of abating (Wilson and McDonald, in press).[6]

The peroxidation of an unsaturated fatty acid is represented schematically in Figure 28. Initiation of the reaction sequence involves abstraction of hydrogen (H•) from a methylene group ($-CH_2-$) adjacent to a double bond. It is unclear how this process normally occurs in desiccated seeds, although studies of isolated oils suggest that metal ions often initiate it by reacting with oxygen to form the superoxide anion:

$$M^+ + O_2 \longrightarrow M^{2+} + O_2^-$$

The addition of a proton produces a peroxy radical ($HO_2^•$), which serves as an effective chain initiator (Frankel, 1980). Superoxides have also been implicated in lipid peroxidation occurring within hydrated, metabolically active systems, where they are formed as a consequence of electron transport or other enzymic processes. Reaction of superoxide with H_2O_2 yields the hydroxyl radical (•OH), which readily abstracts hydrogen from a lipid methylene group, beginning the chain of peroxidative degradation (Leibovitz and Siegel, 1980). The process of initiation may well vary, therefore, depending on the degree of hydration and metabolic status of the seed. Following hydrogen abstraction, and after diene conjugation and addition of oxygen, a peroxy radical (ROO•) is obtained (Fig.28), which, by reaction with another unsaturated fatty acid (RH), forms a lipid hydroperoxide (ROOH) as the

---

6. The earlier literature commonly refers to "seed rancidity" as a cause of deterioration. *Rancidity*, used in this sense, was frequently poorly defined, embracing both the production of free fatty acids (sec. 7.4.2) and lipid peroxidation (Jamieson, 1943). The two processes need not proceed concurrently in stored seed.

primary oxidation product. The lipid hydroperoxides (often inaccurately called "lipid peroxides") decompose readily in the presence of metal ions, generating new free radical species and ultimately leading to highly complex mixtures of olefins, alkanes, alcohols, and carbonyl compounds.

Although atmospheric autoxidation alone is capable of inducing the effects outlined here, it is possible that lipoxygenase plays a significant role in many seeds. Enzyme activity has been suspected of indirectly generating off-flavors in raw peanuts attributable to lipid breakdown products (St. Angelo et al., 1979), for example. Soybean lipoxygenase has also been shown to function in model systems at very low levels of hydration (sec. 2.2), and the evidence of chemiluminescence studies is that it may be active within seeds at 14–16% hydration (sec. 7.4.4). Further, although seed lipids are sensitive to atmospheric autoxidation *in vitro*, they are remarkably resistant to heated oxidizing atmospheres when they are still retained within the seed (Priestley et al., 1985a). It is possible, therefore, that lipoxygenase-sponsored peroxidation is more important than nonenzymic oxidation in some species.[7]

In most seeds, the lipids that are at risk from autoxidation comprise oleate (18:1), linoleate (18:2), or linolenate (18:3) fatty acyl chains. The degree of unsaturation has considerable influence on the rate of degradation. Thus 9,12-linoleate, with a pair of double bonds that are methylene-interrupted, is degraded about 30 to 40 times faster than 9-oleate, which has only one double bond. For 9,12,15-linolenate, with three double bonds, the rate is about 80 to 100 times more rapid than with 9-oleate (Schaich, 1980).

Free radical scavengers in seeds play a key role in preventing or terminating the reaction sequence. The lipophilic tocopherols (vitamin E), for example, are effective in quenching both superoxide and lipid peroxy radicals (Leibovitz and Siegel, 1980). In theory, each tocopherol molecule is inactivated in two steps: a primary oxidation to form a tocopheryl radical, followed by a second oxidation to the tocopheryl quinone. One mole of tocopherol can therefore reduce two moles of free radicals.[8] In similar fashion, ascorbic acid—a free radical scavenger present in some seeds—is first oxidized to an ascorbate free radical and then to dehydroascorbic acid. It is important to bear in mind, though, that these are merely theoretical considerations; the stoichiometry of scavenging reactions *in vivo* is not well understood and may show considerable divergence from the ideal (Pryor, 1976). Other seed constituents that have been implicated in limiting oxidation of the lipid fraction

---

7. Despite considerable attention, the physiological function of lipoxygenase in seeds remains obscure (Galliard and Chan, 1980).

8. Although tocopherols are thought to serve primarily as antioxidants in seeds (e.g., Hove and Harris, 1951), they may also modulate the dynamic properties of membranes (Lucy, 1978). According to one view (Beringer and Northdurft, 1979), their significance as antioxidants in seeds may be secondary to other functions.

include lipid-associated hemoproteins (Butler and Baker, 1965), isoflavone glycosides (Hammerschmidt and Pratt, 1978), and other flavonoids (Whittern et al., 1984). Little is known concerning the role of enzymic defense mechanisms in preventing lipid peroxidation in seeds. Superoxide dismutase is a key enzyme in many cells, protecting against high levels of superoxide radicals; but it is absent in dry soybeans, developing only during the first hours of imbibition (Stewart and Bewley, 1980). On the basis of this evidence, suggestions that the longevity of stored seeds may be directly linked to superoxide dismutase levels (Rabinowitch and Fridovich, 1983) seem rather implausible.

Studies of dried foods have documented the important influence of water on lipid oxidation mediated by free radicals (Karel, 1975; Karel and Yong, 1981). Peroxidation is generally facilitated at low degrees of water sorption, equivalent to zone I (see sec. 2.1). With increasing moisture, the reactions are attenuated for several reasons: water slows the access of oxygen to the sensitive sites; by hydrating metal ions, it lowers their catalytic effectiveness; it renders chelators and antioxidants more efficient by increasing their rate of diffusion; and by hydrogen bonding, it interferes with the decomposition of hydroperoxides. Such considerations may presumably apply to very dry seeds too, a suggestion that has been repeatedly urged (e.g., Harrington, 1972; Ovcharov, 1976). As section 3.2.1 indicated, however, definitive evidence for enhanced deterioration of this kind in highly desiccated seeds has proved to be surprisingly elusive.

The consequences of lipid peroxidation for cellular functioning and survival are likely to be severe. Peroxidation of membrane lipids *in vitro* leads to massive dysfunction: membrane viscosity increases (Dobretsov et al., 1977; Pauls and Thompson, 1981), enhanced bilayer permeability is commonly observed (Van Zutphen and Cornwall, 1973; Smolen and Shohet, 1974), mitochondria swell (Haydar and Hadziyev, 1974), and lysis occurs in severe cases (McKnight and Hunter, 1966; Vladimirov et al., 1980). The rate of transbilayer movement of phospholipids frequently increases (the so-called flipflop phenomenon) (Barsukov et al., 1980; Shaw and Thompson, 1982), and there are indications that nonbilayer phases may form (Galanopoulou et al., 1982).

Lipid oxidation products have pronounced effects on other important cellular systems. DNA, for example, is subject to denaturation (Reiss and Tappel, 1973; Fujimoto et al., 1984) and translation of messenger RNA into protein is hindered (Dhindsa, 1982). Proteins are liable to several forms of perturbation, depending on hydration levels (Gardner, 1979; Pattee et al., 1982b; Ory and St. Angelo, 1982). The interaction of lipid hydroperoxides and proteins at relatively high moisture content leads to protein–protein crosslinking; at lower levels of hydration, scission of the protein is favored. Several amino acids, including histidine, methionine, cysteine, and lysine, are oxidized by hydroperoxides, a phenomenon that can occur in either dry or hydrated conditions. Lipid-protein adducts are also frequently formed, via both hydrogen bonding and covalent linkages. Furthermore, aldehydes are particularly important among lipid breakdown products because of their tendency to form covalent links with proteins in "browning" reactions.

In view of its significant cellular ramifications, many commentators have considered lipid peroxidation to be central to the explanation of seed aging (e.g., Harrington, 1970; Villiers, 1973; Kulka, 1973; Wilson and McDonald, in press). Although there are good grounds for believing that such changes occur in some seeds under certain conditions of storage, however, it seems equally apparent that other forms of seed aging can occur without widespread lipid peroxidation. Evidence of lipid degradation derives from two types of investigation: studies that indicate a loss of susceptible lipids from seeds during storage, and those that report an increase in breakdown products.

Declining levels of lipid unsaturation in deteriorating seeds have generally been followed either as a decrease in iodine value of the extracted oil or, more directly, by analysis of fatty acids by means of gas chromatography.[9] Kazakov and Bolkova (1950), for example, noted a marked downward trend in iodine value for wheat, oats, and rye kept in open storage for periods of up to 16 years. Lowered iodine values were also reported by Scarascia and Scarascia-Venezian (1954) in aged soybeans, by Malysheva (1964) in sunflower, peanut, and perilla over four years of storage, and by Sharma (1977) in deteriorated sesame, castor bean, and cotton. Giovannozzi-Sermanni and Scarascia-Venezian (1956), on the other hand, found no evidence of a decrease in iodine value in tobacco seeds that had lost viability in long-term storage. Physical integrity of the seed evidently exerts some influence: Painter and Nesbitt (1943), working with flaxseed subjected to dry, dark storage for periods up to seven years, found only slight decreases in iodine value with age, although the decline was more obvious in material that had suffered mechanical injury.

Direct analysis of fatty acids by gas chromatography has generally suggested that seeds in dry storage tend to lose polyunsaturation over months or years. Thus for rice stored under laboratory conditions for several months, Lee et al. (1965) found a small decrease in the proportion of linoleate (18:2) in the phospholipids of the whole seed. Priestley and Leopold (1983) likewise reported a fairly meager decline in the proportions of linoleate (18:2) and linolenate (18:3) present in axes and cotyledons of soybean over several years of dry storage, a period during which vigor and viability decreased. Uematsu

---

9. The quantity of iodine that will associate with a lipid mixture under defined conditions—the "iodine value"—is a measure of the sample's average degree of unsaturation.

and Ishii (1981) also recorded a small downward shift in the amount of linoleate in stored peanuts, which was associated with loss of viability. Van Staden et al. (1976), however, found no loss of polyunsaturation in unimbibed embryos of nonviable *Protea compacta* seeds.

Over decades of storage, changes due to lipid peroxidation become much more evident in many seeds. In this respect, a graphic demonstration of the benefits of the hard-seeded condition has been provided by Flood and Sinclair (1981), who divided a sample of 20-year-old subterranean clover into permeable and impermeable seeds for subsequent lipid analysis. On the basis of seed dry weight, the linoleate (18:2) and linolenate (18:3) contents of the permeable seeds had declined 19.7% and 54.1% respectively over those of the harder material. Lipid polyunsaturation is also maintained at high levels in highly impermeable seeds of the Asiatic lotus (*Nelumbo nucifera*), which retain their viability over several centuries (Priestley and Posthumous, 1982; see sec. 6.2); but for maize, a species that remains viable in open storage for no more than a few decades, over half the seed's detectable linoleate (18:2) is degraded within 100 years (Priestley et al., 1981).

Considerable attention has been devoted to the lipids of seeds that undergo accelerated aging at high levels of hydration. In some studies the proportion of polyunsaturated to total fatty acids has remained constant during deterioration. This is true, for example, of lipids within embryonic axes and cotyledons of soybeans (Priestley and Leopold, 1979; Ohlrogge and Kernan, 1982; Priestley et al., (1985b) and of various lipid fractions of peanut seeds (Pearce and Abdel Samad, 1980). Others have reported a decline in polyunsaturation in the axes of soybeans subjected to accelerated aging (Stewart and Bewley, 1980), although much of this decrease was evidently incurred after the seeds had lost viability. Harman and Mattick (1976) noted a loss of polyunsaturated fatty acids in whole peas and embryonic axes during a period of declining viability, which they induced by exposing the seeds to 30°C and 92% relative humidity for several weeks; Yu (1981; Yang and Yu, 1982) reported similar results for peas exposed to 48°C and high relative humidity for nine days. Perl et al. (in press), on the other hand, who exposed seeds to 50°C and saturating humidity, found that the proportion of lipid polyunsaturation during aging remained constant in five species (cucumber, onion, pea, *Papaver orientale*, and *Oenothera erythrosepala*), decreased in pepper (*Capsicum annuum*), increased in *Vicia faba*, and fluctuated rather irregularly in maize. It is difficult to reconcile so many apparently contradictory reports concerning the fate of polyunsaturated lipids in seeds during accelerated aging, save to note that enzymic effects—both anabolic and catabolic—cannot be easily discounted at high levels of water sorption. Differences among species or aging conditions may well influence the equilibrium between degradation and synthesis.

Several investigators have explored the relationship between tocopherol content and seed aging. In model systems comprised of polyunsaturated fatty acids and tocopherol, the autoxidation of the fatty acids becomes significant only following an induction period during which the antioxidant is destroyed (Wu et al., 1979). There is some evidence for a similar induction effect in stored seed: Pattee et al. (1982a), for example, reported that peanut lipids became more vulnerable to artificial oxidation stresses after they had been held for several months at 4°C with 6.3% moisture content; lipid iodine values, however, did not change during the storage period.

Direct measurements of tocopherols in stored seeds have been somewhat contradictory. Nordfeldt et al. (1962) reported that wheat germs lost 5 to 10% of their tocopherols during six months of storage, but provided no indication that this loss was associated with changing viability. Ovcharov (1969) asserted that loss of germinability in stored wheat and maize was associated with declining tocopherol content, but the data he provided for the most part show relatively small differences; Sharma (1977) also noted a decline in tocopherols associated with loss of viability in oilseeds subjected to open storage in India, and Ionesova (1970) suggested that declining tocopherol content was often—but not inevitably—associated with loss of germinabilty in stored seeds of desert plants. Fielding and Goldsworthy (1980), however, could find no consistent evidence of a loss in tocopherols for wheat subjected to various aging conditions, including both high and low humidity. Priestley et al. (1980) also found no change in tocopherols of soybean subjected to accelerated aging at 40°C and 100% relative humidity. Similar examination of seeds that had lost viability in long-term storage revealed considerable variability in tocopherol levels, but no association between antioxidant content and age could be established; yet quite similar storage conditions can lead to small declines in lipid polyunsaturation (Priestley and Leopold, 1983). The role of tocopherols in seed aging consequently remains somewhat unclear. It is possible, for example, that only certain parts of the cell are protected by tocopherols, leaving other areas susceptible to oxidation. Present evidence suggests that seeds can lose viability in dry storage, in part as a result of the effects of lipid peroxidation, yet still retain relatively high levels of antioxidant.[10]

There have been reports that application of antioxidants can prolong seed survival, although for the most part beneficial results have been sporadic and unpredictable. Kaloyereas (1958), for example, found that treatment with starch phosphate, a reputed antioxidant, reduced viability loss in okra and

10. According to Toyoda (1960, 1965), decreases in ascorbic acid and reduced glutathione during storage are much less evident in the exceptionally long-lived seeds of *Nelumbo nucifera* (see secs. 4.2 and 6.2) than they are in other species. Ruge (1947) also suggested that loss of viability in oats was associated with declining ascorbic acid content.

onion, but not in longleaf pine (*Pinus palustris*). Yang and Yu (1982) have questioned the usefulness of this treatment on theoretical and practical grounds. Kaloyereas et al. (1961) also noted that briefly dipping onion seeds in a 1% aqueous emulsion of tocopherol prolonged subsequent viability in storage at about 29°C, although similarly treated okra seeds declined faster than controls. More recently, Woodstock et al. (1983) investigated the effectiveness of tocopherol and butylated hydroxytoluene, a synthetic antioxidant, in preserving onion, pepper, and parsley seeds. Several concentrations and storage protocols were used. In some treatments, the antioxidants apparently helped preserve viability, whereas in others they were inimical to survival; overall, it is difficult to generalize from these data. Similarly, Parrish and Bahler (1983) applied propyl gallate and *tert*-butylhydroquinone to soybean seeds (8–13% moisture content, 20–40°C); small increases in storability were noted in some cases, but the treatments were deleterious in others. Another synthetic antioxidant, *tert*-butyl sulfide, applied to peas has been reported to reduce viability loss and malondialdehyde production (see below) during 48 hours of aging at 60°C (Yu, 1981; Yang and Yu, 1982). Basu and Rudrapal (1979, 1980; Rudrapal and Basu, 1980) also found that treatment with iodine vapor yielded effective protection against various forms of accelerated aging treatment. They suggested that the iodine (which is normally considered an oxidizing reagent) exerted a stabilizing influence by associating with lipid double bonds; it is possible, however, that the treatment conferred some advantage by reducing fungal or microbial effects, either during aging or else in germination assays. According to a subsequent report (Rudrapal and Basu, 1981), chlorine and bromine can be used to similar effect, although both are even stronger oxidants than iodine. Treatment of old seeds with chlorinated or iodinated water has long been recognized as a method for improving their subsequent performance (e.g., Chauvin, 1878), presumably as a consequence of surface sterilization.[11]

A second type of approach to the question of lipid peroxidation and its role in seed aging involves analysis of lipid degradation products. Oxygenated fatty acids (primarily hydroxy and epoxy fatty acids) have been reported to accumulate in stored seeds of sunflower (Mikolajczak et al., 1968), *Cichorium intybus*, and *Crepis* spp. (Spencer et al., 1973), but the significance of these derivatives in the aging process has not been explored. An elevation of "peroxide value," which correlates negatively with loss of germinability, has also been encountered during prolonged storage of peanuts (Mathur et

---

11. Barnes and Berjak (1978) suggested that pretreatment of seeds with antioxidants (sodium thiosulfate and mercaptoethanol) reduced ultrastructural aberrations during accelerated aging. No data on viability were provided.

al., 1956; Uematsu and Ishii, 1981).[12] Kaloyereas (1958) indicated that peroxide values rose for longleaf pine and okra seed as viability was lost in long-term storage, and Sharma (1977) reported similar trends for deteriorated oilseeds after open storage in India.

Other investigators have employed the thiobarbituric acid (TBA) test for malondialdehyde as an index of lipid peroxidation in aged seeds.[13] Stewart and Bewley (1980), for example, investigated TBA reactivity of aqueous extracts from axes of soybeans that had been subjected to accelerated aging (45°C and high humidity). Despite other evidence that lipid peroxidation had occurred during aging, TBA reactivity was higher in unimbibed fresh tissue than in aged axes. Yu (1981; Yang and Yu, 1982), on the other hand, reported that TBA reactivity increased in pea seeds subjected to accelerated aging at 48°C and 92% relative humidity. Elevated lipid peroxidation in wheat and brown mustard seeds aging under conditions of high and low humidity has also been suggested on the basis of the TBA test by Rudrapal and Basu (1982). Using a different approach, Koostra and Harrington (1969) attempted to separate and quantify degraded phospholipids by thin-layer chromatography, and suggested that oxidized lipids increased more than fivefold in cucumber seeds rendered nonviable by prolonged exposure to 38°C and saturating humidity. In general, most assays of seed lipid degradation products have been of a fairly rudimentary character: quantification has been rather uncertain and the relative cytotoxicity of the many derivatives encountered is poorly understood.

Suggestions have appeared in the literature that volatile lipid degradation products may be particularly significant in determining field emergence of aged seeds. Using peas subjected to accelerated aging (30°C, 92% relative humidity), Harman et al. (1978) found that deteriorated seeds produced more than 20 times as many volatile carbonyl compounds during the first hours of imbibition as did fresh seeds. Moreover, the volatiles produced were highly effective in stimulating germination of certain fungal spores. In subsequent work with peas and soybeans, useful correlations were established between volatile release and field emergence (Harman et al., 1982). Recent analyses, however, suggest that acetaldehyde and ethanol are the principal volatiles produced (Gorecki et al., 1985), and it seems probable that they arise from

12. The "peroxide value" of an oil, which can be determined with the aid of a variety of reducing agents, is commonly used to monitor the quality of samples in which oxidative degradation is still in its early stages (Gray, 1978).

13. The condensation of malondialdehyde (a lipid degradation product) with TBA forms the basis of a semiquantitative colorimetric assay that has been widely used by food scientists. The sources and stoichiometry of malondialdehyde production are poorly understood and the assay is notoriously subject to interference (Gray, 1978). Yang and Yu (1982) have discussed some of the difficulties involved in satisfactorily applying this test to the study of aging seeds.

imbalances of metabolism in aged seeds (see sec. 7.7.2) rather than from lipid peroxidation. Using a different form of analysis, Fielding and Goldsworthy (1982) have investigated volatiles produced by wheat seeds upon heating at 60°C for six hours. Although they suggested that old seeds released more volatiles than fresh ones as a result of lipid oxidation, the source and identity of these derivatives were not precisely defined.

7.4.4 The free radical status of seeds

Electron spin resonance (ESR) spectroscopy and low-level chemiluminescence analysis have been increasingly used in recent years to investigate free radical–based processes in stored and hydrating seeds. In ESR, information is obtained by observing the resonance characteristics of an unpaired electron within a free radical when it is exposed to a combination of magnetic field and microwave irradiation. Low-level chemiluminescence (or "ultraweak" chemiluminescence) is to be differentiated from the more effective types of photoemission familiar in luminescent organisms. Low-level chemiluminescence, which has been identified in a variety of plant and animal tissues, is a frequent consequence of lipid peroxidation. Both ESR and low-level chemiluminescence analysis suffer from difficulties of interpretation, but they nevertheless yield potentially useful information for the study of seed aging.

Although organic free radicals are often considered to be highly unstable and reactive, they can be readily trapped by drying. Under these conditions, free radicals are detectable within seeds by ESR spectroscopy. Generally, an increase in hydration leads to enhanced molecular motion (see sec. 2.1) and the free radicals are destabilized; levels detected are then much lower than in the dry state (Randolph et al., 1968; Priestley et al., 1985b; see Fig. 29). Similar considerations also apply to radiation-induced free radicals in seeds (Conger et al., 1969). The nature of the free radicals observed by ESR in unirradiated seeds is poorly understood, although their stability in the dry state suggests a localization in the more polar parts of cells: lipid radicals are unlikely to be sufficiently stable because they would have to exist in a very mobile phase, even in highly desiccated material (sec. 2.1). In dried model systems of lipid and protein, free radical levels increase as a result of lipid peroxidation (Roubal, 1970); in this case, signal enhancement is presumed to arise from trapped protein radicals generated by the degrading lipid.

Despite the uncertainties inherent to ESR observations of seeds, several attempts have been made to investigate changes induced by aging. Conger and Randolph (1968), for example, examined seeds of several species with ages of up to 50 years. Overall, they found some indication that organic free radical levels declined slightly with age. Priestley et al. (1980), who examined

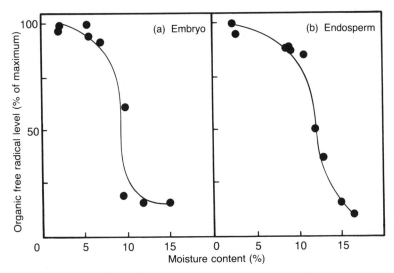

Figure 29. The effect of hydration on the organic free radical content of maize tissues. Powdered seed material was adjusted to the hydrations indicated by means of a humid atmosphere or a desiccator. The amplitudes of organic free radical signals, determined by means of ESR, were adjusted to compensate for microwave absorption by water. Adapted from D. A. Priestley et al., "Organic Free Radical Levels in Seeds and Pollen: The Effects of Hydration and Aging," *Physiologia Plantarum*, 64 (1985):91.

cotyledonary material from soybeans that had been in dry storage for periods of up to ten years, also concluded that free radical signals changed little as a result of this type of aging. Analysis of cotyledons following accelerated aging (40°C and saturating humidity) and redrying demonstrated no obvious alteration in the radical signal. More recently, though, Buchvarov and Gantcheff (1984) reported that the organic free radical content of soybean embryonic axes almost doubled as a result of a similar treatment, and intimated that comparable increases were also associated with aging over several years of storage. Priestley et al. (1985b) confirmed that the organic radicals of soybean axes increase considerably as a result of high humidity aging (but not long-term storage). They found no evidence that the elevated radical content was linked to lipid peroxidation, however, and the source of the increased signal in the embryonic axis remains unexplained.

At present, therefore, the evidence obtained from ESR measurements is incomplete and unsatisfactory; further advances will probably depend on a better understanding of the identities of the radicals concerned. Nevertheless, the information gathered so far serves to emphasize yet again the marked differences that exist between accelerated aging at high humidities (sorption zone III as defined in sec. 2.1), when free radicals are destabilized and presumably reactive, and aging in drier conditions (zones I and II), when "free" water is absent and radicals are trapped. Recent work on *Hibiscus cannabinus* with respect to free radical–mediated radiation injury suggests that the mechanism of seed deterioration may be exquisitely sensitive to hydration (Mahama and Silvy, 1982; see Fig. 3). In sorption zone I, seeds were most susceptible to radiation damage; sensitivity was somewhat lower in zone III, but in a narrow window of hydration between 8 and 11% moisture content (zone II), the seeds were much less affected by radiation than in the other two sorption zones. Although the mechanisms of radiation-induced injury in seeds may well be rather different from those involved in aging, these findings serve to emphasize that very small shifts in seed hydration can exert great influence on degradation processes induced by free radicals.

Low-level chemiluminescence, an index of free radical activity, has been increasingly subjected to analysis in seeds. Singlet oxygen represents one of the more distinctive products resulting from lipid peroxidation, and as this species decays to a state of lower energy, photons are emitted.[14] This radiation is characterized by wavelengths in the red part of the spectrum (about 630 to 700 nm). Another important species within the context of lipid degradation is the excited carbonyl group, which emits around 420–450 nm (Cadenas and Sies, 1984). Atmospheric and enzymic peroxidation of lipids, either *in vitro* or *in vivo*, leads to an increase in low-level chemiluminescence, which can be detected fairly simply by means of the photomultiplier of a scintillation counter. Intact seed tissues of soybean characteristically give rise to several hundred counts per minute per gram. In terms of spectral quality, seeds, seed particles, and lipoxygenase-plus-lipid systems show clear affinities to each other (Boveris et al., 1980). By incubating soybean seeds (14–16% hydration) at various temperatures, Boveris et al. (1983) derived an activation energy for spontaneous light emission, and were able to discriminate in this way between enzymic and nonenzymic oxidation. At room temperature, lipoxygenase was evidently responsible for most light emission from the seed (activation energy, 20 kJ/mol), but at temperatures in excess of about 40 to 45°C the activation energy was much higher (68 kJ/mol), indicative of a nonenzymic process. Changes in steady-state chemiluminescence from unimbibed soybeans as a function of age have been explored by Kiyashko (1981; Likhatchev et al., 1984), who reported that light emission increased severalfold as seeds lost germinability over 20 weeks in an incubator at 37°C. It is unclear whether enzymic activity was significant under these conditions. Perelberg et al. (1981), in contrast, briefly noted that spontaneous chemiluminescence from soybeans declined approximately 20% per year during long-

---

14. Singlet molecular oxygen is an excited form in which one electron occupies an orbital of higher energy than it would in the ground state.

term storage, but it increased as a result of accelerated aging at high humidity. Additional studies of these age-induced phenomena are desirable.

Hydration leads to an abrupt alteration in spontaneous light emission. In soybeans, there is a four- to eightfold increase during the first minute of imbibition, but thereafter changes are relatively slow (Boveris et al., 1983). According to Buchvarov et al. (1983), chemiluminescence of aged soybeans (wavelength unspecified) is lower than that of fresh material during the first minutes of hydration. Grabiec et al. (1968) also reported on the chemiluminescence (300–600 nm) of aqueous rye seed homogenates, noting a decrease with age. The signals in this case were presumably derived from a variety of reactions, some of which may not have proceeded at all readily within the intact seed. In part, though, the differences observed may be attributable to a decline in enzyme activity with age. More recently, Kolesnikov et al. (1980) found that the chemiluminescence of rice homogenates (primarily at 380–420 nm) increased during the early stages of seed aging but declined thereafter. They ascribed this phenomenon to changes in peroxidase activity.

Low-level chemiluminescence studies are appealing because the technique can be employed nondestructively and noninvasively. There is at least one major limitation, however, in applying it to intact seed tissue: light is emitted only from the most peripheral cells (Boveris et al, 1984). As the outermost cell layers are often subject to unusual stresses during imbibition (Simon and Mills, 1983), and may also be atypical compositionally or enzymically, it remains questionable whether their performance can be considered representative of the remainder of the seed.

The widespread preconception that seed aging is linked to free radical reactions has led to the development of protocols that—nominally, at least—limit the damage incurred by such activity. Some of the most intriguing work of this nature has concerned so-called cathodic protection of maize seed (Pammenter et al., 1974). Seeds at 13.6% hydration placed on aluminum foil maintained at a negative potential of 300 V declined in viability more slowly at 40°C than controls. In other tests, fully viable seeds that had suffered some deterioration in storage were subjected to "cathodic therapy" at 20°C for ten days. Rate of emergence in germination tests was reported to be enhanced by such treatments (Berjak, 1978). In explanation of these effects it has been proposed that the cathode nullifies free radicals by providing a source of electrons. However, there is as yet no direct evidence to show either that free radical–mediated injury is important during aging of this species or that the cathodic treatments depress free radical levels.

The subjection of seeds to a brief wetting-and-drying cycle has also received considerable advocacy as a method for slowing the effects of aging.

According to some workers, the period of treatment can be very short: Sahadevan and Narasinga Rao (1947), for example, noted that deterioration in rice could be checked by soaking for five minutes and then drying. Others have found that 18 or 24 hours of imbibition, followed by drying, confers considerable resistance to subsequent accelerated aging treatments (Savino et al., 1979). Similar procedures have been extensively applied to a variety of species by Basu and his co-workers, and a voluminous literature attests to their success: wheat (Basu, 1976; Basu and Das Gupta, 1974, 1978; Das Gupta et al., 1976; Rudrapal and Basu, 1979; Ray, 1982; Mandal and Basu, 1982, 1983), rice (Basu, 1976; Basu et al., 1974; Basu and Pal, 1978, 1979, 1980), barley (Punjabi et al., 1982), soybean (Saha and Basu, 1981, 1982, 1984), cotton (Dharmalingam and Basu, 1978), jute (Basu et al., 1974, 1978), sunflower (Pathak and Basu, 1980; Dey and Basu, 1982; Basu and Dey, 1983), beet (Basu and Dhar, 1979), carrot (Kundu and Basu, 1981; Pan and Basu, 1985), lettuce (Basu et al., 1979; Pan et al., 1981; Pan and Basu, 1985), tomato (Mitra and Basu, 1979), and several other species (Basu et al., 1975) have been shown to benefit from treatment. The wetting-and-drying cycles routinely employed in these studies extend over several hours, although much briefer treatments have proven effective in some cases; a period of imbibition in a humid atmosphere is also often sufficient to slow subsequent aging in storage. In general, partially deteriorated seeds have yielded the best results, whereas highly vigorous material may even suffer from the treatment. Consequently, the technique is best employed midway through a normal period of storage, before serious losses in germinability are evident. From these findings Basu (1977) has developed a practical method of treating stored seeds for use in India.

Explanations for the beneficial effects of the wetting-and-drying procedure on storability have been sought in terms of free radical metabolism; in particular, attention has been focused on analogous systems in which seed deterioration is promoted by ionizing radiation. When dry seeds are artificially irradiated, an accumulation of organic free radicals is usually detectable by ESR spectroscopy. Gradual decay of these free radicals in storage leads to loss of seed performance, but a wetting-and-drying cycle soon after irradiation serves to eliminate injury, apparently by encouraging radical–radical recombination (Haber and Randolph, 1967; Conger et al., 1968). Basu and his co-workers have pointed to similarities between age- and irradiation-induced seed deterioration, suggesting that wetting-and-drying treatments may serve similar roles in both systems (Punjabi and Basu, 1982). In support of this contention, they argue that wetting-and-drying treatments are even more effective if radioprotective agents are incorporated during the hydration phase (e.g., Basu, 1976; Das Gupta et al., 1977). Recent ESR work on soybean axes has also demonstrated that organic free radical levels can be reduced about 25% by a wetting-and-drying experience (Priestley et al., 1985b); but although it is possible that the free radical status of seed tissues may be effectively modulated by such treatments, it is unclear whether any direct physiological benefit necessarily accrues. It is possible, for example, that some advantages are to be gained for partially deteriorated seed through priming of germinative processes or by enzymic repair (sec. 7.11.1), although on the basis of work with the inhibitor cycloheximide, Pan et al. (1981) have claimed that cytoplasmic protein synthesis is not implicated in the enhanced resistance to aging. It is relevant to note also that Bagchi (1974) suggested that a correlation existed between resistance to gamma radiation (undoubtedly a free radical–mediated stress) and long-term storability in seeds of various rice strains. Fresh seeds of genetic lines that possessed poor storage characteristics were also found to be particularly radiosensitive. Unfortunately, only two susceptible and two resistant stains were investigated; further corroboration of this interesting finding is desirable.

Despite much work, it is still difficult to define the role of lipid peroxidation in seed aging, and a satisfactory understanding of the importance of free radicals in deterioration is even more elusive. Almost any generalization that may be made can be controverted by at least one significant observation. Seed lipids do tend to undergo peroxidative degradation in long-term dry storage, for example, but not always before viability is lost. Further, when seeds age under conditions of high humidity, they may or may not exhibit signs of lipid peroxidation; to compound the complexity, free radicals have apparently increased as a result of aging in some studies, but not in others. Despite these confusions and ambiguities, one conclusion is apparently unassailable: seeds will not retain viability if their lipids are subjected to extensive peroxidation. That fact, at least, attests to the importance of this component for seed survival.

7.5 Changes in the structure and chemistry of proteins

Proteins serve two major functions in seeds, acting both as storage reserves and as enzymes. Most work concerning alterations in protein structure with seed age has addressed the storage fraction. Much less is known concerning enzymic structure, although changes in enzyme activity with age have been quite extensively cataloged (see sec. 7.6). Deleterious changes in storage proteins could presumably be indicative of similar alterations in enzymes.

One hypothesis that formerly enjoyed some favor suggested that seed aging was determined by the rate of protein denaturation. Crocker and Groves (1915), for example, pointed to evidence that both seed aging and protein denaturation *in vitro* were similarly influenced by temperature and moisture. Several investigators have analyzed seed protein characteristics (primarily those of the storage fraction) as a function of storage time. Solubility properties have commonly been found to change over several months or years, indicating that alterations in protein structure certainly occur. Such effects have been noted, for instance, in wheat (Jones and Gersdorff, 1941), maize (Jones et al., 1942), peanut (Moorjani and Bhatia, 1954), *Phaseolus vulgaris* (Grange, 1980), and soybean (Echigo, 1965; Saio et al., 1980). Nikolova and Dencheva (1984) found that decreased solubility of maize endosperm proteins was associated with declining viability, and Grzesiuk and Kulka (1971b) reported much lower levels of extractable albumins (water-soluble proteins) from unimbibed oat seeds that had lost viability in long-term dry storage. Such pronounced trends were evidently absent, though, from soybean seeds deteriorated at 15% moisture content (Grzesiuk and Łuczyńska, 1972). Most investigators, however, have not attempted to relate such changes to loss of germinability, nor is it clear by what means the proteins in seeds become denatured.

Proteolysis has not been widely linked to seed aging, although Ovcharov and Genkel' (1973) claimed that declining levels of protein in the embryo and endosperm fractions of maize (14% hydration) were related to loss of viability in long-term storage; they also suggested that the number of protein components that could be resolved electrophoretically diminished considerably with age. González-Juliá et al. (1983) also described alterations in the electrophoretic profile of albumins and globulins (salt-soluble proteins) from pine seed (*Pinus pinea*) subjected to accelerated aging at 45°C and saturating humidity. According to data of Krishtofovich (1974), however, such changes are seen in wheat only at relatively high moisture content, and they are absent from seeds aging at 10–11% moisture content. Certainly, at high sorption levels, the effects of microbial or fungal proteinases on stored seeds cannot be discounted entirely (Cherry, 1983).

As section 7.4.3 noted, changes in protein structure during storage may arise from interactions with lipid peroxidation products: scission, polymerization, and other forms of deterioration are possible. Crowe (1971) has suggested that the oxidation of the sulfhydryl groups of cysteine residues may be a particularly important factor in limiting the longevity of dried organisms, leading to protein malfunction. An oxygen-dependent loss of sulfhydryls has been shown to occur in wheat flour during storage (Yoneyama et al., 1970), and similar changes have been suggested to take place within seeds. Buch (1960), for example, reported that sulfhydryl levels were unusually high in some types of tree seeds with poor storage characteristics, and he intimated that aging in these species may be linked to protein lability. According to Sutulov (1965), sulfhydryl oxidation is associated with loss of viability in such crop seeds as wheat and sunflower, and Lazukov (1969a) made similar claims for some species of crucifer. Semerdzhyan et al. (1973)

likewise noted that thiol levels decreased in wheat embryos with age, and Stefanov and Dencheva (1984) have suggested that the disulfide/sulfhydryl ratio in maize storage proteins increases as viability is lost. Miranenko et al. (1981) also reported a loss of protein sulfhydryls accompanying declining germinability in yellow lupine. In these experiments, though, seeds hydrated to 34% or 43% moisture content were "aged" by drying in an oven at temperatures of 40 to 65°C. Despite some initial indications of this nature, the mechanisms by which proteins are altered during aging still awaits detailed elucidation. Nevertheless, from the limited information available it is clear that age-induced changes in seed proteins are relatively commonplace.

One aspect of deterioration that has received considerable attention from cereal chemists is the so-called sick wheat syndrome, in which seeds stored at high sorption levels (zone III as defined in sec. 2.1) suffer discoloration of the embryo. Elevated fat acidity and fungal activity are often associated phenomena, although it is likely that neither is essential for darkening to occur (Cole and Milner, 1953). The discoloration probably arises as the result of a condensation reaction between lysine or methionine residues of proteins and reducing sugars—the Maillard reaction (McDonald and Milner, 1954; Feeney and Whitaker, 1982). Elevated levels of reducing sugars frequently arise in seeds stored at high sorption levels (see sec. 7.7.1), providing a convenient stimulus for the reaction (Linko et al., 1960). Germ damage of this type may be a late event in deterioration, as visibly "sick" wheat is usually not viable.

7.6 Enzymes

7.6.1 Changes in enzyme activity induced by aging

When dry seed material is placed in water, many previously latent enzymes become immediately detectable. These activities are often considerably reduced in aged seeds, presumably as a consequence of deleterious structural changes of the type characterized in section 7.5. MacLeod (1952), for example, assayed several enzymes in barley seed of various ages up to 81 years, and found that all declined in activity within a few decades. Proteinase, β-amylase, phosphatase, and catalase decreased very slowly and were detectable at fairly high levels even after germinability had been lost, whereas peroxidase activity declined more abruptly and was closely linked to deteriorating viability. Perl et al. (1978) similarly found that the levels of total amylase, acid phosphatase, alanine aminotransferase, glutamate decarboxylase, and ribonuclease decreased in sorghum seeds subjected to prolonged accelerated aging at elevated humidity.[15]

15. Enzyme nomenclature has been regularized to conform to the recommended forms listed

Some enzymes are absent or inactive in newly hydrated seeds, and they reach readily detectable levels only after a few hours of development. The rate at which new enzymes appear is often very slow in deteriorated seeds, probably because of deficiencies in *de novo* protein synthesis (sec. 7.9). Thus, in the case of alcohol dehydrogenase in peas, for example, prolonged storage leads to a lower extractable activity from the unimbibed seed as well as slower development of additional activity during imbibition (Modenesi, 1977, 1979). On the other hand, diminished levels of isocitrate lyase in aged soybeans (Rao and Wagle, 1983) and cauliflower seeds (Houlberg and Madsen, 1985) during imbibition are due solely to ontogenetic deficiencies (Rao and Wagle, 1983). Many studies have indicated that a range of enzymic activities are present at abnormally low levels in aged seeds during the first 24 hours of germination, although it is not always clear whether these shortcomings are due to degradation of latent enzymes in the stored seed or arise less directly from a failure of proper development. The catalog of deficiencies includes amylases (Paul et al., 1970; Krishtofovich, 1975; Boshnakov and Ivanov, 1980; Saxena and Maheshwari, 1980), lipase (Łuczyńska, 1973; Boshnakov and Ivanov, 1980), proteinases (Mitra et al., 1974; Nowak and Mierzwinski, 1978), ascorbate oxidase (Nagata et al., 1964), cytochrome c oxidase (Ching, 1972), glyceraldehyde-phosphate dehydrogenase (Harman et al., 1976), succinate dehydrogenase (Nagata et al., 1962), succinate-semialdehyde dehydrogenase (Galleschi and Floris, 1978), alcohol dehydrogenase (Sutamihardja et al., 1969), pyruvate decarboxylase (Nagata and Hayashi, 1966), and various phosphatases (Sircar, 1970; Ching, 1972; Krishtofovich, 1975; Bukhtoyarova et al., 1979a), as well as activities that degrade ATP, phytin, and RNA (Nagata et al., 1964; Paul et al., 1970; Łuczyńska, 1973; Sedenko, 1975; Bukhtoyarova et al., 1979a). The effectiveness of the trypsin inhibitor of rye seeds at early stages of germination is also lowered by several years of storage (Sobieraj and Kulka, 1983). The depression of enzymic activity in old seeds has sometimes been considered useful for testing purposes, and is discussed further in section 7.6.2.

A few studies have suggested that enzymic activities may increase in seeds as a consequence of aging. Rao and Wagle (1981), for example, found that β-amylase activity was higher in soybeans following accelerated aging (40°C and 94% relative humidity). Proteinase has also been noted to increase in wheat (Shvetsova and Sosedov, 1958) and sorghum (Perl et al., 1978) following aging at fairly high sorption levels. In addition, Mazzini et al. (1980)

---

by Dixon and Webb (1979). Enzyme Commission (EC) numbers for the various enzymes discussed can be determined from the same listings. It should be noted that terms such as "proteinase" or "phosphatase" are generic in character, rather than referring to specific enzymes of well-defined function.

claimed that proteinase activity increased in stored rice during the late summer months (when water content was 10–12%), although losses in germinability were relatively minor under these conditions. González et al. (1982) also reported that accelerated aging of *Pinus pinea* seed (45°C, saturating humidity) resulted in the detection of increased levels of proteinase during germination. Grange et al. (1980), who examined seeds of *Phaseolus vulgaris* exposed to humid storage, found that endopeptidase activity declined in the early stages of aging but increased thereafter; Zelenskii and Zelenskaya (1983) recorded a similar pattern for proteinase activities in soybean seeds aged at 13% hydration and 37°C. Others have suggested that ribonuclease activity is much more evident during imbibition of peas that have previously been subjected to accelerated aging (De Leo et al., 1973; Dell'Aquila and De Leo, 1974). Rice embryos that had lost viability during prolonged storage in India have also been reported to possess elevated ribonuclease levels (Ghosh and Chaudhuri, 1984), and Agrawal (1981) similarly associated enhancement of IAA-oxidase with aging in chickpea and broad bean.

Reports of increased enzyme activity during seed deterioration almost always derive from studies in which aging has been accelerated at high humidity. Increased enzyme activity as a result of prolonged dry storage has been noted in the case of rye embryonic DNAse (Cheah and Osborne, 1978), however, and it seems possible that declining levels of the enzyme's natural inhibitor may be responsible for this phenomenon. Schimpff et al. (1978) also argued that ribonucleoside-diphosphate reductase activity was present in higher amounts in imbibed, aged wheat embryos as a result of deficiencies in an oligosaccharide inhibitor.

Many enzymes exist in multiple forms, or isozymes, that vary slightly in composition and are coded by different genes. Frequently, separate compartments within a cell will each possess a distinct isozyme of a particular enzyme. A number of electrophoretic studies have documented changes in isozyme profiles in aged seeds. In rice, aging in storage is associated with the preferential loss of certain forms of peroxidase (Shutova et al., 1973; Yarosh and Antonova, 1978; Kolesnikov et al., 1980; Shi et al., 1982), acid phosphatase (Bukhtoyarova et al., 1982), malate dehydrogenase, glutamate dehydrogenase, and alcohol dehydrogenase (Morgunova et al., 1975; Krasnook et al., 1976). Similar results have been reported for succinate dehydrogenase and alcohol dehydrogenase in maize (Ovcharov et al., 1978), malate dehydrogenase and alcohol dehydrogenase in barley (Yarosh and Antonova, 1978), esterase in cabbage seed (Ostromecki, 1977), and peroxidase in barley (Yarosh and Antonova, 1978), maize (Klisurka and Dencheva, 1984), and wheat (Shi, 1981), although isozymes of alcohol dehydrogenase in the latter species were reported to show minimal change with age, and alterations in malate dehydrogenase were not marked (Krishtofovich, 1980). Pitel (1982) also found fairly minor quantitative and qualitative changes in the esterase, glutamate dehydrogenase, leucine aminopeptidase, and peroxidase isozyme profiles of jack pine (*Pinus banksiana*) seed subjected to accelerated aging at 43°C and saturating humidity. Limited data are available concerning the cellular location of the more susceptible isozymes in stored seeds. According to Krasnook et al. (1979), degradation of malate dehydrogenase isozymes was evident in both the cytosol and a crude mitochondrial fraction from aged rice embryos. Changes in isozyme profiles are often very marked during the germination of vigorous seeds because of *de novo* protein synthesis, and such developmental transformations are slowed or absent in badly deteriorated seeds (Akhmedov and Ovcharov, 1974; Krasnook et al., 1976, 1979).

Several factors are likely to complicate investigations of enzyme performance as a function of age. One obvious concern, particularly relevant when seeds deteriorate at relatively high levels of hydration, is that some reported increases in hydrolytic activity may be due to fungal invasion. A second problem relates to the simplifying character of most sampling protocols. Thus, although lipase in castor beans has been reported to decrease with age (Guillemet, 1931), it evidently does so more rapidly in the peripheral layers of the endosperm than deeper within the tissue (Kunert, 1966). In most investigations, differences within seeds (and among seeds) have been averaged before any analysis is made. Other uncertainties can arise from the mechanics of assay. The use of centrifugation to prepare "soluble" supernatant fractions for analysis, for example, may introduce bias in some cases: Eldan et al. (1974) found that aging in lettuce was associated with a much greater pelletability of isocitrate lyase activity from the imbibed seed. There have also been suggestions that the soluble fraction of aged seeds may contain potent inhibitors of enzyme function (Modenesi, 1978; Riggio-Bevilacqua, 1979). Centrifugation, followed by resuspension, could obviously minimize such influences on membrane-bound activities. Additionally, seed enzymes are often assayed under physiologically unrealistic conditions, so that an enzyme's activity *in vitro* may yield an inaccurate indication of its performance *in vivo*. Particular complexity is introduced when aging tissues are examined. From a study of oats, for example, Kulka (1971a) suggested that the optimal pH for embryonic proteinase shifted downward with age; concurrently, the activation energy for reaction decreased. More recently, Zelenskii (1982) investigated the pH optimum of proteinase activity in soybean seeds of different ages, noting a variability of about one unit of pH between samples; however, he observed no trend in this parameter with increasing length of storage. Clearly, it may sometimes be misleading to draw conclusions about enzyme functioning *in vivo* on the basis of a single arbitrary set of assay conditions. Nevertheless, it appears to be beyond dispute that many enzymes suffer serious deterioration as a consequence of seed aging and that others fail to develop to normal levels during germination.

It has frequently been suggested that depletion of coenzymes may be responsible for reduced enzymic activity within aged seeds. Particular attention has been focused on the assay of thiamin (vitamin B_1), which is active in cells as thiamin pyrophosphate (TPP), and nicotinic acid (niacin), which is functional as nicotinamide adenine dinucleotide (NAD^+) or its phosphate ($NADP^+$). Analyses of stored grain have been somewhat contradictory in establishing the rate at which these coenzyme precursors are lost in storage (Burton, 1982); nevertheless, several attempts have been made to link coenzyme deficiencies to the seed aging process. Kondō and Okamura (1932/33) reported that conditions that favored loss of germinability in rice were also conducive to decreased thiamin content, and Sutamihardja et al. (1970) suggested that loss of thiamin and nicotinic acid from wheat seeds was particularly evident under conditions of humid storage, when viability decreased rapidly. Ovcharov and Murashova (1973) noted very briefly that depressed levels of riboflavin (vitamin B_2) in maize embryos were linked to lowered viability; two of its derivatives, flavin mononucleotide (FMN) and flavin adenine dinucleotide (FAD), are important coenzymes in mitochondrial electron transport. Decreased pyridoxine (vitamin B_6) has been associated with loss of viability in wheat (Oe and Nagata, 1969), although changes in the embryo were found to be particularly slight; phosphorylated derivatives of pyridoxine are especially significant in transamination reactions. Depressed levels of pantothenic acid (vitamin B_3) and biotin (vitamin B_7) in wheat also occur under conditions that lead to diminished germinability (Sutamihardja et al., 1970); pantothenic acid is an integral component of coenzyme A, which is particularly important in fatty acid metabolism, and bound biotin serves as coenzyme in some carboxylation reactions.

Despite this rather extensive list linking diminished levels of coenzymes or their precursors to loss of germinability, there is at present little support for the contention that such phenomena represent the primary cause of seed deterioration. The mechanism and kinetics of coenzyme degradation during storage are poorly understood, and it seems likely that destruction of coenzymes is merely one of many symptoms constituting the aging syndrome in seeds.

7.6.2 The use of enzyme assays in testing for vigor and viability

A few enzymic activities have received particular attention because of their possible use as rapid indicators of seed vigor or viability. "Phenolase" (Davis, 1931) and peroxidase (Brucher, 1948) assays were once regarded in this light; both activities tend to decline as seeds age, but they can also yield somewhat unreliable estimates of seed performance. The measurement of catalase activity was formerly popular, but it can also be unpredictable in identifying older, less vigorous seeds (Gračanin, 1927; Knecht, 1931). Reliability may be improved by a comparison of activities before and after imbibition: in vigorous seeds, catalase levels increase during germination, but soaking of nonviable material leads to a loss of enzymic activity (Davis, 1926; Singh et al., 1938). More recently, the assay of glutamate decarboxylase as an indicator of aging has received some support (e.g., Cseresnyes, 1971; Mora and Echandi, 1976), although it, too, is plagued by problems of variability, at least in some species (James, 1968; Azizul Islam et al., 1973).

By far the most important form of enzymic seed testing has involved the assay of respiratory dehydrogenase activities, functions that are usually presumed to reside in mitochondria. Historically, a number of redox reagents have been employed for this purpose, including selenium and tellurium salts, methylene blue, and malachite green (Delouche et al., 1962). These substances have now been almost universally succeeded by triphenyltetrazolium chloride (Lakon, 1942), which is reduced by vigorous seed tissues to a colored, nondiffusable formazan product over one or more hours (Fig. 30). It has sometimes been suggested that the predictive success of tetrazolium testing provides presumptive evidence that the respiratory chain is the site of a principal lesion in seed deterioration (discussed by Woodstock, 1973); however, the physiological basis of tetrazolium testing is not always very clear: a negative response need not necessarily point to enzymic deterioration.

The redox potential of triphenyltetrazolium chloride is approximately -0.08 V. In theory, pyridine-linked dehydrogenase systems (those requiring NAD^+ or $NADP^+$ as coenzyme) may reduce it to the formazan, since they are considerably more negative in redox value (Jensen et al., 1951). Much of this reducing activity is doubtless associated with the respiratory chain, although suitable dehydrogenases may also be available outside of the mitochondrial compartment. Moreover, formazan production is not solely dependent on enzymic integrity, as lack of coenzymes or substrates in leaky cells can also lead to a negative response. Working with partially deteriorated poplar seeds (*Populus nigra*), Zheng et al. (1964) reported that dehydrogenase activity could be restored by the addition of NAD^+ or $NADP^+$, although completely nonviable seeds did not respond (Tang et al., 1964). Similarly, Powell and Matthews (1981a) were able to restore tetrazolium-reducing activity to deteriorated pea seeds by supplying succinate as substrate.[16] A failure to reduce tetrazolium is not inevitably indicative of fundamental lesions in

---

16. The redox potential of the succinate-fumarate couple is slightly less negative than that for tetrazolium reduction, but in practice succinate dehydrogenase will serve to produce formazan (Kun and Abood, 1949).

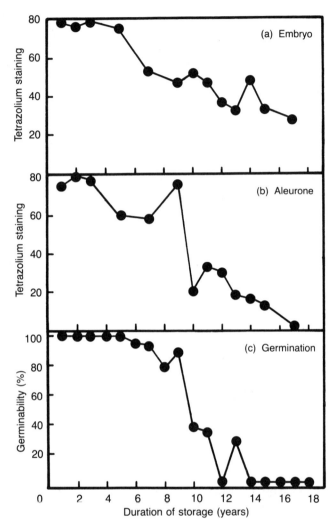

Figure 30. Tetrazolium staining of wheat seeds. Seeds were harvested in successive years from wheat (cv. Seewari) grown in Glen Osmond, South Australia. For the tetrazolium test, samples of 20 grains from each lot were used. Each grain was halved, stained, and scored: 2 for intense staining, 1 for light staining, and 0 for no coloration. A score of 80 consequently indicates that all 40 half-grains within a sample stained intensely. Adapted from D. Aspinall and L. G. Paleg, "The Deterioration of Wheat Embryo and Endosperm Function with Age," *Journal of Experimental Botany* 22 (1971): 926, 931, by permission of Oxford University Press.

mitochondrial function, therefore. An additional complexity in the use of this reagent concerns its penetrability. Triphenyltetrazolium chloride sometimes enters more readily into aged seeds, presumably because of their declining cellular integrity (Byrd and Delouche, 1971; Powell and Matthews, 1977). When seeds are equally permeated, however, old seed tissue usually produces less formazan than younger material.

7.7 Respiration

Age-induced alterations in stored seeds are ultimately revealed as metabolic deficiencies during germination. The respiratory characteristics of deteriorated seeds during imbibition have received particular attention from investigators. The earlier literature on this topic has been reviewed at some length by Abdul-Baki and Anderson (1972) and Anderson (1973). Briefer surveys have been given by Abdul-Baki (1980) and Anderson and Baker (1983).

7.7.1 Quantitative and qualitative changes in metabolic reserves

It has been well documented that respiratory changes in stored seeds (or the activity of the seed microflora) can lead to reductions in dry weight. In most cases, the losses incurred are quite small, even at relatively high levels of sorption. Chudoba (1972), for example, who investigated rape seed stored at 20°C for ten weeks, found no weight loss at 5% hydration, a decrease of 0.4% at 11% hydration, and a 1% loss at 13% hydration. Similarly, Kurdina (1963) indicated that the losses of dry weight in various vegetable seeds stored at 11–12% hydration and 20–22°C were generally much lower than 1% over a year. On the basis of gravimetric analyses, therefore, respiratory depletion of nutrient reserves within seeds is usually trivial during long-term storage.

Direct chemical analysis of seed reserves leads to similar conclusions: lipids (see sec. 7.4.1), proteins (e.g., Ching and Schoolcraft, 1968; Gryzesiuk and Łuczyńska, 1972; Srivastava and Sareen, 1974), and starch (e.g., Kondō and Okamura, 1934b; Ching and Schoolcraft, 1968; Agrawal, 1978) are usually present at relatively high levels in aged seeds. Stored sugars, however, may constitute a more important reserve during the first hours of imbibition, and there are some indications that substantial declines in sugar content are associated with aging (e.g., Ching and Schoolcraft, 1968; Edje and Burris, 1970; Mitra et al., 1974; Ovcharov and Koshelev, 1974; Tsvetkov et al., 1975; Agrawal, 1977; Likhachev et al., 1978), although others have noted the opposite trend (Teixeira et al., 1980). There may also be important qualitative changes in this fraction. At high sorption levels (zone III as defined in sec. 2.1), the ratio of reducing sugars to nonreducing sugars tends to increase (Zeleny and Coleman, 1939; Ramstad and Geddes, 1942). An elevation of the levels of reducing sugars in seeds during storage is probably a destabilizing factor, as these compounds are prone to participate in deleterious Maillard reactions with some amino acids (see sec. 7.5).

Although large-scale metabolite depletion does not accompany seed aging in storage, subtler changes have been proposed as a possible explanation for loss of viability. Harrington (1973), for example, noted that an insufficiency of metabolites within a few key embryonic cells may account for declining germinability (see also sec. 3.1.1). Transport of reserves to the embryo probably also suffers as a result of aging: Abdul-Baki and Srivastava (1973) pointed to deficiencies in the transfer of amino acids from the cotyledons to the embryonic axis of deteriorated lima beans during the first days of germination. Others have suggested that depletion of one important intermediate metabolite may account for decreased performance. Täufel and Pohloudek-Fabini (1955), for example, found that accelerated aging at 32°C and an elevated (but unspecified) level of hydration led to a decrease in citrate levels in six species. This depletion was closely linked to loss of germinability. Linko and Milner (1959a, 1959b) similarly proposed that seed viability may be imperiled by depletion of other tricarboxylic acid cycle metabolites such as α-ketoglutarate and oxalacetate, which they suggested may result from limited enzymic activity at relatively high sorption levels. Wyttenbach (1955), on the other hand, intimated that declining viability in alfalfa was associated with accumulation of lactate, presumably indicative of glycolytic activity in the absence of mitochondrial oxidation. Although shifts in metabolic intermediates undoubtedly occur in seeds stored at relatively high levels of hydration, the significance of such changes for seed deterioration remains unclear.

7.7.2 Alterations in respiratory pathways

Low rates of oxygen uptake have been commonly observed during imbibition of aged seeds (e.g., Woodstock and Feeley, 1965; Woodstock and Grabe, 1967; Abdul-Baki, 1969; Filipenko, 1983; see Fig. 31), and both embryonic and nonembryonic parts of seeds may exhibit depressed oxygen consumption (Wahab and Burris, 1971; Kulka, 1971b; Paul and Mukherji, 1972; Parrish and Leopold, 1977). In wheat, however, Anderson and Abdul-Baki (1971) found that embryonic oxygen consumption decreased as a result of deterioration, whereas the endosperm (presumably the aleurone) was unaffected. Carbon dioxide evolution is often unnaturally high during the first hours of imbibition for slightly deteriorated seeds (Anderson 1970; Anderson and Abdul-Baki, 1971), although the cause of this increase is uncertain. Woodstock and Taylorson (1981a) have suggested that imbalances between glycolysis and the tricarboxylic acid cycle may lead to an accumulation of glycolytic products such as pyruvate, and that these products may in turn become subjected to decarboxylation. Whatever the complete explanation

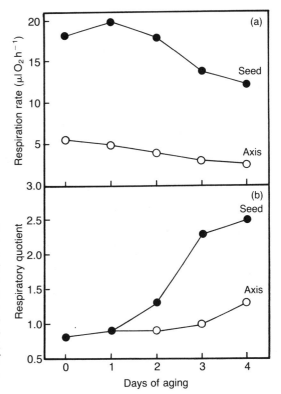

Figure 31. Respiration of soybean embryonic axes and whole seeds. Respiration (per seed or per axis) and respiratory quotients were measured after 4 hours of imbibition. Seeds had previously been subjected to up to 4 days of accelerated aging at 41°C and saturating humidity. A 4-day exposure to the aging treatment was sufficient to decrease seedling length from 140 mm to 71 mm (recorded after 4 days of germination). Data of Woodstock et al., 1984.

may be, an elevated respiratory quotient (the molar ratio of carbon dioxide produced to oxygen consumed) is often characteristic of aged seeds early in imbibition (e.g., Woodstock and Grabe, 1967; Anderson 1970; Byrd and Delouche, 1971; Woodstock et al. 1984; Fig. 31).

Depressed rates of oxygen consumption have usually been ascribed to mitochondrial deterioration. In vigorous seeds, the first hours following imbibition are characterized by considerable mitochondrial differentiation, either by synthesis of new organelles or by modification of preexisting ones (Nakayama et al., 1980; Morohashi et al., 1981). During this period, the ability of mitochondria to oxidize exogenous substrates *in vitro* is greatly enhanced, and coupling between oxidation and phosphorylation becomes increasingly evident (Pradet, 1982).[17] Rates of oxygen consumption during imbibition of aged seeds may therefore be depressed by two factors: direct lesions in

---

17. Although this developmental scenario has been accepted for many years, work on hydrating pollen indicates that mitochondrial deficiencies early in germination may arise partly as an artifact of isolation (Hoekstra and van Roekel, 1983).

mitochondrial structure induced by aging (which may be subject to repair in the later stages of imbibition) and sluggish mitochondrial development.

Evidence for mitochondrial lesions was adduced by Leopold and Musgrave (1980) when they worked on soybean seeds that had been subjected to accelerated aging (40°C and saturating humidity) and that displayed relatively low levels of oxygen uptake. On the basis of the response of intact tissues to inhibitors, these workers proposed that aging of the embryonic axis causes a deterioration of the cyanide-sensitive electron transport chain through cytochrome c oxidase; in compensation, a less efficient cyanide-insensitive alternative electron transport pathway is engaged. In unaged axes, the alternate pathway is apparently inactive, and its normal function remains obscure. If we assume that the inhibitors used in this study were specific for the two mitochondrial pathways, the results suggest that deficiencies in electron transport are detectable well within the first hour of imbibition. Working on the same tissue, Woodstock et al. (1984) have similarly noted that oxygen consumption is lower in aged embryonic axes within the first minutes of imbibition.

Evidence suggestive of inferior mitochondrial development during germination of aged seeds has also been reported. Woodstock et al. (1984), for example, compared mitochondria from axes of vigorous and long-stored soybeans. After 4 or 24 hours of imbibition, mitochondria from aged seeds displayed depressed rates of oxidation with succinate as substrate, lower respiratory control, and a decrease in the amount of inorganic phosphate esterified into ATP per unit oxygen consumption (the ADP:O ratio). Abu-Shakra and Ching (1967), who investigated mitochondria from aged and unaged pea embryonic axes at a relatively late stage in germination (four days), concluded that the organelles were fewer and less efficient in seedlings from aged material, presumably because of improper development. Axes from aged seeds exhibited ADP:O ratios that were about half those from unaged ones, and application of the respiratory uncoupler dinitrophenol (DNP) confirmed that mitochondria from aged axes were inherently less well coupled. Similar conclusions were reached by Tłuczkiewicz (1980) with mitochondria from rye after one to two days of germination, and by Sojka and Zaidan (1983) with mitochondria from wheat embryos after two days. Bukhtoyarova et al. (1979b) have also argued that respiratory uncoupling is characteristic of seed deterioration on the basis of DNP application to intact germinating rice seeds of various ages.

The pentose phosphate pathway represents a second glucose-utilizing respiratory sequence in seeds, in addition to glycolysis. An estimate of the importance of this pathway in seed metabolism has usually been achieved by means of glucose labeled at the C6 or C1 position. The quantity of $^{14}CO_2$ respired following application of glucose-6-^{14}C is compared with that released from similar tissue supplied with glucose-1-^{14}C; for simplicity, the results are usually expressed as a ratio (the C6/C1 ratio). If glycolysis alone is significant, the C6/C1 ratio is theoretically equal to 1, because both labeled carbons (C1 and C6) in glucose are metabolized to carbon dioxide. When the pentose phosphate pathway is operational, the C6/C1 ratio decreases, as only C1 is involved in a decarboxylation step. Takayanagi (1977) found direct evidence for a change in emphasis between the glycolytic and pentose phosphate pathways in aging rape seed. For fresh material (99% germinable) the C6/C1 ratio in the first 3 hours of imbibition was 0.5 to 0.6, rising in excess of 0.9 after 24 hours; the pentose phosphate pathway appeared to play an important role, therefore, in the early stages of imbibition of unaged seeds, although glycolysis was almost completely predominant after 24 hours. For seeds previously subjected to accelerated aging at 30°C and 74% relative humidity (82% germinable), the C6/C1 ratio was about 0.8 during the initial stages of imbibition, but had dropped to 0.4 after 24 hours. Moreover, in aged seeds, release of $^{14}CO_2$ from both labels was several times less than that from fresh material. Recent work of Kharlukhi and Agrawal (1984) appears to corroborate these findings. After 24 hours of imbibition, the C6/C1 ratios determined for artificially aged mung bean, chickpea, and wheat seeds were in the order of 0.5 to 0.6, whereas values were close to unity in fresh material. Findings of this nature serve to emphasize yet again that aging promotes unusual metabolic imbalances in the imbibing seed. It is important to bear in mind, though, that deterioration does not affect all tissues and metabolic pathways in the same manner. In cereal grains in particular, metabolic alterations induced by aging may be quite dissimilar in the embryo and in the endosperm (Anderson, 1973).

7.7.3 ATP content and energy charge

The content of adenosine triphosphate (ATP) in seeds has received increasing scrutiny in recent years, in part because of its potential as a vigor test (Ching, 1982). In view of the central role of ATP in metabolism, it is self-evident that vigorous seed tissue must maintain an appropriate supply. Levels of ATP and energy charge tend to be very low in dry seeds, but they increase substantially during early imbibition.[18] Some researchers have suggested that oxidative phosphorylation develops far too slowly to account for this sudden surge (Mayer, 1977), although at least one report has intimated that oxidative phosphorylation may well be responsible for ATP generation within minutes of imbibition (Raymond et al., 1982). Perl (1980), on the

18. Energy charge $= \dfrac{[ATP] + \frac{1}{2}[ADP]}{[ATP] + [ADP] + [AMP]}$

An energy charge of 0.8 or more is usually considered to be characteristic of actively metabolizing cells.

other hand, has argued that ATP accumulates rapidly in newly hydrating seeds through the agency of a specialized enzymic system utilizing AMP, phosphoenolpyruvate, and inorganic orthophosphate, but the extent to which such a system may operate *in vivo* remains unclear.

Several investigators have reported that ATP levels increase only slowly during the hydration of aged seeds. Thus Ching (1973), who examined crimson clover seeds after four hours of imbibition, found that ATP content ranged from about 2.2 nmol per seed in fully viable materials down to 0.02 nmol per seed in aged, nongerminable seeds. Negative correlations between total ATP content and aging (at 30°C and high moisture content) were also established by Ching and Danielson (1972) for lettuce seeds. Van Onckelen et al. (1974) also found that ATP levels were depressed after six hours of imbibition in both embryos and endosperms of barley that had previously been subjected to accelerated aging (43°C and 85% relative humidity). Diminished ATP accumulation in imbibing soybean axes likewise reflects the effects of aging at 20% hydration and 35°C (Anderson, 1977). Lunn and Madsen (1981) observed that total ATP content increased more slowly during imbibition of aged cauliflower, rape, and sugar beet seeds (see Fig. 32); Banarjee et al. (1981), working with rice, and Zheng et al. (1982), using cabbage seed, similarly noted that ATP levels reflected age-induced deterioration. There is also evidence that the accumulation of other nucleoside triphosphates (such as uridine triphosphate) is hindered during the imbibition of low-vigor seeds, although aged material has not been specifically studied (Standard et al., 1983).

Despite the apparently quite widespread success of ATP measurements in predicting seed quality, some workers have questioned their utility. At a practical level, Styer et al. (1980), studying maize, radish, onion, and cucumber stored under various conditions (5–25% hydration, -10 to 35°C), reported that ATP content at four hours of imbibition usually did not correlate with changes in germination, germination rate, or radicle length of most seeds. Further, Harman et al. (1976) have cautioned that the time during imbibition at which measurements are made can have a profound effect on the result. They reported that the ATP content of their pea embryo system oscillated during the first 24 hours of imbibition. Differences between aged and unaged axes were most evident at 9 hours, less so at 3 hours, and absent at 15 hours. At least one report has suggested that the ATP content of soybean axes may increase as a result of accelerated aging at high humidity (Buchvarov and Alekhina, 1984).

The use of ATP measurements in vigor testing has also been criticized on theoretical grounds: Mazor et al. (1984) have argued that ATP is subject to very rapid turnover and that there is no reason to assume that steady-state levels necessarily reflect rates of synthesis or utilization. Future clarification

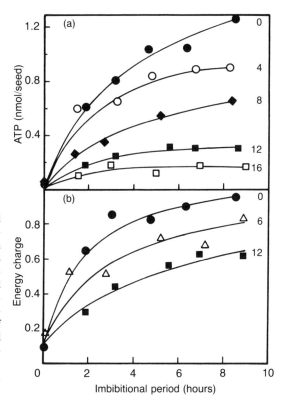

Figure 32. Accumulation of ATP and development of energy charge in cauliflower seeds during imbibition as a function of deterioration. The seeds were subjected to accelerated aging (0–16 weeks) at 16–17% hydration and 25°C. Unaged seeds were of high germinability (about 93%) and were rendered nonviable by 16 weeks of treatment. Adapted from G. Lunn and E. Madsen, "ATP-Levels of Germinating Seeds in Relation to Vigor," *Physiologia Plantarum* 53 (1981): 165, 167.

of these issues is obviously required. It should be evident, however, that if age-induced lesions prevent ATP generation, the synthesis of many cellular constituents will be overwhelmingly restricted (Abdul-Baki, 1969; Anderson and Abdul-Baki, 1971; sec. 7.9).

7.8 Chromosomal aberrations and deterioration of DNA

It has long been evident that seedlings grown from aged seeds show a high incidence of morphological abnormality. Nobbe (1876) drew attention to this phenomenon in his classic manual on seeds, and Burgerstein (1895) produced quantitative data: in maize, abnormalities increased steadily, from about 4% of seedlings produced from one-year-old seed to 54% of seedlings grown from nine-year-old material. In many cases these aberrations no doubt arise from localized necroses or from hindered cell division in meristematic tissues (Gori, 1956; Ching, 1958; Hunter and Presley, 1963; Triolo et al., 1975), but it is certain that genetic mutation is also an important concomitant of seed aging. Several reviews have addressed the genetic aspects of seed deterioration: Roberts (1972b, 1973a, 1978), Orlova (1972), Roos (1980, 1982), Roberts and Ellis (1984).

7.8.1 Chromosomal aberrations

Cytological observations of mitotic cells provided the first concrete evidence that seed deterioration is linked to changes in genetic material. Navashin (1933a, 1933b; Nawaschin, 1933) was the first to demonstrate that aging of *Crepis tectorum* seeds was associated with an increase in chromosomal aberrations in the plants grown from them. Less than 1% of plants grown from fresh seeds were affected, but after five years of storage (when germinabilty had slipped to about 40%) the great majority displayed irregularities.[19] A similar increase in chromosomal aberration due to age was recorded almost simultaneously by Peto (1933), working with maize; Peto also reported that heating the seeds to 95°C for 25 minutes (at 9% moisture content) induced similar effects, and found that mutant cells were gradually eliminated as the seedling developed. Navashin likewise established that the effects of aging could be mimicked by heating seeds at 50–60°C for several weeks (Navashin and Shkvarnikov, 1933; Schkwarnikow and Nawaschin, 1934), and in a series of papers he and his co-workers documented in great detail that increases in temperature and relative humidity during seed storage led to a subsequent enhancement of aberrations in both *Crepis* and wheat.[20] These early observations have since been repeatedly confirmed; storage conditions that favor seed deterioration also promote chromosomal aberrations (e.g., Murín, 1961; Sax and Sax, 1962; Harrison, 1966; Abdalla and Roberts, 1968; Zelenov, 1968; Slesaravichyus, 1969a, 1969b, 1969c; Lazukov, 1970; Villiers, 1974; Orlova and Soldatova, 1975; Ghosal and Mondal, 1978; Murata et al., 1981; see Fig. 33). It should be noted, though, that in some particularly stressful aging environments, the frequency of irregularities may be poorly correlated with decreasing germinability (Harrison, 1966; Abdala and Roberts, 1968; McLeish et al., 1968). Furthermore, response can vary widely within an aged seed lot. According to Orlova and Rogatykh (1970), whereas in one seedling virtually all divisions may be aberrant, another may be completely free of deficiencies. Dourado and Roberts (1984a) have also emphasized that seeds that have already accumulated a few deficient cells subsequently tend to accumulate more aberrations than would be expected by chance.

---

19. It has been commonly asserted that the earliest relevant observations linking genetic alteration to seed age were made by De Vries (1901) and Nilsson (1931), who purportedly noted an increase in mutant phenotypes developing from old *Oenothera* seeds. As I have argued elsewhere (Priestley, 1985), however, these data have been regularly misconstrued, and are probably irrelevant to the question of genetic changes induced by seed aging.

20. Much of the prolific output of this group was summarized by Shkvarnikov (1939).

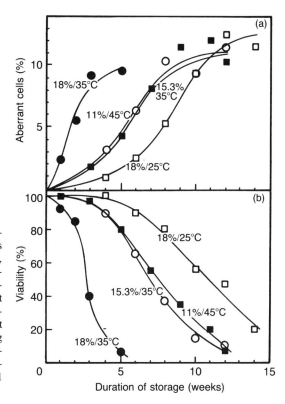

Figure 33. Increase in mitotic aberrations in broad beans with age. Seeds were stored at 11% hydration at 45°C, 15.3% hydration at 35°C, and 18% hydration at 25 and 35°C. For each treatment, 300 anaphase figures from about 10 seeds were assessed for abnormalities. Adapted from E. H. Roberts et al., "Nuclear Damage and the Ageing of Seeds, with a Model for Seed Survival Curves," *Symposium of the Society for Experimental Biology* 21 (1967): 74.

Aging entails profound consequences for mitotic divisions in the germinating seed. The onset of mitosis is delayed (Orlova et al., 1975; Murata et al., 1980), the cycle is prolonged (Tagliasacchi and Vocaturo, 1977), and various abnormal alterations in mitotic figures become evident. The structural abnormalities that are induced can be broadly divided into two groups: "chromosome" types, in which both chromatids are broken at the same locus, and "chromatid" types, involving fragmentation of a single chromatid. The distinction is potentially significant, as it may relate to the time during rehydration at which the lesion is incurred. Chromosome-type aberrations arise before the chromosome is duplicated, and hence affect both chromatids identically following duplication. Lesions of this type are thought to be induced in the resting stage of the cell cycle (G_1) and may accumulate in storage or during early imbibition. Chromatid-type aberrations, on the other hand, occur after duplication of the chromosome—following the S stage of the cell cycle—and probably indicate a lesion that develops somewhat later in imbibition.

Opinions have been divided on the question of whether chromosome or chromatid breaks are more significant in old germinating seeds. A considerable body of evidence suggests that chromosome-type aberrations predominate at anaphase in aged seeds, including onion (both *Allium cepa* and *A. fistulosum*) (Nichols, 1941; Kato, 1954; Jackson and Barber, 1958; Orlova, 1967; Orlova and Nikitina, 1968; Morikawa and Inomata, 1978), *Crepis capillaris* (Protopopova et al., 1970; Kagramanyan, 1971), lettuce (Harrison and McLeish, 1954), tobacco (both *Nicotiana tabacum* and *N. rustica*) (Scarascia and Venezian, 1953; Scarascia, 1957), and various cereals and legumes (D'Amato, 1951; Scarascia and Di Guglielmo, 1953; Gunthardt et al., 1953; Nuti Ronchi and Martini, 1962; Abdalla and Roberts, 1968). Other studies have similarly indicated an increase in the importance of chromosome-type aberrations during storage, although chromatid-type irregularities may still be prevalent in the early stages of aging (Dubinin et al., 1965; Dubinin and Dubinina, 1968). In durum wheat, however, several workers have suggested that chromatid-type breaks are much more characteristic of deteriorated material (Corsi and Avanzi, 1969; Avanzi et al., 1968; Innocenti and Avanzi, 1971; Floris and Anguillesi, 1974). One explanation of the results obtained with this species may be that some nuclei in the unimbibed seed are already in the postsynthetic (G_2) phase of the cell cycle (Avanzi et al., 1963), and that these nuclei are in some way more susceptible to the stresses imposed by aging than are presynthetic (G_1) nuclei. In an extensive study of the first mitotic cycle of aged barley seeds, Murata et al. (1982) found that although some of the fragments observed could be ascribed to chromatid breaks, other frequently noted irregularities could not be assigned to either the chromosome or chromatid types. Chromatid-type lesions tend to be favored in aging maize (Sevov et al., 1973; Khristov and Khristova, 1978), and Dourado and Roberts (1984a), working with barley and pea, have also noted a predominance of chromatid-type aberrations in aged seeds. As Roos (1982) has emphasized, however, the assignment of an irregularity to the chromosome type or the chromatid type can sometimes be quite doubtful. Fragments and broken chromatids may fuse in a variety of permutations, and some aberrations may justifiably be referred to both the chromosome and chromatid types. In addition, interpretations may vary depending on whether the cells observed are in the first or second mitotic cycle following imbibition. In some cases, a chromosome-type aberration in the second cycle will arise from a lesion at the chromatid level in the first cycle; confusion of the two cycles on the part of the observer would consequently lead to erroneous conclusions. Other uncertainties may arise from the effects of hydration. Romanov (1980), for example, found that when old *Crepis capillaris* seeds were subjected to extreme dehydration before germination, the frequency of irregularities decreased overall, but the ratio of chromatid to chromosome aberrations was apparently enhanced. Dourado and Roberts (1984a) have also argued that chromatid-type aberrations may arise following replication as a consequence of improper repair. Clearly, despite much work, it is still difficult to generalize about the nature of the lesions induced by aging, although studies of DNA integrity (see sec. 7.8.3) imply that some breaks in the genetic material are present before imbibition.

Many of the chromosomal aberrations noted in germinating seeds may be lethal to the cell, and it has been suggested that loss of a critical number of cells can compromise viability (Roberts et al., 1967). It may be significant, therefore, that the proportion of aberrant cells found in the surviving seeds of a deteriorated seed lot has generally been 15% or less, although figures considerably in excess of 15% have sometimes been reported (Orlova and Rogatykh, 1970; Orlova, 1973; Soldatova, 1976), and in old seeds treated with inhibitors of DNA synthesis, values approaching 40% have been recorded (Orlova and Ezhova, 1974). Even if some aberrant cells die, many others continue to divide and transmit their chromosomal irregularities to the tissues of the mature plant. Observations by Murata et al. (1984) on barley, however, confirm earlier suggestions that the incidence of irregularity greatly decreases as the plant develops, presumably because the afflicted cells fail to compete with normal ones. Although it has been asserted that chromosomal aberrations may be transmitted at low levels through gametes to the next generation (Gerasimova, 1935a, 1935b), other workers have failed to find any evidence of such phenomena (Harrison, 1966; Murata et al., 1984); genetic irregularities in the haploid generation (discussed below) are apt to prove particularly lethal. Roos (1982) has consequently argued that age-induced lesions of the type visualized in mitotic figures are unlikely to threaten the genetic integrity of stored germ plasm.[21]

7.8.2 Other genetic irregularities

There are several indications that "point" mutations accumulate as seeds age. Two types of study have been commonly pursued in this context: pollen abortion analysis and the observation of chlorophyll mutant phenotypes.

Pollen abortion occurs at a relatively high frequency in A_1 plants grown from aged seed.[22] In a classic series of observations, Cartledge and Blakeslee (1933, 1934; Cartledge et al., 1935) demonstrated that pollen infertility in A_1 plants of jimson weed (*Datura stramonium*) arose because of point mutations as well as chromosomal aberrations; moreover, they were also able

---

21. Despite the widespread assumption that deficiencies in a chromosome present in single dose are especially deleterious, it is relevant to note that most of the seeds of wheat monosomics studied by Piech and Supryn (1979) were not inferior in storability to the disomic norm. (Monosomic wheat lines are genetic strains that lack one chromosome from the normal complement.) Wheat may be unusual in this respect because of its homeologous chromosomes.

22. In this discussion, the convention adopted by Roberts (1972b) is employed. Seeds that have undergone aging—and the plants grown from them—are designated the A_1 generation. The progeny of A_1 plants constitute the A_2 generation.

to show that these defects were promoted by increased temperature and hydration during storage of the seeds from which the A_1 plants were grown (Cartledge et al., 1936). These conclusions have generally been confirmed by later workers, although it has not always been evident whether the decreased pollen viability observed in more recent studies arose from deficiencies at the level of the gene or the chromosome. Pollen fertility has been found to be greatly reduced as a result of aging in perennial ryegrass seeds (Griffiths and Peglar, 1964), for example, and small reductions in pollen viability were noted by Abdalla and Roberts (1969a) in A_1 plants of barley, broad bean, and pea. Other workers have noted this effect in tobacco (Gisquet et al., 1951), pea and wheat (Purkar and Banerjee, 1979, 1983), and soybean and barley (Chauhan and Swaminathan, 1984), although Murata et al. (1984) were unable to confirm that pollen infertility was linked to seed aging in the latter species. In addition to pollen studies, there are reports that rates of seed setting are deficient in A_1 plants, even when effective pollen is supplied (Griffiths and Peglar, 1964), and this finding may indicate similar lesions in the female haploid generation.

Recessive chlorophyll mutations induced by seed aging have been a convenient subject of study because of their distinctive appearance in the A_2 generation (Stubbe, 1935; Floris and Meletti, 1972). Presumably a host of other characters are affected in similar fashion, although the consequences are less visually dramatic (Purkar et al., 1979, 1981, 1982; Chauhan and Banerjee, 1983a, 1983b). Abdalla and Roberts (1969a), working with barley, broad bean, and pea, have presented extensive analyses of age-induced chlorophyll mutations. In this study the A_1 plants grown from aged seeds were selfed to produce a second generation (A_2), and this progeny was screened for defectively pigmented individuals. Under conditions leading to a 50% loss of viability in A_1 seeds, approximately 1 to 3% of the surviving A_1 plants yielded A_2 seeds segregating for chlorophyll mutations. Recent work by Dourado and Roberts (1984b) on peas and barley has confirmed and extended these conclusions by examining segregation of A_3 progenies. Such observations provide clear evidence that heritable mutations are induced by seed aging, even when viability loss is relatively minor; this is obviously a matter of substantial concern for germ plasm curators. It has even been suggested that artificial seed aging may be substituted for other mutagenic procedures in breeding programs. Khristov (1979; Khristov and Khristova, 1980), for example, exposed maize seed to 50°C (over saturated NaCl solutions) in an attempt to create new breeding lines. It is possible that a similar increase in genetic diversity due to seed aging may have played a significant role in the evolution of new genotypes through natural selection.[23]

23. A further practical problem for those concerned with the purity of stored germ plasm

7.8.3 Changes in the integrity of DNA

Amounts of DNA have been reported to alter little (Ching, 1972; Hallam et al., 1973; Abdul-Baki and Chandra, 1977; Kulka and Konopka, 1978; Osborne et al., 1980/81) or to decline (Mitra et al., 1974; Samshery and Banerji, 1979; Banerjee et al., 1981) as seeds age. Even if levels of DNA are unaltered, important qualitative changes take place: amounts of spoolable DNA from nonviable rye embryos decline—an index of fragmentation (Hallam et al., 1973; Roberts and Osborne, 1973a, 1973b; Osborne et al., 1980/81)—and separation of total nuclear DNA on a sucrose density gradient or polyacrylamide gel reveals that randomly sized fragments of low molecular weight accumulate in the stored seed (Cheah and Osborne, 1978). Moreover, the number of fragments—some as small as 200 base pairs—increases during imbibition of nonviable embryos, presumably as a result of endodeoxyribonuclease activity. Location of the 3'OH termini of single-stranded breaks in fixed sections of rye embryos by autoradiography confirms that "nicked" DNA is more prevalent in the nuclei of nonviable embryos and embryos of low viability (Cheah and Osborne, 1978; Osborne et al., 1980/81). A sufficiently large number of breaks would presumably impair DNA template function and limit RNA transcription (sec. 7.9).

There is evidence that ^{3}H-methyl-thymidine is incorporated into double-stranded DNA as part of a repair function during the first hour of imbibition, and that this "unscheduled" DNA synthesis is more intense in partially deteriorated seeds (Osborne, 1982a, 1982b, 1983; see Fig. 34). Presumably the defective region is excised and the missing section resynthesized on the template of the complementary DNA strand. Under some circumstances, however, the repaired DNA in aged seeds is evidently unstable, as much of the label incorporated early in imbibition may subsequently be lost; the chromosome-type abnormalities witnessed at mitosis provide additional evidence that repair is not completely efficient. In most vigorous seeds, scheduled DNA synthesis prior to mitosis is a relatively late event in germination, but it is considerably reduced and further delayed in older material (Hecker, 1974; Sen and Osborne, 1977; Dell'Aquila et al., 1980; Bernhardt et al., 1981; see Fig. 35).

---

is the possibility that genetic shifts may occur within heterogeneous accessions as a result of differences between individual genotypes in resistance to aging. Such effects have been found to be significant in some studies (Roos, 1984), but not in others (Roos and Rincker, 1982).

Work on the soybean variety T219 has yielded indications that some aging regimes are more effective than others in producing mutations as vigor is lost (Murata and Vig, 1985). Spotting of A_1 plants, a phenomenon that has been ascribed in this variety to chromosome aberrations or point mutations, is promoted by deterioration of the dry seeds (moisture content 5.6%) at 50°C, but not at 37°C.

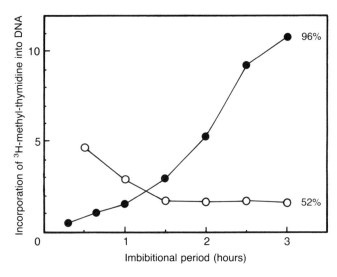

Figure 34. Incorporation of thymidine by rye embryos of different ages. The seed lot of lower viability (52%) showed a higher level of thymidine incorporation (into double-stranded DNA) during the first hour of imbibition than did high-quality embryos (96% viable). This finding provides presumptive evidence of DNA repair. Thymidine incorporation is expressed in arbitrary units. Based on data of A. Dell'Aquila; adapted from D. J. Osborne, "Deoxyribonucleic Acid Integrity and Repair in Seed Germination," in *The Physiology and Biochemistry of Seed Development, Dormancy, and Germination*, ed. A. A. Khan, p. 438 (New York: Elsevier, 1982).

Although considerable attention has been focused on alterations to DNA as a result of aging, some studies have implied that DNA transcription in deteriorated seeds may be affected by alterations in the histone fraction, proteins that are assumed to play a role in regulating gene expression. Pitel (1982) indicated that aging of *Pinus banksiana* seeds under conditions of high humidity was associated with a minor decrease in the lysine-rich F1 histone, for example. It has also been claimed that incorporation of labeled orthophosphate into histones of rice seed is greatly reduced as a result of aging (Banerjee et al., 1981); on the basis of cytophotometric evidence, however, others have asserted that the ratio of histones to DNA in presynthetic stage (G_1) nuclei is higher in older seeds of durum wheat (Innocenti and Bitonti, 1977, 1979, 1981) and *Haynaldia villosa* (Innocenti and Bitonti, 1980) than it is in fresh material. Such changes are apparently most marked in the embryonic radicle tip (Innocenti and Bitonti, 1984). The significance of these findings is uncertain, but alterations in chromatin structure (i.e., the nuclear histone-DNA complex) have previously been deduced from ultrastructural studies (see sec. 7.2). On the basis of evidence that histone mRNA

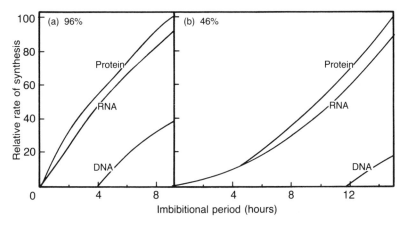

Figure 35. Development of protein and nucleic acid synthesis in rye embryos of high (96%) and low (46%) viability. All major classes of RNA are included. Adapted from D. J. Osborne, "Deoxyribonucleic Acid Integrity and Repair in Seed Germination," in *The Physiology and Biochemistry of Seed Development, Dormancy, and Germination*, ed. A. A. Khan, p. 453 (New York: Elsevier, 1982).

species are not present in the dry seed, but rather are transcribed *de novo* during imbibition (Kato et al., 1982), Osborne (1983) has pointed out that the production of new histones during germination would be particularly sensitive to disruption of the histone-coding cistrons.

The sources of the defects in genes and chromosomes associated with seed aging have engendered much speculation, although satisfactory evidence is scarce. Cheah and Osborne (1978) have suggested that endodeoxyribonuclease activity, triggered by age-induced deficiencies of the enzyme's natural inhibitor, may be responsible for lesions in genetic material. Much more common, however, has been the idea that genetic alterations are provoked by mutagens. Natural background radiation has usually been discounted, as it is insufficient in magnitude to account for the changes observed (Navashin and Gerasimova, 1935; Nawaschin and Gerassimowa, 1936; Gunthardt et al., 1953), and it does not explain the dependence of aberrations on temperature and hydration. Automutagenesis as a result of physiological or biochemical activity within the seed has long seemed to present a more satisfactory hypothesis. Lipid oxidation products, for example, are known to degrade DNA (sec. 7.4.3), and their involvement in the genetic aberrations of old seeds has been suspected for a long time (e.g., Navashin, 1934); unfortunately, definitive evidence of such interactions within seeds is lacking, although oxygen treatments of unimbibed seeds are known to promote the appearance of chromosomal aberrations during germination (Ehrenberg et al., 1957; Kronstad et al., 1959; Berg et al., 1965; Abdalla and Roberts, 1968). There are also indications that the imbibition of old seeds in oxygenated water leads to a great increase in the aberration rate (Protopopova et al., 1974, 1979).

In hopes of shedding light on these issues, many investigators have applied aqueous or lipophilic extracts from old seeds to healthy tissue in the expectation of identifying mutagenetic aging products. Although apparent success has been reported by some workers (e.g., Marquardt, 1949; Gisquet et al., 1951; Keck and Hoffman-Ostenhof, 1952; Kato, 1954; D'Amato and Hoffman-Ostenhof, 1956; Wittmer, 1958, 1960, 1961a), others have raised serious doubts concerning the means by which the purported mutagens have been assayed in many of the studies (Abdalla and Roberts, 1968; Roberts, 1972b). A satisfactory resolution of these questions must await more definitive methods of analysis.

7.9 RNA and protein synthesis

Protein synthesis, which is usually evident well within the first hour of imbibition, is reduced in aged seeds (Marcus et al., 1966; Hallam et al., 1973; Roberts and Osborne, 1973a, 1973b; Osborne et al, 1974; Hecker, 1974; Bray and Chow, 1976a; Bray and Dasgupta, 1976; Abdul-Baki and Chandra, 1977; Anderson, 1977; Khekker et al., 1977). Moreover, polysomes are often conspicuously absent from the cytosol of imbibed aged seeds (Bray and Chow, 1976b; Zalewsky, 1982; see also sec. 7.2), confirming that limited protein translation occurs. The declining rate of protein synthesis may be related in part to an inability to produce adequate levels of nucleoside triphosphates, particularly ATP and GTP (sec. 7.7.3); but it is also evident that seed aging entails severe deficiencies in the cellular mechanisms responsible for the translation of mRNA into protein. This field has been authoritatively reviewed by Bray (1979) and Osborne (1980, 1982b, 1983).

7.9.1 Deficiencies in messenger RNA

The mRNA responsible for protein synthesis during the earliest stages of seed imbibition may be of two sorts. Some mRNA survives desiccation within the seed and is capable of translation soon after rehydration (Spiegel and Marcus, 1975). In addition to this long-lived mRNA, newly transcribed mRNA is available within a few hours after the start of imbibition (Osborne, 1983). Deficiencies in either type of mRNA may presumably have serious consequences for germination.

There are indications that long-lived mRNA is lost to some extent during extended storage. Osborne and her colleagues (Osborne et al., 1977; Osborne, 1983) have isolated polyadenylated RNA (potential mRNA species) from unimbibed rye embryos of varying viability (90% or 0%), and have found that both preparations sustain protein synthesis in a wheat germ cell–free system, producing a similar variety of polypeptides in both cases. Nonviable embryos, however, yielded only about half as much polyadenylated RNA as the material that was 90% germinable, and specific template activity was also reduced by aging. The effects of five years of storage on wheat are highly comparable (Peumans and Carlier, 1981), but in this species viability was not impaired; aging in durum wheat (Grilli et al., 1982) and rice (Ghosh and Choudhuri, 1984) has also been associated with loss of polyadenylated RNA. It is possible that much of the stored RNA is not essential for successful germination, since it is normally subject to extensive degradation during the first two hours of imbibition, even in highly vigorous material (Smith and Bray, 1982). This deleted RNA may code for polypeptides that are important during seed maturation but less relevant during germination. Indeed, it appears that the degradation of stored polyadenylated RNA during rehydration progresses more slowly in partially deteriorated embryos than it does in vigorous ones (Smith and Bray, 1984). Consequently, the significance of declining levels of long-lived mRNA during seed aging is still uncertain.

Newly synthesized mRNA may also be lacking in aged seeds during the early imbibitional phase. It has generally been suggested that seed deterioration leads to a depression in the rate of RNA synthesis following hydration (Grzesiuk and Kulka, 1971a; Osborne et al., 1974; Van Onckelen et al., 1974; Bray and Dasgupta, 1976; Sen and Osborne, 1977; Anderson, 1977; Bernhardt et al., 1981; Fernández-García de Castro and Martínez-Honduvilla, 1982), and it is apparent that all the major classes of RNA are affected. This failure is probably related in part to a general synthetic debility, but deficient transcription may be of greater significance. In aged peas (Bray and Dasgupta, 1976), rye (Sen and Osborne, 1977), and wheat (Weidner and Zalewski, 1982) only low-molecular-weight RNA (4S to 5S) is transcribed. It is possible that messages of greater length cannot be readily synthesized because of fragmentation or other irregularities in the DNA template (sec. 7.8.3).

7.9.2 Deficiencies in translation

There is evidence that ribosomes suffer some degree of degradation in seeds during prolonged storage. Ribosomes from nonviable embryos of rye (Roberts et al., 1973; Osborne et al., 1974) and peas (Bray and Chow, 1976a) are sufficiently intact to support polypeptide synthesis *in vitro* when polyuridylic acid is used as a synthetic mRNA template, although in the case of rye this process occurs with considerably reduced efficiency. Structural deficiencies in rRNA with seed age have been well documented. In rye embryos the 18S and 28S rRNA subunits may suffer loss of integrity (Roberts and Osborne, 1973a, 1973b; Osborne et al., 1974; Wisñiewski and Kulka, 1979), which has been attributed to ribonuclease activity during storage. Bray and Chow (1976b) have likewise described limited breakdown of rRNA subunits extracted from dry nonviable pea embryos, although fragmentation became more evident upon prolonged imbibition. Partial degradation of rRNA has also been implicated in the aging of seeds of tobacco (Brocklehurst and Fraser, 1980), durum wheat (Grilli et al., 1982), and rice (Ghosh and Chaudhuri, 1984). In addition to modifications in rRNA, there has been one report that the electrophoretic properties of ribosomal proteins from wheat embryos are altered with age (Zalewski and Weidner, 1982). Despite deficiencies in structure, the sedimentation characteristics of ribosomes from nonviable rye embryos suggest that all major components are still present (Osborne et al., 1977), although in another species (*Pinus densiflora*) complete disintegration of ribosomes from aged seeds has been indicated (Sasaki, 1977). The conclusion to be drawn from most of these studies, however, is that ribosomes are still partially functional even in severely aged seeds, although their synthetic capacity may well be reduced.

At least one major age-induced lesion is associated with the soluble factors that mediate peptide bond formation at the ribosome. In particular, the capacity for GTP-dependent binding of amino acyl–tRNA to ribosomes by elongation factor 1 (EF1) is significantly diminished in deteriorated embryos of rye (Roberts and Osborne, 1973a), pea (Bray and Chow, 1976a), and wheat (Dell'Aquila et al., 1976). Decline in EF1 activity closely parallels the loss of protein synthetic capacity *in vivo*. Presumably the polypeptide aggregate comprising EF1 is destablized or inhibited, although details remain obscure (Dell'Aquila et al., 1978). In imbibed embryonic axes of peas, deterioration of elongation factor 2 (EF2)—responsible for translocation of the mRNA through the ribosome—and phenylalanyl-tRNA synthetase have both been noted (Bray and Chow, 1976a), although in rye embryos these functions are less deficient (Roberts and Osborne, 1973a).

The evidence currently available clearly indicates that each of the principal elements of the protein synthesizing system in aged seeds may suffer from deficiencies: long-lived mRNA is lost or less efficient, newly synthesized messenger may be absent or incoherent, ribosome structure and effectiveness are altered to a certain degree, and some of the soluble factors responsible for peptide formation are incapacitated. There is, in addition, some question of a sufficient and continuing supply of the nucleoside triphosphates required by the synthetic apparatus (sec. 7.7.3). Several or all of these factors probably serve to reduce the overall ability of aged seeds to synthesize proteins. This deficiency, in turn, has severe consequences for the effectiveness of cellular repair mechanisms that operate during the early stages of imbibition (sec. 7.11.1).

7.10 Hormones

Seeds are quite abundantly supplied with plant hormones. Although the functions of these growth substances are often incompletely understood, it

Morphological, Structural, and Chemical Changes

is clear that they play a key role in dormancy, in reserve metabolism (the effects of gibberellin in cereal grains are well studied), and probably in several other areas of seed physiology.

A scattered and sometimes rather unconvincing literature attempts to document an important role for hormones and other growth factors in seed aging. Two types of report are prevalent: studies that suggest that aging is related to a deficiency in a growth-promoting factor (a situation that can sometimes be remedied by exogenous supply) and those that point to inhibitory levels of growth factors that are thought to prevent normal germination.

In recent years the use of rigorous chemical techniques to identify and quantify plant growth substances has expanded considerably (Hillman, 1978). These techniques have supplanted many of the older, less certain methods, such as bioassay. It is important to bear in mind that few of the studies reviewed in this section have employed definitive analytical procedures. In consequence, a substantial degree of uncertainty must surround much of the work discussed.

7.10.1 Hormonal deficiencies

Juel (1941/42) was probably the first to investigate hormone content as a function of seed age (by means of bioassay), and concluded that auxins in maize and oats declined with several years of storage. Köves et al. (1965) also reported decreased auxin content in deteriorated rice seeds, although according to Sircar and Biswas (1960), aging in this species is accompanied by a substantial increase in auxin. Skrabka (1964) detected no significant difference in auxin levels following deterioration of wheat, but more recently Sreeramulu (1983c) associated lowered auxin content with loss of viability in bambara nut (*Voandzeia subterranea*). Few substantive conclusions can be drawn from these reports; it is evident that a rigorous analysis of the effects of seed age on the levels of indole acetic acid and its conjugates has yet to be published. Further, since auxin is not known to affect germination even in vigorous seeds, it is questionable whether an age-induced alteration in this hormone must be considered particularly consequential.

More attention has focused on the possibility that failing levels of gibberellins may be implicated in seed aging. Aspinall and Paleg (1971) found that production of α-amylase by wheat aleurone layers in response to added gibberellin decreased with seed age, suggesting that they were inherently less efficient in translating the hormonal stimulus. Mierzwińska (1977) noted similar findings in aged rye endosperms. She also found, however, that the application of gibberellin (10 mg/l) to old wheat seeds promoted their rate of germination, presumably an indication that the aged embryo was incapable of providing a sufficient level of the hormone. Deficiencies in both embryonic gibberellin synthesis and aleurone responsiveness were consequently implicated in the reduced rate of germination characteristic of old seeds. The production of α-amylase by aged wheat seed has received some recent rescrutiny by Artsruni and Panosyan (1984), who concluded that the failure to produce appropriate levels of the enzyme during germination was due more to the inadequacy of the hormonal stimulus than to a decrease in the aleurone cells' capacity to synthesize amylase.

Several reports indicate a small improvement in the performance of deteriorated seeds following gibberellin application. These include studies of Cuban pine (*Pinus caribaea*) (Venator, 1972), celery (Harrington, 1973), eggplant (Takahashi and Suzuki, 1975), rice (Kwun, 1980), and tobacco (Yamaguchi et al., 1983). Mostafa et al. (1982), however, noted more impressive stimulation of old sweet pepper (*Capsicum annuum*) seeds by gibberellin treatment. In addition, Galleschi et al. (1977) suggested that gibberellin could stimulate the activation or synthesis of glutamate decarboxylase in aged seeds, and Mierzwińska (1973, 1975) noted that use of 0.5 mM gibberellic acid reduced the number of abnormal seedlings produced from old lupine seeds. Harrison (1977) also reported that imbibition of aged barley seed in the presence of gibberellin (0.09%) increased mean shoot length of the seedlings; however, the same treatment applied to unaged seeds occasioned even bigger increases. Preapplication of gibberellin (0.3 mM in acetone) to durum wheat seeds has been reported to decrease sensitivity to aging at 14.5% hydration and 30°C (Petruzzelli and Taranto, 1985). Perhaps understandably, there are few reports that gibberellins were found to be ineffective in promoting the performance of old seeds. Huber and McDonald (1982), however, analyzed in considerable detail the effects of an application of 10μM gibberellic acid on germinating aged barley seeds (a standard protocol used by seed testers to remove cereal dormancy), and found their vigor and viability to be minimally affected.

Cytokinins have received rather less consideration in the context of seed aging. Khan et al. (1976) applied the synthetic cytokinin kinetin (0.5 mM in acetone) to lettuce seed and suggested that it slowed the rate of deterioration during subsequent accelerated aging (43°C and 85% relative humidity); the benefit achieved was fairly small, however. Styer and Cantliffe (1977) also claimed that prestorage application of kinetin (0.5 and 1 mM) to onion seed delayed loss of germinability in one cultivar examined; however, the treatment was without effect on two other cultivars. The results of a mixed hormonal application were reported by Puls and Lambeth (1974), who found that an imbibitional cocktail of gibberellic acid (10 mg/l), kinetin (1–50 mg/l), and KNO_3 (0.01 M) improved the rate of germination of ten-year-old tomato seeds.

Ethylene has also received some attention. Takayanagi and Harrington (1971) maintained that application of this hormone (60 ppm) to aged rape seed accelerated the rate of radicle emergence, without improving germinability; as deteriorated seeds also produced less ethylene during the emergence of radicles, these authors suggested that the degeneration of an ethylene-producing system in the seed may be one of the factors responsible for aging. Depressed ethylene production is also characteristic of the germination of aged seeds of rice (Kwun, 1980) and *Phaseolus vulgaris* (Samimy and Taylor, 1983). Dell'Aquila and Uggenti (1974) found that germinability was greatly increased in pea seeds (previously aged at 38°C and 70% relative humidity) when they were treated with 1 ppm chloroethylphosphonic acid as an ethylene source. As the seeds were assessed after four days of growth, the ethylene presumably increased the rate of germination rather than restored viability. Application of the same compound (500 ppm) to seeds of durum wheat has been reported to decrease their deterioration at 14.5% hydration and 30°C (Petruzzelli and Taranto, 1985).

Abscisic acid has received much less attention than the four other principal hormones. Yamaguchi et al. (1982) report that treatment with abscisic acid may promote germination of aged rice seed, but others have suggested that abscisic acid may have a depressing influence on the germination of old seeds (see sec. 7.10.2).

The evidence currently available indicates that hormonal deficiencies in aged seeds probably do not exert a very marked influence on their germination. Although application of exogenous hormones can in some cases increase their vigor or viability, the gains achieved are mostly fairly marginal.

7.10.2 Growth inhibitors and seed aging

A number of investigators have been drawn to the suggestion that slowed or delayed germination in aged seeds is in some way related to an increase in inhibitory compounds (e.g., Went, 1957), and some have attempted to identify these putative inhibitors with substances well known to regulate plant growth. Loss of germinability in rice during storage, for example, has been ascribed to supra-optimal concentrations of indole acetic acid (Sircar and Dey, 1967; Dey and Sircar, 1968a), elevated levels of an abscisic acid–like material (Dey and Sircar, 1968b; Ghosh et al., 1978), or accumulation of phenolic substances (Dey et al., 1968; Sircar and Sircar, 1971). A subsequent report implicated caffeic acid, kaempferol, and quercetin and its methyl ester in the aging of this species (Chatterjee et al., 1976). As Roberts (1973b) has aptly remarked, much of this work seems to indicate that aging rice seeds are particularly unfortunate in the number of lethal inhibitors they accumulate. Loss of viability of jute (Chakraverty, 1976; Bhaumik and Mukherji, 1981) and wheat (Monin and Girard, 1973) in storage has also been ascribed to the accumulation of rather ill-defined "growth inhibitors." More recently,

Narasimhareddy and Swamy (1977) have implied that loss of germinability in peanuts is in part associated with an increase in abscisic acid–like material.

Other investigators have suggested a role for phenolics in the aging of seeds other than rice. Sreeramulu (1983c) implied that loss of viability in bambara groundnut was associated with an increase in such "inhibitory" phenolics as vanillic, ferulic, and hydroxybenzoic acids. Gesto and Vazquez (1976) also reported that phenolic acids accumulate in embryos of *Phaseolus vulgaris* over several years of storage and are linked to slower germination. When Volynets and Pal'chenko (1980, 1982) surveyed a variety of free phenols and phenolic conjugates in seeds of yellow lupine, flax, and barley during long-term storage, however, they found that they changed little and were probably not responsible for deteriorating vigor and viability in these species. Generally, the suggestion that loss of viability is caused by phenolics or any other type of inhibitory plant growth regulator is not compelling.

The accumulation of polyamines, considered by some researchers to be important regulators of plant growth, has also been implicated in seed aging; on the basis of a rather small survey of durum wheat samples, Anguillesi et al. (1974) reported that spermidine and putrescine levels increased in the embryos with age, and suggested that this may have impaired metabolic regulation. Subsequently, Mukhopadhyay et al. (1983) also noted that loss of viability in rice was associated with a great increase in the polyamine content of the imbibed embryo. The physiological consequences of these changes remain largely unexplored, however.

7.11 Repair and reinvigoration

With the exception of hard-seeded species, it is probable that many seeds become hydrated fairly soon after dispersal. Consequently, the storage of seeds of temperate zone plants in a comparatively desiccated state for months or years imposes a relatively unnatural stress. According to Darwin (1855a), "the power of retaining vitality in a dry and artificial condition must be an indirect, and in one sense accidental, quality in seeds of little or no use to the species." This is not to deny, though, that unconscious artificial selection for longevity in the dry state may have been significant in the development of cultivated forms.

Many seeds of wild plants are adapted to persist in the soil for years or decades, and for much of this time they are presumably hydrated (chap. 5). In such cases longevity depends on a different set of criteria from those that apply in the dry condition. In the imbibed state, dormancy must certainly be maintained for the seed to persist, but it has long seemed probable that some extra factor is at work to help sustain viability. From Darwin's perspective, "the power in seeds of retaining their vitality when buried in damp soil may well be an element in preserving the species, and, therefore, seeds may be specially endowed with this capacity'' (1855a). In recent years it has become increasingly obvious that part of this special endowment consists of an ability for cellular repair and maintenance (Villiers, 1973).

7.11.1 Cellular repair .

When lettuce seeds are held in an imbibed state at about 30°C, they become thermodormant and will not germinate. Using a range of storage conditions, Toole and Toole (1953b) were able to demonstrate that although increasing temperature and humidity normally accelerate the deterioration of lettuce seeds (Toole et al., 1948; sec. 3.2), thermodormant seeds are exceptional in that they lose viability far more slowly than the usual trends would predict. It has since been well established that imbibed dormant seeds possess a highly developed metabolism. For example, respiration is evident (Powell et al., 1983), membrane lipids and proteins are subject to turnover (Cuming and Osborne, 1978a, 1978b), RNA is synthesized (Hecker and Bernhardt, 1976), and active polysomes are present (Fountain and Bewley, 1973). It is presumably the existence of such metabolic functions in the hydrated seed that permits extensive repair to take place.

Recent studies on lettuce have also shown that longevity is promoted when water content is increased above 15 to 20% (Ibrahim and Roberts, 1983), and that this extension in viability is strictly dependent on the availability of oxygen (Ibrahim et al., 1983). An increase in hydration of onion seeds above about 15% has similarly beneficial effects on their longevity (Ward and Powell, 1983; see Fig. 36). In addition, Villiers (1974, 1975) has shown that storage of lettuce in an imbibed dormant condition reduces chromosomal aberrations (Fig. 37). Moreover, if lettuce seeds are subjected to occasional brief interludes of wetting-and-drying during conventional dry storage, the usual age-induced accumulation of chromosomal injury is largely eliminated (Villiers and Edgcumbe, 1975). For seeds in the soil, continuous or intermittent hydration is obviously commonplace. Evidence that genetic abnormalities accumulate relatively slowly in buried seeds—even over several decades—has been presented by Cartledge and Blakeslee (1935) and by Avery and Blakeslee (1943).

Despite this capacity for maintenance and repair, seeds will eventually show the effects of aging after a prolonged period of storage in the hydrated state. Thus in European ash seeds (*Fraxinus excelsior*), ultrastructural aberrations gradually become apparent (Villiers, 1972); in particular, mitochondria are swollen, and nuclei show lobing. Dormant Norway maple (*Acer platanoides*) seeds stored in an imbibed state show increased electrolyte leakage as aging progresses and viability is lost (Pukacka, 1983), and it is possible that depletion of nutritional reserves over many years may limit the

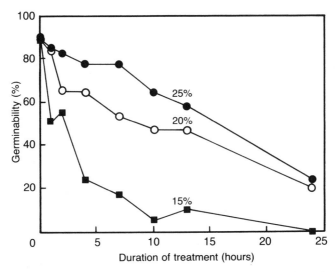

Figure 36. Survival curves for onion seeds (cv. Robusta) subjected to deterioration at 15–25% hydration and 45°C. Adapted from F. H. Ward and A. A. Powell, "Evidence for Repair Processes in Onion Seeds during Storage at High Seed Moisture Contents," *Journal of Experimental Botany* 34 (1983): 279, by permission of Oxford University Press.

viability of some species (Powell et al., 1983). There are also indications that deteriorated seeds are inherently less able to maintain dormancy: McCoy and Harrington (1970) found that lettuce seeds that had aged in dry storage were subsequently less likely to display thermodormancy. Similarly, Foard and Haber (1966), who studied lettuce seeds that had previously been subjected to several years of dry storage and that were almost completely nonviable, reported mitotic activity in the radicle tips when the seeds were imbibed at high temperatures. In young seeds, mitotic activity was suppressed, and thermodormancy was almost complete.

It is probable that for most seeds the first hours of imbibition are marked by the activation of cellular repair processes that gradually eliminate the lesions acquired in dry storage. Berjak and Villiers (1972), for example, demonstrated that in maize seed that had previously been subjected to accelerated aging (14% moisture content and 40°C), ultrastructural abnormalities present upon rehydration (see sec. 7.2) gradually disappeared over the next 24 to 48 hours. Further, the rate of ^{3}H-uridine and ^{3}H-leucine incorporation was unusually high; one possible explanation is that extensive repair demanded elevated levels of RNA and protein synthesis. Biochemical evidence for increased DNA synthesis during early imbibition of deteriorated rye seeds is presented in section 7.8.3.

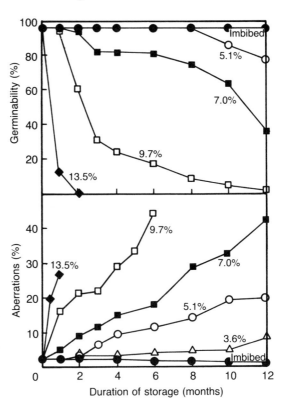

Figure 37. Germination and percentage of aberrant nuclear divisions in radicle tips of lettuce (cv. Arctic King). Seeds were stored at 3.6–13.5% hydration, or fully imbibed, at 30°C. Upon germination, the percentage of aberrant nuclear divisions was scored for late anaphase figures in the radicle tip. Adapted from T. A. Villiers, "Seed Aging: Chromosome Stability and Extended Viability of Seeds Stored Fully Imbibed," *Plant Physiology* 53 (1974): 876, 877.

Considerable attention has been given to the possibility that the effects of aging during storage could be reversed or ameliorated if seeds were soaked briefly and then allowed to dry. "Priming" treatments of this sort are known to accelerate the germination of many types of seed (Heydecker et al., 1975; Hegarty, 1978); some form of metabolic advancement is thought to take place upon hydration and to be retained even after redrying. It is probable that when wetting-and-drying procedures are applied to partially deteriorated seeds, cellular repair becomes possible during the brief imbibitional treatment; metabolic priming may also be significant. If such seeds are returned to storage, their longevity is often extended; if instead they are immediately germinated, their vigor is apparently increased. Some workers have considered that the benefits noted may arise from strictly physical rather than metabolic interactions; they claim that wetting-and-drying treatments applied to partially deteriorated seeds are effective because they dispel debilitating free radicals (see sec. 7.4.4).

As the reinvigoration of aged seeds by wetting-and-drying treatments has usually been defined in terms of an enhanced rate of radicle emergence, it is not completely clear whether the seeds have been genuinely invigorated through repair of cellular lesions or whether they are merely metabolically advanced in comparison with untreated control material. Savino et al. (1979), for example, reported that imbibing aged pea seeds for 18 hours at 20°C, followed by drying, led to a significant increase in radicle length in subsequent germination assays. Using similar criteria, Goldsworthy et al. (1982) suggested that soaking for 30 minutes or less is sufficient to reinvigorate aged wheat seeds. Coolbear et al. (1984) have reported that tomato seeds previously subjected to accelerated aging (70% moisture content, 45°C) showed faster germination after being imbibed at 10°C for 21 days. Brocklehurst and Dearman (1983) and Dell'Aquila et al. (1984) have found that germination rate is improved in aged seeds following priming with the aid of polyethylene glycol solutions; and Simak (1976), using 14-year-old Scots pine seed primed in polyethylene glycol, noted not only faster germination, but also a small increase in germinability. Wetting-and-drying treatments applied to aged soybeans have been reported to suppress leakage during subsequent imbibition (Tilden and West, 1985).

There are suggestions that seeds can be primed by very brief exposure to conditions that would normally accelerate aging. Sorghum seeds subjected to 17% hydration and 30°C show some improvement in rate of emergence following six days of "aging," although performance slipped dramatically over the next 42 days of treatment (Gelmond et al., 1978). After exposure to elevated temperature and humidity for six days, many enzymes were more active than they were in untreated material; further aging led to a decline in activity for all the enzymes examined except proteinase (Perl et al., 1978). Kole and Gupta (1982) have argued that similar changes can be identified in safflower subjected to accelerated aging, and Likhachev (1977, 1980; Likhatchev et al., 1984) noted that short periods of artificial aging may stimulate seed performance. It may be reasonable, therefore, to view accelerated aging at high humidity as "overpriming," a situation that is far removed from the fate of seeds with minimal metabolism stored at low levels of hydration.

The general experience with priming treatments has been that the vigor of old seeds—or perhaps more precisely, their rate of germination—can be accelerated. Sánchez and Miguel (1983), however, have argued that hydration-dependent repair processes can restore viability to aged embryos of *Datura ferox*. In this species, the isolated embryo will not grow if seeds have been in storage for eight to ten months, although whole seeds have been reported to germinate after much longer periods, once dormancy is removed (Soriano et al., 1964). Sánchez and Miguel ascribed the incapacity of the embryos to aging on two grounds: older embryos displayed increased leakage during imbibition, and after only a few months of storage large numbers of seedlings with anthocyanin-deficient phenotypes were produced (reportedly 100% of seedlings in some cases). It is possible, though, that the effects of aging and dormancy have been confounded in this case; certainly, it would be very unusual if repair processes alone were responsible for the restoration of complete germinability to otherwise nonviable embryos.

Numerous studies have suggested that wetting-and-drying treatments may promote seed longevity. The work of Basu and his colleagues has already been discussed in section 7.4.4; several other investigators have reported comparable effects. Savino et al. (1976, 1979) found that after seeds of durum wheat, pea, tomato, and maize were imbibed for 16 to 24 hours and then redried, they were much more resistant to accelerated aging (38°C, 13–17% hydration). In this case it is unclear whether metabolic priming was in some way responsible or whether all the seeds used for the experiment were already partially deteriorated, so that they were subject to repair during the imbibitional treatment. Burgass and Powell (1984) found that soaking deteriorated cauliflower seed for two hours not only accelerated germination but also led to higher retention of germinability in an artificial aging treatment (20% hydration, 45°C), and Baboth (1978) also indicated that presoaking of onion seeds in trace element solutions led to better preservation of viability over 30 months of storage under room conditions; Simancik (1968), however, found that soaking of larch seeds (*Larix decidua*) for eight hours and then drying had no effect on subsequent viability during three years of laboratory storage. It has been the experience of Basu and his colleagues (sec. 7.4.4) that such treatments are most effective when they are applied to seeds that have already undergone a certain amount of deterioration, and this finding may explain some apparent contradictions in the literature.

One important but poorly explored consideration is that repair processes that normally operate during hydration presumably themselves become subject to increased inefficiency with age. Such incapacity may well be responsible for the increased radiosensitivity of aged seeds (Gustafsson, 1937; Nilan and Gunthardt, 1956; Sax and Sax, 1961; Matsumura and Fujii, 1966; Ando and Vencovsky, 1967; Semerdzhyan et al., 1969; Nechitailo, 1969; Lebedeva et al., 1970; Sizova, 1976; Salmonson, 1977) and their greater susceptibility to mutagens (Kaul, 1969). According to Tarasenko et al. (1965), equal numbers of free radicals are detectable by ESR in young and old potato seeds irradiated to the same extent, indicating that equivalent injuries are sustained, but chromosomal damage is far more consequential in the aged material. It may well be relevant, therefore, that DNA polymerase activities have been shown to decrease as seeds age (Yamaguchi et al., 1978; Dell'Aquila et al., 1980), as this enzyme has a central role to play in genetic repair.

7.11.2 Reinvigoration by chemical and physical treatments

Attempts to reinvigorate seeds have not been limited to the wetting-and-drying treatments discussed in sections 7.4.4 and 7.11.1. Indeed, experimental treatments to improve seed vigor have been an abiding concern for centuries. Although such treatments have been refined somewhat over the "spirit of urine mixed with phlegm of elderberries" prescribed by Sharrock (1672), a few of the more recently proposed invigoration treatments for aged seeds have been unusually imaginative.

Many remedies that have been claimed to reinvigorate old seeds (or to prevent deterioration of fresh seeds in storage) have probably produced their effects by reducing fungal invasion: conventional fungicides are often particularly beneficial in improving the performance of aged seeds (Wallen et al., 1955).[24] Sometimes it may not be immediately apparent that a fungistatic effect is involved in protocols that purportedly reduce or alleviate aging effects. Treatments to promote seed longevity by irradiation with lights that partially emit in the ultraviolet range (Jensen, 1941, 1942), for example, may fall into this category.

In the case of some putative invigoration treatments, overoptimism on the part of the investigator has no doubt been quite pervasive. A graphic example is provided by the aftermath to Acton's discovery that old wheat seeds were deficient in proteinase and amylase activities (Acton, 1893). Within a few years, no fewer than three publications independently suggested that deteriorated seeds could be reinvigorated by enzyme solutions (Thomson, 1896; Waugh, 1897; Stone and Smith, 1901); thereafter, others reported poor responses to this protocol, and enthusiasm dwindled. With modern hindsight, the reasoning behind such a treatment seems hopelessly unsophisticated; but it would be naive to believe that recent work in the field of seed invigoration has always been free of similarly optimistic biases.

Chemical treatments have been frequently employed. Ruge (1952), for example, claimed that extremely dilute solutions of 2-chloroethanol (10 ng/l) boosted the germinability of old spinach seeds considerably. Gumińska (1958) and Gumińska and Sulej (1964) suggested that chelating agents could invigorate deteriorated seed, although Skrabka and Szuwalska (1969) applied a variety of chelators to old seeds with little effect. The use of nitrates has been recommended, apparently independently, in two studies: Řetovský (1934) stimulated aged barley seeds with uranyl nitrate ($UO_2(NO_3)_2$), and Chakraverty (1975) found that KNO_3 or thiourea promoted the germination of deteriorated jute seeds. On the other hand, Harty et al. (1983), who investigated applications of KNO_3 to *Panicum maximum* seeds—a treatment designed to assist in breaking dormancy—found that the germinability of aged material was depressed.

A purely physical invigoration was attempted by Jakovuk et al. (1970), who claimed that ultrasonic waves (1 MHz) stimulated the germinability of

---

24. *Prolonged* exposure to sterilizing agents, however, may be especially stressful to older seeds, leading to diminished germinability (Lomagno Caramiello and Montacchini, 1976).

old tobacco seeds. Their data indicate, however, that not all old seed lots responded similarly, and the general utility of the technique is therefore questionable. According to Biswas and Sircar (1976), red light stimulated the germinability of badly deteriorated rice seed; these workers intimated that red irradiation in some way corrected hormonal imbalances, which they regarded as responsible for loss of viability in this species (see sec. 7.10). Promotion of aged rice seeds by red light has also been claimed by Banerjee and Ghosh (1982). A third type of physical technique, "cathodic therapy," has been reported to increase the germination rate of partially deteriorated maize seeds (Berjak, 1978; see sec. 7.4.4).

Invigoration by biochemicals has perhaps been most frequently attempted. Such efforts have usually involved the application of recognized plant hormones (see sec. 7.10), although success has also been claimed for other procedures. Skrabka (1965) noted small but significant increases in the germinability of aged wheat as a result of imbibition in glutathione (10 mg/l), for example. More recently, Petruzzelli et al. (1982) reported that fusicoccin (5 μM)—a fungal toxin that stimulates H^+/K^+ exchange of plant cells—promoted the viability of durum wheat seeds that had aged under conditions of laboratory storage; fusicoccin-treated seeds were about 80% germinable, whereas untreated controls were limited to 50%. Application of K_2SO_4 or Na_2SO_4 (2 mM) was reported to be similarly effective. In explanation, it has been proposed that the H^+-extrusion/K^+-uptake mechanism on the plasmalemma is inactivated by aging, and that fusicoccin and the monovalent cations are responsible for its stimulation or restitution. Fusicoccin and salt treatments are ineffective, however, when applied to seeds that have been subjected to accelerated aging (12.5% or 14.5% moisture content and 30°C), although both naturally aged and artificially aged seeds show a decreased capacity for proton extrusion (Petruzzelli and Carella, 1983). The cause of this discrepancy in response is unclear, although it has been suggested to point to yet another difference between the types of seed aging that occur at high and low humidity.

7.12 Synopsis

The evidence reviewed here indicates clearly that very few components of a stored seed are immune to degradation, and that most cellular processes within aged seeds are found to be hindered after rehydration. Unfortunately, because the majority of studies have been concerned with the effect of aging on one component or process, we have little knowledge of the way in which these manifold deficiencies interact to induce cellular debility.

Figure 38 outlines some of the pathways by which deterioration may progress in unimbibed seeds. It is possible that many or all of these forms

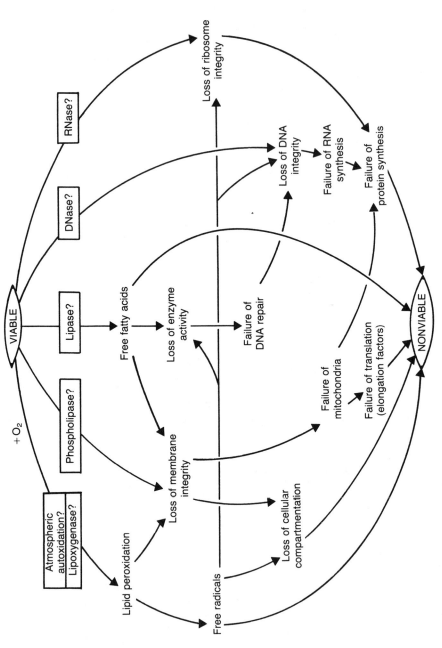

Figure 38. A flow diagram illustrating some of the major paths by which seed viability may be lost during aging. Adapted by permission from D. J. Osborne, "Senescence in Seeds," in *Senescence in Plants*, ed. K. V. Thimann. Copyright © 1980 by

of degradation can occur to some extent within a single seed; but no doubt certain deteriorative processes will be more important than others, depending on species and aging environment. In some cases, lipid peroxidation and its many ramifications may be of greatest significance. In other conditions, autolytic degradation of nucleic acids may be the key to debilitation. A dry, largely ametabolic seed that lacks the capacity for immediate cellular repair (characteristic of sorption zones I and II, as defined in sec. 2.1) will eventually reach some level of deterioration beyond which restoration is impossible; the seed will fail because the cumulative effect of the infidelities and irregularities in all its components is too great to be countered.

Figure 38 may be less appropriate for an assessment of the causes of aging in seeds stored under conditions of high relative humidity (sorption zone III) or present in an imbibed state in the soil. We know even less about the debilitation of seeds stored under such conditions than about the deterioration of seeds stored in a dry state, but presumably some equilibrium between degradation and repair is attained. With time, this balance must inevitably shift in favor of deterioration, although the factors responsible are still very poorly understood. As with seeds in dry storage, the challenge is to understand how the integrated system becomes subject to disarray and to trace the overall effect of alterations to nucleic acids, membrane integrity, respiratory activity, and repair functions. The theoretical and practical value of these questions should no doubt ensure that satisfactory solutions will eventually be forthcoming.

Appendix

Selected Tabulations of Seed Longevity

Table A.1
Mean number of years that horticultural and agronomic species can be expected to maintain high germinability under favorable conditions of commercial storage in Germany, compiled by W. Ullman ("Über die Keimfähigkeitsdauer [Lebensdauer] von landwirtschaftlichen und gartenbaulichen Samen," *Saatgutwirtschaft* 1 [1949]:174–75, 195–96) from his own observations and those of others, in Germany and elsewhere. The nomenclature of the original list has been extensively revised to conform to the standards of *Hortus Third* (Liberty Hyde Bailey Hortorium, 1976). A few entries have been omitted because of taxonomic synonymy; others have been eliminated because of uncertain identity. The chief value of this tabulation probably lies more in its comprehensiveness than in its precision.

Species	Years	Species	Years
Abelmoschus esculentus	3–5	*Amaranthus caudatus*	4–5
Abutilon hybridum	3–4	*Ampelopsis* spp.	2
A. indicum	3–4	*Anchusa capensis*	3
Acanthus latifolius	2–3	*Anemone coronaria*	3
A. mollis	2–3	*Anethum graveolens*	2–3
Achillea millefolium	2–4	*Angelica archangelica*	2–3
Aconitum napellus	4	*Anthriscus cerefolium*	2–4
Adlumia fungosa	2–3	*Anthyllis vulneraria*	3–4
Ageratum houstonianum	2–4	*Antirrhinum majus*	3–4
Agrostis nebulosa	2–3	*Apium graveolens*	3–6
Albizia distachya	8–10	*Aquilegia* spp.	2
Alcea rosea	3–4	*Arabis alpina*	2–3
Allium ampeloprasum	2–3	*Arctotis stoechadifolia*	3
A. cepa	1–3	*Argemone polyanthemos*	4–5
A. schoenoprasum	1–2	*Armeria pseudarmeria*	2–3
Alopecurus pratensis	2–3	*Armoracia rusticana*	5
Althaea officinalis	3–4	*Arnica montana*	2–3

Table A.1—*continued*

Species	Years	Species	Years
Arrhenatherum elatius	3	*Chamaemelum nobile*	2–3
Artemisia absinthium	3–4	*Chasmanthium latifolium*	2–3
A. dracunculus	2	*Cheiranthus cheiri*	3–5
A. vulgaris	4–5	*Chrysanthemum coccineum*	1–4
Asarina sp.	2–3	*C. coronarium*	3–5
Asclepias syriaca	2–3	*C. parthenifolium*	1–4
Asparagus densiflorus	1	*C. segetum*	2–5
A. officinalis	2–3	*Cichorium endiva*	4–5
A. setaceus	1	*C. intybus*	3–5
Asphodeline liburnica	2–3	*Citrullus lanatus*	4–10
Asphodelus microcarpus	2–3	*Clarkia unguiculata*	2–3
Aster sp.	1–3	*Clematis paniculata*	1
Atriplex hortensis	4–6	*Cleome* sp.	2–3
Avena sativa	2–3	*Cobaea scandens*	2
Baptisia australis	3–4	*Coix lacryma-jobi*	2–3
B. tinctoria	3–4	*Coleus* × *hybridus*	2–5
Begonia cucullata	2–3	*Consolida* sp.	1–2
B. gracilis	2–3	*Convolvulus tricolor*	3–5
B. rex	2–3	*Cordyline indivisa*	1
Bellis perennis	1–3	*Coreopsis basalis*	3–4
Beta vulgaris	4–6	*C. grandiflora*	2–3
Borago officinalis	2–3	*C. lanceolata*	2
Brachycome iberidifolia	3–4	*Coriandrum sativum*	3
Brassica hirta	3–5	*Cortaderia selloana*	2–3
B. napus	4–5	*Cosmos* sp.	3–4
B. nigra	4–5	*Cryptostemma calendulaceum*	3–4
B. oleracea	3–5	*Cucumis melo*	5–6
B. rapa	4–6	*C. sativus*	5–7
Bromus arvensis	5	*Cucurbita pepo*	4–6
Browallia speciosa	2–3	*Cyclamen persicum*	2–3
Calceolaria tripartita	2–3	*C. purpurascens*	2–3
Calendula officinalis	3–5	*Cynara cardunculus*	2–3
Callistephus chinensis	2–3	*C. scolymus*	2–3
Caltha palustris	4–5	*Cynosurus cristatus*	3–4
Campanula medium	3–5	*Cyperus alternifolius*	1–2
C. persicifolia	3	*Dactylis glomerata*	2–3
Canna indica	3–5	*Dahlia imperialis*	3
Cannabis sativa	4–6	*D. pinnata*	3
Capsicum annuum	2–4	*Daucus carota*	3–5
Carum carvi	2	*Delphinium hybridum*	3–4
Catharanthus roseus	2–3	*Dianthus barbatus*	2–4
Celosia cristata	4–5	*D. caryophyllus*	2–5
Centaurea cyanus	2–5	*D. chinensis*	3–5
C. moschata	2–4	*D. plumarius*	3–5
Centaurium erythraea	3–4	*Digitalis purpurea*	2–3
Centranthus ruber	3–4	*Dimorphotheca pluvialis*	2
Cerastium tomentosum	2–4	*Dolichos lablab*	3–4
Chaerophyllum bulbosum	1	*Dorotheanthus tricolor*	3–5

Table A.1—*(continued)*

Species	Years	Species	Years
Dracaena draco	1	*Impatiens balsamina*	5–8
Echinocystis lobata	4–5	*I. wallerana*	2–6
Echinops ritro	2	*Inula helenium*	1–2
Emilia spp.	2–3	*Ipomoea nil*	3–4
Ensete ventricosum	1–2	*I. purpurea*	3–4
Erica spp.	3–4	*Iris* ×*germanica*	2
Eryngium alpinum	2–3	*I. kaempferi*	2
Erysimum hieraciifolium	5	*Kniphofia hybrida*	1–2
E. pulchellum	3–4	*Kochia scoparia*	2
Eschscholzia californica	2–3	*Lactuca sativa*	4–5
Eupatorium cannabinum	2	*Lagenaria siceraria*	5–6
Euphorbia marginata	3–4	*Lantana camara*	1–2
Fagopyrum esculentum	2–3	*Lathyrus latifolius*	3–4
F. tartaricum	2–3	*L. odoratus*	2–5
Festuca ovina	3–4	*L. sativus*	4–5
F. pratensis	2–4	*Lavandula angustifolia*	2–5
F. rubra	3–5	*Lavatera trimestris*	3–4
Foeniculum vulgare	2–3	*Lens culinaris*	3
Fuchsia spp.	2	*Leontopodium alpinum*	3
Gaillardia ×*grandiflora*	2–3	*Lepidium sativum*	3–5
G. pulchella	2–3	*Levisticum officinale*	1–2
Gerbera jamesonii	0.25–0.5	*Liatris spicata*	2
Geum coccineum	2–3	*Lilium* spp.	1–3
Gloxinia spp.	2–3	*Limonium latifolium*	2–3
Glycine max	2	*L. sinuatum*	1–2
Gomphrena globosa	3–5	*Linaria maroccana*	2–5
Grevillea robusta	4	*Linum grandiflorum*	5–6
Gypsophila elegans	5	*L. perenne*	5
G. paniculata	5	*L. usitatissimum*	2–3
Helenium autumnale	3–4	*Lobelia cardinalis*	2–3
H. hoopesii	3–4	*L. erinus*	4–5
Helianthus spp.	4–5	*Lobularia maritima*	3–4
H. annuus	2–5	*Lolium multiflorum*	3–4
H. debilis	2–5	*L. perenne*	3–4
Helichrysum spp.	2–3	*Lotus corniculatus*	3–5
Heliopsis helianthoides	2–3	*L. pedunculatus*	3–5
Heliotropium arborescens	1–3	*L. tetragonolobus*	4–5
Helipterum spp.	2–4	*Lunaria annua*	3–4
H. roseum	2–3	*Lupinus albus*	2–4
Hesperis matronalis	3–4	*L. angustifolius*	3–5
Heuchera spp.	3	*L. hartwegii*	2–5
Hibiscus trionum	2–4	*L. hirsutus*	3–5
Hippeastrum spp.	1–3	*L. luteus*	3–5
Hordeum vulgare	2–4	*L. mutabilis*	3–5
Hunnemannia fumariifolia	2	*L. nanus*	3–5
Hyssopus officinalis	2–4	*L. polyphyllus*	4–5
Iberis sempervirens	2–3	*Lychnis chalcedonica*	3–4
I. umbellata	2–3	*L. coronaria*	3–4

Table A.1—*continued*

Species	Years	Species	Years
L. viscaria	3–4	*P. zonale*	2–3
Lycopersicon esculentum	4–6	*Pennisetum setaceum*	2–3
Lythrum flexuosum	2	*P. villosum*	2–3
Macleaya cordata	1–3	*Penstemon* spp.	2–5
Malva moschata	2–5	*Petrorhagia saxifraga*	2–3
Matricaria spp.	2–3	*Petroselinum crispum*	1–3
Matthiola longipetala	4–5	*Petunia* × *hybrida*	2–5
Medicago lupulina	3–4	*Phacelia tanacetifolia*	4–5
M. sativa	3–5	*Phalaris arundinacea*	3–5
Melilotus alba	3–4	*P. canariensis*	3–5
M. officinalis	3–4	*Phaseolus coccineus*	3–4
Melissa officinalis	2–3	*P. vulgaris*	3–4
Mentzelia lindleyi	3	*Phleum pratense*	2–4
Mesembryanthemum crystallinum	3–5	*Phlox cuspidata*	1–2
Mimosa pudica	4–7	*P. drummondii*	1–2
Mimulus × *hybridus*	6	*P. paniculata*	1–2
M. moschatus	3–6	*Physalis alkekengi*	4–5
Mirabilis jalapa	3–5	*P. peruviana*	4–5
Momordica spp.	3–4	*Pimpinella anisum*	2–3
Myosotis alpestris	2–3	*Pisum sativum*	2–5
M. scorpioides	2–3	*Platycodon* spp.	2–3
Myrrhis odorata	1–2	*Poa pratensis*	1–3
Nasturtium spp.	2–3	*P. trivialis*	2–3
Nasturtium officinale	4–5	*Portuclaca grandiflora*	3–4
Nemesia strumosa	2–3	*P. oleracea*	3–4
Nicotiana alata	3–5	*Poterium sanguisorba*	1–2
N. glauca	3–4	*Primula auricula*	2–3
N. sylvestris	3–5	*P.* × *kewensis*	2–3
N. tabacum	3–5	*P. malacoides*	2
Nigella damascena	2–4	*P. obconica*	1–2
N. sativa	3–4	*P. sinensis*	2–3
Ocimum basilicum	4–5	*P. veris*	2–5
Oenothera biennis	2–4	*P. vulgaris*	2
O. drummondii	2–4	*Psylliostachys suworowii*	2–3
O. missouriensis	2–4	*Pueraria lobata*	3–4
O. rosea	2–4	*Raphanus sativus*	3–5
Onobrychis viciifolia	2–3	*Reseda odorata*	2–4
Origanum majorana	1–3	*Rheum* spp.	1–2
Ornithopus sativus	3	*Ricinus communis*	3
Paeonia suffruticosa	3–4	*Rosmarinus officinalis*	2–3
Papaver dubium	4–5	*Rudbeckia* spp.	3–5
P. nudicaule	3–5	*Rumex patientia*	3–4
P. orientale	3–5	*Ruta graveolens*	2–3
P. somniferum	4–5	*Saintpaulia ionantha*	4–5
Pastinaca sativa	1–2	*Salpiglossis sinuata*	4–5
Pelargonium carnosum	3	*Salvia azurea*	4–5
P. × *hortorum*	3	*S. farinacea*	4–5
		S. officinalis	1–3

Table A.1—*continued*

Species	Years	Species	Years
S. patens	2–3	*Tragopogon porrifolius*	1
S. sclarea	1–3	*Trifolium alexandrinum*	4–5
S. splendens	1–2	*T. arvensis*	3–4
Saponaria ocymoides	3–5	*T. hybridum*	2–3
S. pyramidata	2–3	*T. incarnatum*	1–3
Satureja hortensis	1–2	*T. pratense*	3–4
Scabiosa atropurpurea	3	*T. repens*	2–4
S. caucasia	3	*Trigonella caerulea*	3–4
Schizanthus spp.	4	*Trisetum flavescens*	2–3
Scirpus cernuus	2	*Triticum aestivum*	2–4
Scorzonera hispanica	1–2	*Tropaeolum* spp.	2–3
Secale cereale	1–3	*Tropaeolum majus*	3
Sedum reflexum	3–4	*Valeriana officinalis*	3–4
Setaria italica	2–5	*Valerianella locusta*	3–4
Sinningia speciosa	2–3	*Verbascum thapsiforme*	3–4
Smilax spp.	1–3	*Verbena canadensis*	1–2
Solanum spp.	4–5	*V.* ×*hybrida*	2–3
Solanum melongena	5–6	*V. rigida*	2–3
Spergula arvensis	3–5	*Veronica spicata*	2–3
Spinacia oleracea	3–5	*Vicia faba*	3–5
Stevia serrata	2–3	*V. sativa*	3–4
Stipa pennata	2–3	*V. villosa*	3–4
Stokesia laevis	2	*Vinca major*	2–3
Tagetes spp.	3–4	*V. minor*	2–3
Taraxacum officinale	2	*Viola cornuta*	1–2
Tetragonia tetragonioides	4–5	*V. odorata*	1–2
Thunbergia alata	2–3	*V. tricolor*	1–3
T. coccinea	2–3	*Xeranthemum annuum*	2–3
T. fragrans	2–3	*Zea mays*	1–3
T. grandiflora	2–3	*Zinnia elegans*	3–7
Thymus vulgaris	2–3	*Z. haageana*	3–4
Torenia fournieri	2–5		

Table A.2
Number of years that horticultural and agronomic species may be stored under favorable conditions in a Leningrad seed collection before regrowing of the accession is required (Khoroshailov and Zhukova, 1973).

Species	Years	Species	Years
Agrostis sp.	5	*Brassica oleracea*	6
Allium cepa	3	*Bromus erectus*	5
Alopecurus sp.	5	*Cannabis sativus*	3
Anethum graveolens	5	*Cicer arietinum*	7
Avena sativa	8	*Citrullus lanatus*	6

Table A.2—*continued*

Species	Years	Species	Years
Cucumis melo	7	*Onobrychis viciifolia*	8
Cucurbita sp.	4	*Oryza sativa*	4
Dactylis glomerata	5	*Panicum miliaceum*	10
Daucus carota	5	*Perilla frutescens*	1
Festuca sp.	5	*Phaseolus vulgaris*	7
Glycine max	3	*Phleum pratense*	5
Helianthus annuus	5	*Poa palustris*	5
Hordeum vulgare	6	*Ricinus communis*	6
Lactuca sativa	3	*Secale cereale*	4
Linum usitatissimum	8	*Sorghum bicolor*	8
Lolium sp.	5	*Trifolium* sp.	8
Lycopersicon esculentum	8	*Triticum aestivum*	6
Medicago sativa	8	*Zea mays*	10

Table A.3
Average number of years required for germinability of seeds to fall below 50% when stored in Cambridge, England. Seeds that had reached equilibrium with laboratory humidity were placed in cans with close-fitting lids and stored at room temperture. Each can was opened annually for testing purposes. The total data set, most of which is summarized here, was based on "some 10,000 readings" obtained over 25 years. Adapted from D. B. Mackay, J. H. B. Tonkin, and R. J. Flood, "Experiments in Crop Seed Storage at Cambridge," *Landwirtsch Forsch, Sonderheft* 24 (1970): 189–96.

Species	Years	Species	Years
Allium cepa	4	*L. perenne*	7
Alopecurus pratensis	6	*Medicago lupulina*	8
Arrhenatherum elatius	8	*M. sativa*	>12
Avena sativa	15	*Onobrychis viciifolia*	7
Beta vulgaris	13	*Pastinaca sativa*	4
Brassica napus	9	*Phleum pratense*	9
B. oleracea	9	*Pisum sativum*	12
B. rapa	11	*Poa nemoralis*	7
Cynosurus cristatus	7	*P. pratensis*	9
Dactylis glomerata	9	*P. trivialis*	7
Daucus carota	9	*Secale cereale*	6
Festuca elatior	6	*Trifolium hybridum*	7
F. ovina	5	*T. incarnatum*	6
F. pratensis	9	*T. pratense*	6
F. rubra	3	*T. repens*	9
F. tenuifolia	4	*Triticum aestivum*	8
Hordeum vulgare	8	*Vicia sativa*	9
Lolium multiflorum	11		

Appendix

Table A.4
Number of years of storage required for germinability of horticultural and agronomic seeds in Hungary to fall below 50% (Szabó and Virányi, 1970). Seeds of various ages (1–10 years) stored at 5–25°C and 60–70% relative humidity were tested for germinability. Data for each species were collected from 10 to 15 varieties.

Species	Years	Species	Years
Allium cepa	3–4	*Lycopersicon esculentum*	>9
Arachis hypogaea	3	*Nicotiana tabacum*	8–10
Avena sativa	5	*Oryza sativa*	4–5
Brassica hirta	>10	*Panicum miliaceum*	5–8
Capsicum annuum	5	*Papaver somniferum*	3–4
Cucumis melo	9	*Phaseolus vulgaris*	4–7
Datura stramonium	6–10	*Pisum sativum*	5
Glycine max	4	*Raphanus sativus*	8
Hordeum vulgare	5	*Setaria italica*	4–5
Lactuca sativa	4–8	*Sorghum bicolor*	6–8
Lathyrus sativus	>10	*Spinacia oleracea*	3–4
Lens culinaris	>10	*Triticum aestivum*	5
Linum usitatissimum	8	*Vicia faba*	>8
Lupinus albus	>5	*V. sativa*	>5
L. luteus	5	*V. villosa*	5–7

Table A.5
Number of years of storage required for germinability of flower seeds in Sacramento, California, to fall below half-maximum (Goss, 1937). Seeds were tested annually during 10 years of open storage. From one to six seed lots were used for each species. In several cases, maximum germinability was not attained until after some years of storage, and about a quarter of the stocks investigated had a germinability of less than 60% at the first test.

Species	Years	Species	Years
Alcea rosea	10	*Iberis umbellata*	2
Arctotis stoechadifolia	6	*Kochia scoparia*	3
Calendula sp.	10	*Lathyrus odoratus*	6
Callistephus chinensis	4	*Lobularia maritima*	9
Centaurea cyanus	10	*Matthiola* sp.	10
C. gymnocarpa	8	*Nigella damascena*	6
C. moschata	10	*Papaver* sp.	>10
Chrysanthemum carinatum	10	*Petunia* ×*hybrida*	6
C. leucanthemum	3	*Phlox drummondii*	4
C. segetum	10	*Salpiglossis* sp.	>10
Coreopsis sp.	7	*Scabiosa atropurpurea*	6
Cosmos sp.	6	*Schizanthus* ×*wisetonensis*	>10
Delphinium grandiflorum	4	*Tagetes erecta*	6
Dianthus caryophyllus	>10	*T. patula*	6
D. heddewigii	9	*Tropaeolum majus*	>10
Eschscholzia californica	8	*Verbena* sp.	6
Gilia capitata	>10	*Viola* sp.	6
Helichrysum bracteatum	3	*Zinnia* sp.	>10

Table A.6
Number of years of storage required for germinability of flower seeds in Brno, Czechoslovakia, to fall below half-maximum (Nádvorník, 1949). Germination tests were conducted on seeds of various ages (1–8 years) stored under laboratory conditions.

Species	Years	Species	Years
Adelocaryum coelestinum	>3	Gaillardia hybrida	>5
Ageratum houstonianum	6	Helichrysum bracteatum	4
Ammobium alatum	6	Helipterum manglesii	>5
Anchusa capensis	>8	H. roseum	>4
Antirrhinum majus	7	Hesperis matronalis	6
Asperula orientalis	>3	Iberis imperialis	6
Aurinia saxatilis	>8	Impatiens balsamina	>8
Begonia cucullata	7	Lathyrus odoratus	6
Bellis perennis	6	Limonium sinuatum	3
Calendula officinalis	>8	Linaria maroccana	>8
Callistephus chinensis	3	Linum grandiflorum	>8
Campanula medium	>7	Malope trifida	7
Celosia cristata	>8	Matthiola annua	7
Centaurea americana	6	Mimulus ×hybridus	>8
Cerastium tomentosum	7	Myosotis alpestris	>4
Cheiranthus allionii	4	Nemesia grandiflora	6
Chrysanthemum coronarium	>8	Osteospermum sp.	7
C. maximum	>5	Papaver rhoeas	>8
C. parthenium	5	Petunia ×hybrida	4
C. segetum	>8	Phlox drummondii	4
Clarkia amoena	>8	Portulaca grandiflora	5
C. unguiculata	>8	Rudbeckia hirta	>8
Consolida regalis	5	Salpiglossis sinuata	>4
Convolvulus tricolor	>5	Salvia splendens	>3
Coreopsis coronata	>3	S. viridis	>8
C. tinctoria	>8	Sanvitalia procumbens	5
Cuphea llavea	>8	Scabiosa atropurpurea	7
Dianthus barbatus	6	Tagetes erecta	8
D. caryophyllus	8	Thelesperma burridgeanum	>8
D. chinensis	>8	Ursinia anthemoides	8
Digitalis purpurea	3	Venidium fastuosum	>5
Doreanthus bellidiformis	>8	Verbena ×hybrida	7
Echium lycopsis	>5	Viola tricolor	5–6
Emilia javanica	7	Xeranthemum annuum	3
Eschscholzia californica	>7		

Table A.7
Number of years of storage required for germinability of flower seeds in Leningrad, USSR, to fall below half-maximum (Pidotti, 1952). Germination tests were conducted on seeds of various ages (1–8 years) stored under laboratory conditions.

Species	Years	Species	Years
Aquilegia caerulea	4–5	Arabis alpina	5
A. vulgaris	5	Armeria juncea	2

Table A.7—(continued)

Species	Years	Species	Years
Aster alpinus	4	G. septemfida	3
Astragalus cicer	2	Hedysarum alpinum	5
Bellis perennis	6–7	Heuchera americana	3
Bergenia crassifolia	2–3	Lupinus polyphyllus	5
Cephalaria gigantea	3	Nepeta grandiflora	5
Dianthus barbatus	5–6	Papaver persicum	4
D. discolor	6	Polemonium caeruleum	4
D. giganteus	6	Prunella grandiflora	5
D. seguieri	6	Saxifraga caespitosa	4
Digitalis ciliata	4	S. rotundifolia	4
D. grandiflora	4	Sedum kirilowii	4
Dodecatheon meadia	6	Tellima grandiflora	6
Dracocephalum ruyschiana	3	Veronica gentianoides	6
Gentiana lagodechiana	3		

Table A.8
Number of years of storage for germinability of flower seeds in Moscow, USSR, to fall below half-maximum (Nesterenko, 1960). Germination tests were conducted on seeds of various ages (1–6 years) stored under room conditions.

Species	Years	Species	Years
Ageratum houstonianum	>3	Helichrysum bracteatum	4
Antirrhinum majus	>6	Hesperis matronalis	>5
Arctotis stoechadifolia	>6	Iberis amara	>5
Calendula officinalis	>5	I. umbellata	>5
Campanula medium	>4	Impatiens balsamina	>6
C. persicifolia	>5	Lilium regale	3
Celosia cristata	>6	Linum altaicum	>4
Coreopsis grandiflora	4	Matthiola incana	>4
Dahlia pinnata	>6	Papaver dubium	>5
Dianthus barbatus	>4	P. nudicaule	>5
D. fischeri	>4	P. somniferum	>6
D. fragrans	>5	Penstemon angustifolius	>4
D. plumarius	>4	P. barbatus	>4
Dimorphotheca pluvialis	>5	P. gracilis	3–4
Gypsophyla elegans	>5	Potentilla recta	>5

Table A.9
Estimated half-viability periods (P50) of seeds of cultivated species at 13 storage locations (Priestley et al., *Plant Cell Environ.*, in press).

Previously documented data from 15 storage stations were selected for analysis on the basis of the following criteria: (1) They described loss of germinability under open storage conditions in a temperate climate; (2) they provided the results of a number of tests on seed lots over several years, so that a well-defined deteriorative trend could be established; (3) they covered several species stored in a similar fashion at the same locality. Data from the following storage stations were used: South Australia (Pritchard, 1933); Ottawa, Canada (Sifton, 1920); Děčín,

Table A.9—*continued*

Czechoslovakia (Gross, 1917); Brno, Czechoslovakia (Nádvorník, 1947); Denmark (Dorph-Peterson, 1924); England (Carruthers, 1911); France (Bussard, 1935); Germany (Filter, 1932); Ireland (Lafferty, 1931); Poland (Lityński and Chudoba, 1964); Fort Collins, Colorado (Robertson et al., 1943); Yonkers, New York (Barton, 1935a, 1939a, 1953, 1966a, 1966b); and Leningrad, USSR (Adamova, 1964; Gvozdeva, 1966, 1970, 1971; Gvozdeva and Yarchuk, 1969; Gvozdeva and Zhukova, 1971).

These data were assessed by probit analysis. Values of P50 (the half-viability period) were calculated for all the species represented at each storage location by means of the FORTRAN IV program developed by Moore et al. (1983). Data were rejected if initial germinability (or germinability after one year if no initial figure was given) was less than 90%; the average initial value was about 98%. When several sets of data were available for a single species at the same locality, a mean response was calcualted before probit analysis. Legume seeds classed as "hard" were regarded as viable. In these species, tailing of the survival curve owing to the presence of hard seed caused the P50 to be overestimated slightly.

Seed viabilty rankings were calculated by a least square means procedure under the assumption of a two-way crossed classification model without interactions. As a result, each species can be assigned a value that indicates its relative longevity in comparison with all other species in the list. This value (in years) represents the estimated half-viability period (P50) averaged over all 13 storage locations. These estimates are likely to be far more reliable for species present at several localities than they are for those observed at only one or two.

Species	Estimated P50 value (years)	Number of localities providing data
Achillea millefolium	5.75	1
Agrostis gigantea	9.69	1
Allium ampeloprasum	5.30	1
A. cepa	5.43	3
Alopecurus pratensis	6.19	2
Anthriscus cerefolium	3.37	1
Anthyllis vulneraria	9.24	2
Apium graveolens	4.11	1
Arrhenatherum elatius	5.31	3
Asparagus officinalis	3.92	1
Avena sativa	12.96	8
Beta vulgaris	16.51	3
Brassica hirta	13.71	2
B. napus	13.94	5
B. oleracea	7.15	6
B. rapa	8.74	4
Bromus inermis	3.38	1
B. mollis	4.90	1
Cannabis sativa	5.20	3
Carum carvi	4.19	1
Cicer arietinum	15.29	1
Cichorium intybus	5.42	1
Cucumis sativus	4.92	1
Dactylis glomerata	6.61	3
Daucus carota	6.63	2
Dipsacus sylvestris	4.99	1

Table A.9—*continued*

Species	Estimated P50 value (years)	Number of localities providing data
Fagopyrum esculentum	7.46	2
Festuca elatior	4.98	5
F. ovina	4.84	1
F. rubra	4.70	1
Glycine max	3.43	2
Helianthus annuus	5.42	2
Holcus lanatus	10.71	1
Hordeum vulgare	7.19	9
Isatis tinctoria	9.82	1
Lactuca sativa	6.42	4
Lathyrus sativus	13.63	1
Lens culinaris	10.65	2
Lepidium sativum	5.09	2
Linum usitatissimum	8.75	4
Lolium multiflorum	9.36	4
L. perenne	7.19	4
Lotus corniculatus	6.72	2
L. pedunculatus	20.59	1
Lupinus angustifolius	3.81	1
L. luteus	6.20	1
Lycopersicon esculentum	24.52	3
Medicago lupulina	8.76	2
M. sativa	10.56	5
Melilotus alba	33.40	1
Nicotiana tabacum	10.30	1
Onobrychis viciifolia	6.43	1
Ornithopus sativus	8.24	2
Panicum miliaceum	11.90	1
Papaver somniferum	7.28	1
Pastinaca sativa	4.04	2
Perilla frutescens	2.33	1
Petroselinum crispum	3.41	1
Phalaris canariensis	10.96	2
Phaseolus acutifolius	10.39	1
P. coccineus	7.99	1
P. lunatus	13.12	1
P. vulgaris	15.97	3
Phleum pratense	5.73	7
Pisum sativum	15.86	4
Poa nemoralis	3.82	1
P. pratensis.	6.63	1
P. trivialis	4.81	2
Raphanus sativus	13.82	2
Rheum sp.	8.68	1
Ricinus communis	13.31	1

Table A.9—continued

Species	Estimated P50 value (years)	Number of localities providing data
Scorzonera hispanica	6.74	1
Secale cereale	4.51	7
Solanum tuberosum	8.92	1
Spergula arvensis	11.83	1
Spinacia oleracea	12.76	1
Tragopogon porrifolius	2.58	1
Trifolium hybridum	6.16	5
T. incarnatum	5.25	3
T. pratense	5.36	6
T. repens	8.21	4
Triticum aestivum	7.59	9
T. turgidum	8.59	1
Valerianella locusta	6.74	1
Vicia ervilia	11.42	1
V. faba	15.58	4
V. narbonensis	14.95	1
V. sativa	7.33	2
V. villosa	20.82	2
Vigna angularis	11.76	1
V. radiata	19.54	1
Zea mays	9.60	4

Bibliography

ABDALLA, F. H., and E. H. ROBERTS. 1968. Effects of temperature, moisture, and oxygen on the induction of chromosome damage in seeds of barley, broad beans, and peas during storage. *Ann. Bot.* (Lond.) 32:119–36.

ABDALLA, F. H., and E. H. ROBERTS. 1969a. The effects of temperature and moisture on the induction of genetic changes in seeds of barley, broad beans, and peas during storage. *Ann. Bot.* (Lond.) 33:153–67.

ABDALLA, F. H., and E. H. ROBERTS. 1969b. The effects of seed storage conditions on the growth and yield of barley, broad beans, and peas. *Ann. Bot.* (Lond.) 33:169–84.

ABDELMAGID, A. S., and A. M. OSMAN. 1975. Influence of storage period and temperature on viability and chemical composition of cotton seeds. *Ann. Bot.* (Lond.) 39:237–48.

ABDUL-BAKI, A. A., 1969. Relationship of glucose metabolism to germinability and vigor in barley and wheat seeds. *Crop Sci.* 9:732–37.

ABDUL-BAKI, A. A., 1980. Biochemical aspects of seed vigor. *HortScience* 15:765–71.

ABDUL-BAKI, A. A., and J. D. ANDERSON. 1970. Viability and leaching of sugars from germinating barley. *Crop Sci.* 10:31–34.

ABDUL-BAKI, A. A., and J. D. ANDERSON. 1972. Physiological and biochemical deterioration of seeds. In T. T. Kozlowski, ed., *Seed Biology*, 2:283–315. New York: Academic Press.

ABDUL-BAKI, A. A., and G. R. CHANDRA. 1977. Effect or rapid ageing on nucleic acid and protein synthesis by soybean embryonic axes during germination. *Seed Sci. Technol.* 5:689–98.

ABDUL-BAKI, A. A., and A. K. SRIVASTAVA. 1973. Transport of arginine and γ-aminoisobutyric acid from cotyledon to axis in germinating *Phaseolus lunatus*. *J. Am. Soc. Hortic. Sci.* 98:181–85.

ÅBERG, E. 1950. Barley and wheat from the Saqqara pyramid in Egypt. *K. Lantbruksho-egsk. Ann.* (Swed.) 17:59–63.

ABOU-HUSSEIN, M. R., and M. S. EL-BELTAGY. 1977. Effect of seed storage period on flowering and sexuality of squash plants. *Egypt. J. Hortic.* 4:175–80.

ABU-SHAKRA, S., G. AKL, and S. SAAD. 1969. Seed longevity of field and vegetable crops under natural conditions of storage in Lebanon. *Publ. Fac. Agric. Sci. Am. Univ. Beirut* 39:1–19.

ABU-SHAKRA, S. S., and T. M. CHING. 1967. Mitochondrial activity in germinating new and old seeds. *Crop Sci.* 7:115–18.

ACKER, L. 1962. Enzymic reactions in foods of low moisture content. *Adv. Food Res.* 11:263–330.

ACKER, L. W. 1969. Water activity and enzyme activity. *Food Technol.* 23:1257–70.

ACKER, L., and H. KAISER. 1959. Über den Einfluss der Feuchtigkeit auf den Ablauf enzymatischer Reaktionen in wasserarmen Lebensmitteln. II. Mitteilung. *Z. Lebensm. Unters. Forsch.* 110:349–56.

ACKER, L., and R. WIESE. 1972. Verhalten der Lipase im wasserarmen Milieu. I. Einfluss des Aggregatzustandes des Substrates auf die enzymatische Lipolyse. *Lebensm. Wiss. Technol.* 5:181–84.

ACKIGOZ, E., and R. P. KNOWLES. 1983. Long-term storage of grass seeds. *Can. J. Plant Sci.* 63:669–74.

ACTON, E. H. 1893. Changes in the reserve materials of wheat on keeping. *Ann. Bot.* (Lond.) 7:383–87.

ADAMOVA, O. P. 1964. Prodolzhitel'nost' zhizni semyan bobovykh rastenii v estestvennykh usloviyakh. [Longevity of legume seeds under natural conditions.] *Sb. Tr. Vses. Inst. Rastenievod. im. N. I. Vavilova* (Leningr.) 8:3–12.

AGRAWAL, B. R. 1981. A comparison of viable and non-viable seeds of *Cicer* and *Vicia* with respect to some biochemical characters. *Plant Biochem. J.* 8:117–21.

AGRAWAL, P. K. 1977. Germination, fat acidity, and leaching of sugars from five cultivars of paddy (*Oryza sativa*) seeds during storage. *Seed Sci. Technol.* 5:489–98.

AGRAWAL, P. K. 1978. Changes in germination, moisture, and carbohydrate of hexaploid triticale and wheat (*Triticum aestivum*) seeds stored under ambient conditions. *Seed Sci. Technol.* 6:711–16.

AGRAWAL, P. K. 1979. Genotypic variation in germination and membrane permeability in wheat (*Triticum aestivum*) seeds during storage under ambient conditions. *Seed Res.* (New Delhi) 7:120–27.

AGRAWAL, P. K. 1980. Relative storability of seeds of ten species under ambient conditions. *Seed Res.* (New Delhi) 8:94–99.

AGRAWAL, P. K., and M. N. SIDDIQUI. 1973. Influence of storage temperature and seed moisture on germination, free fatty acid content, and leaching of sugars of soybean seeds during storage. *Seed Res.* (New Delhi) 1:75–82.

AGRAWAL, P. K., and S. K. SINHA. 1980. Response of okra seeds (*Abelmoschus esculentus* L.) of different chronological ages during accelerated aging and storage. *Seed Res.* (New Delhi) 8:64–70.

AIGRET, C. 1909. Note sur la conservation multiséculaire de la propriété germinative des graines de certaines plantes annuelles. *Bull. Soc. R. Bot. Belg.* 46:295–99.

AKAMINE, E. K. 1943. The effect of temperature and humidity on viability of stored seeds in Hawaii. *Hawaii Agric. Exp. Stn. Bull.* 90:1–23.

AKHMEDOV, A., and K. E. OVCHAROV. 1974. Izmenenie aktivnosti izofermentov malat- i alkogol'degidrogenazy v semenakh kukuruzy raznoi zhiznesposobnosti. [Changes in the activity of the isozymes of malate and alcohol dehydrogenase in maize seeds of differing viability.] *Uzb. Biol. Zh.*, no. 1, pp. 12–14.

AKSENOV, S. I., N. A. ASKOCHENSKAYA, and E. A. GOLOVINA. 1977. Izuchenie sostoyaniya vody v semenakh raznogo kachestvennogo sostava i ego izmeneniya pri temperaturnykh vozdeistviyakh. *Fiziol. Rast.* (Mosc.) 24:1251–60. [State of water in

seeds of different qualitative composition and changes of it under temperature influences. *Sov. Plant Physiol.* 24:1007–15.]
AKSENOV, S. I., N. A. ASKOCHENSKAYA, and N. S. PETINOV. 1969. O fraktsiyakh vody v semenaka pshenitsy. *Fiziol. Rast.* (Mosc.) 16:71–77. [The fractions of water in wheat seeds. *Sov. Plant Physiol.* 16:58–63.]
ALLEN, G. S. 1962. The deterioration of Douglas-fir seed under various storage conditions. *For. Chron.* 38:145–47.
ALLEN, R. H. 1899. *Star Names and Their Meanings.* New York: Stechert. [*Star Names: Their Lore and Meaning.* New York: Dover. 1963.]
ALLERS. 1922. 40jährige Keimfähigkeit der gelben Lupine. *Forstl. Wochenschr. Silva* 10:319.
ALMEIDA, L. D., and S. M. P. FALIVENE. 1982. Efeito da trilhagem e do armazenamento sobre a conservação de sementes de feijoeiro. *Rev. Bras. Semen.* 4:59–67.
AMSDEN, C. A. 1949. *Prehistoric Southwesterners from Basketmaker to Pueblo.* Los Angeles: Los Angeles Southwest Museum.
ANDERSON, J. D. 1970. Physiological and biochemical differences in deteriorating barley seed. *Crop Sci.* 10:36–39.
ANDERSON, J. D. 1973. Metabolic changes associated with senescence. *Seed Sci. Technol.* 1:401–16.
ANDERSON, J. D. 1977. Adenylate metabolism of embryonic axes from deteriorated soybean seeds. *Plant Physiol.* 59:610–14.
ANDERSON, J. D., and A. A. ABDUL-BAKI. 1971. Glucose metabolism of embryos and endosperms from deteriorating barley and wheat seeds. *Plant Physiol.* 48:270–72.
ANDERSON, J. D., and J. E. BAKER. 1983. Deterioration of seeds during aging. *Phytopathology* 73:321–25.
ANDERSON, J. D., J. E. BAKER, and E. K. WORTHINGTON. 1970. Ultrastructural changes of embryos in wheat infected with storage fungi. *Plant Physiol.* 46:857–59.
ANDERSON, R., R. OSREDKAR, and V. ŽAGAR. 1977. Diffusion properties of lipids—pure and in seeds. *J. Am. Oil Chem. Soc.* 54:487–89.
ANDO, A., and R. VENCOVSKY. 1967. The effect of γ-irradiation on a polygenically inherited character in *Nicotiana tabacum* L. in relation to selection after treatment of aged seeds. *Mutat. Res.* 4:605–14.
ANGUILLESI, M. C., N. BAGNI, and C. FLORIS. 1974. Polyamines and RNA content in wheat embryos from seeds of different age. *G. Bot. Ital.* 108:305–9.
Anonymous. 1805. *Geoponika.* Trans. T. Owen. London: White.
Anonymous. 1824. Wirkung eines Erdhebens auf das Wachstum von Maizen. *Not. Gebiete Nat. Heilkunde* 7:167.
Anonymous. 1830. Botanische Notizen. *Flora* (Jena) 13:583–84.
Anonymous. 1835. Erste Sitzung. *Flora* (Jena) 18:3–7.
Anonymous. 1840. Extraordinary vitality of seeds. *Times* (Lond.), Sept. 21, p. 7.
Anonymous. [Lindley, J.] 1843. [Editorial.] *Gard. Chron.* pp. 787–88.
Anonymous. [Lindley, J.] 1849. [Editorial.] *Gard. Chron.* p. 115.
Anonymous. 1852. Growth and vitality of seeds. *Phytologist* 4:776–77.
Anonymous. [Lindley, J.] 1855. [Editorial.] *Gard. Chron.* pp. 739–40.
Anonymous. 1860. British Association at Oxford. *Gard. Chron.* pp. 713–14.
Anonymous. 1894. Germination of old seeds. *Gard. Chron.* ser. 3, 15:406.
Anonymous. 1931. News and views. *Nature* (Lond.) 127:675.
Anonymous. 1932. Some interesting statistics on the longevity of flower seeds. *Seed World* 31 (11):17.
Anonymous. 1933. Seeds undrowned after 23 years under water. *Sci. Newsl.* 24:131.

Anonymous. 1934. Mummy wheat. *Nature* (Lond.) 134:730.
Anonymous. 1935. Some physical and chemical characteristics of a sample of "mummy" wheat. *Annu. Rep. Dom. Grain Res. Lab.* (Winnipeg) 9:63–64.
Anonymous. 1942a. Duration of viability in seeds. *Gard. Chron.* 111:234.
Anonymous. 1942b. Recent work on germination. *Nature* (Lond.) 149:658–59.
Anonymous. 1945. Peas from King Tut's tomb flourish in Florida. *Turtox News* 23:107–8.
ANTHONY, K. R. M., and S. A. J. TARR. 1952. The causes of deterioration of cotton in the Equatoria Province of the Anglo-Egyptian Sudan. *Emp. J. Exp. Agric.* 20:56–65.
ARNOLD, R. E. 1963. Effects of harvest damage on the rate of fall in viability of wheat stored at a range of moisture levels. *J. Agric. Eng. Res.* 8:7–16.
ARRHENIUS, S. 1908. *Worlds in the Making*. Trans. H. Borns. New York: Harper.
ARTHUR, J. C. 1882. Prolonged vitality of seeds. *Bot. Gaz.* 7:88.
ARTSRUNI, I. G., and G. A. PANOSYAN. 1984. α-amilaznaya aktivnost' aleironovykh sloev khronologicheski starykh i molodykh semyan pshenitsy pri prorastanii i gormonal'nom vozdeistvii. *Fiziol. Rast.* (Mosc.) 31:32–39. [α-amylase activity of the aleurone layers of chronologically old and young wheat seeds during germination and under hormonal influence. *Sov. Plant Physiol.* 31:23–28.]
ASHWORTH, L. J., and J. L. MCMEANS. 1966. Association of *Aspergillus flavus* and aflatoxins with a greenish yellow fluorescence of cotton seed. *Phytopathology* 56:1104–5.
ASKOCHENSKAYA, N. A. 1978. Vodnyi rezhim semyan pri khranenii. [The water regime of seeds during storage.] *Byull. Vses. Nauchno-Issled. Inst. Rastenievod. im. N. I. Vavilova* (Leningr.) 77:49–53.
ASKOCHENSKAYA, N. A. 1982. Sostoyanie vody i ee biologicheskaya rol' v nizkoovodnennoi rastitel'noi tkani na primere semyan. [The state of water and its biological role in plant tissue with low water content as exemplified by seeds.] *Fiziol. Biokhim. Kul't. Rast.* 14:29–41.
ASKOCHENSKAYA, N. A., and S. I. AKSENOV. 1970. Ob osobennostyakh gidratsii semyan raznokachestvennogo sostava. *Fiziol. Rast.* (Mosc.) 17:116–22. [Characteristics of the hyrdration of seeds of different composition. *Sov. Plant. Physiol.* 17:95–100.]
ASKOCHENSKAYA, N. A., S. I. AKSENOV, and N. S. PETINOV. 1970. Sostoyanie vody v semenakh bobovykh. *Dokl. Akad. Nauk SSSR* 192:927–29. [On the state of water in leguminous seeds. *Dokl. Akad. Nauk SSSR, Bot. Sci. Sect.* 192:29–31.]
ASKOCHENSKAYA, N. A., and E. A. GOLOVINA. 1981. State of water in seeds and its alterations under various actions. *Stud. Biophys.* 85:19–20.
ASPINALL, D., and L. G. PALEG. 1971. The deterioration of wheat embryo and endosperm function with age. *J. Exp. Bot.* 22:925–35.
Associated Seed Growers. 1954. The preservation of viability in vegetable seed. *Asgrow Monogr.* 2:1–32.
Association of Official Seed Analysts. 1978. Rules for testing seeds. *J. Seed Technol.* 3 (3):1–126.
Association of Official Seed Analysts, Seed Vigor Test Committee. 1983. *Seed Vigor Testing Handbook*. N.p.
ATABEKOVA, A. I., and V. E. ERMAKOVA. 1973. Vskhozhest' raznovozrastnykh semyan lyupina. [Germinability of lupin seeds of various ages.] *Izv. Timiryazevsk. S'kh. Akad.* (Mosc.), no. 1, pp. 62–67.
AUFHAMMER, G., AND U. SIMON. 1957. Die Samen landwirtschaftlicher Kulturpflanzen im Grundstein des ehemaligen Nürnberger Stadttheaters und ihre Keimfähigkeit. *Z. Acker-Pflanzenb.* 103:454–72.

AVANZI, S., A. BRUNORI, F. D'AMATO, V. NUTI RONCHI, and G. T. SCARASCIA MUG-NOZZA. 1963. Occurrence of 2C (G_1) and 4C (G_2) nuclei in the radicle meristems of dry seeds in *Triticum durum* Desf.: Its implications in studies on chromosome breakage and on developmental processes. *Caryologia* 16:553–58.

AVANZI, S., A. M. INNOCENTI, and A. M. TAGLIASACCHI. 1968. Comportamento alla germinazione e mutabilità cromosomica spontanea in *Triticum durum* Desf. durante i primi due anni di vita del seme. *G. Bot. Ital.* 102:381–95.

AVERY, A. G., and A. F. BLAKESLEE. 1943. Mutation rate in *Datura* seed which has been buried 39 years. *Genetics* 28:69–70.

AZIZUL ISLAM, A. J. M., J. C. DELOUCHE, and C. C. BASKIN. 1973. Comparison of methods for evaluating deterioration in rice seed. *Proc. Assoc. Off. Seed Anal.* 63:155–60.

BABAYAN, R. S. 1971. O deistvii subletal'nykh temperatur na raznovozrastnye semena pshenitsy. *Genetika* 7 (2):174–75. [On the action of sublethal temperatures on wheat seeds of different ages. *Sov. Genet.* 7:267–68.]

BABOTH, E. 1978. A mikroelemes nedves magcsávázás hatása a hagymavetömagvak tárolhatóságára. [Effect of wet dressing with microelements on the stability of onion seeds.] *Zoldsegtermesztesi Kut. Intez. Bull.* 13:77–83.

BACHTHALER, E. 1983. Pelargonien: Skarifizierte Saat. *Gartnerborse Gartenwelt (Gb + Gw)* 83:1103.

BADDELEY, M. S. and J. B. HANSON. 1967. Uncoupling of energy-linked functions of corn mitochondria by linoleic acid and monomethyldecenylsuccinic acid. *Plant Physiol.* 42:1702–10.

BAGCHI, S. 1974. A possible correlation between ageing and irradiation damage in rice. *Radiat. Bot.* 14:309–13.

BAIRD, L. A. M., A. C. LEOPOLD, W. J. BRAMLAGE, and B. D. WEBSTER. 1979. Ultrastructural modifications associated with imbibition of the soybean radicle. *Bot. Gaz.* 140:371–77.

BAKER, D., M. H. NEUSTADT, and L. ZELENY. 1957. Application of the fat acidity test as an index of grain deterioration. *Cereal Chem.* 34:226–33.

BAKIR, Ö., A. ERAÇ, and M. TOKLUOĞLU. 1970. Bazi önemli yem bitkisi tohumlarinin çimlenme güçlerini muhafaza süreleri üzerinde araştirmalar. [Studies on the longevity of some important range plant seeds.] *Cayir-Mer'a Yem Bitkileri ve Zootekni Arastirma Enstitusu Yayinlari [Publ. Grassl. Anim. Husb. Res. Inst. (Ankara)]* 8:1–10.

BALDWIN, H. I. 1942. *Forest Tree Seed of the North Temperate Regions.* Waltham, Mass.: Chronica Botanica.

BANERJEE, A., M. M. CHOUDHURI, and B. GHOSH. 1981. Changes in nucleotide content and histone phosphorylation of ageing rice seeds. *Z. Pflanzenphysiol.* 102:33–36.

BANERJEE, A., and B. GHOSH. 1979. Glucose metabolism of rice embryos during storage under light conditions. *Indian J. Plant Physiol.* 22:57–60.

BANERJEE, A., and B. GHOSH. 1982. Effect of red light on water content of ageing rice seeds. *Indian J. Plant Physiol.* 25:303–5.

BANERJEE, S. K. 1978. Observations on the initiation of seed deterioration and its localization in barley and onion. *Seed Sci. Technol.* 6:1025–28.

BARBER, S. 1972. Milled rice and changes during aging. In D. F. Houston, ed., *Rice: Chemistry and Technology,* pp. 215–63. St. Paul, Minn.: American Association of Cereal Chemists.

BARCLAY, A. S., and F. R. EARLE. 1974. Chemical analyses of seeds. III. Oil and protein content of 1253 species. *Econ. Bot.* 28:179–236.

BARNER, H., and F. DALSKOV. 1954. Erfaringer med opbevaring af douglasfrø. [Experience with the storage of Douglas fir seed.] *Dan. Skovforen. Tidsskr.* 39:570–75.

BARNES, G., and P. BERJAK. 1978. The effect of some antioxidants on the viability of stored seeds. *Electron Microsc. Soc. South. Afr. Proc. [Elektronmikroskopiever. Suidelike Afr. Verrigt.]* 8:95–96.

BARNETT, J. P., and B. F. MCLEMORE. 1970. Storing southern pine seeds. *J. For.* 68:24–27.

BARRINGTON, R. M. 1905. The vitality of seeds. *Ir. Nat.* 14:69–70.

BARSUKOV, L. I., A. V. VICTOROV, I. A. VASILENKO, R. P. EVSTIGNEEVA, and L. D. BERGELSON. 1980. Investigation on the inside-outside distribution, intermembrane exchange, and transbilayer movement of phospholipids in sonicated vesicles by shift reagent NMR. *Biochim. Biophys. Acta* 598:153–68.

BARTHOLOMEW, D. P., and W. E. LOOMIS. 1967. Carbon dioxide production by dry grain of *Zea mays. Plant Physiol.* 42:120–24.

BARTON, L. V. 1932. Effect of storage on the vitality of *Delphinium* seeds. *Contrib. Boyce Thompson Inst.* 4:141–54.

BARTON, L. V. 1935a. Storage of vegetable seeds. *Contrib. Boyce Thompson Inst.* 7:323–32.

BARTON, L. V. 1935b. Storage of some coniferous seeds. *Contrib. Boyce Thompson Inst.* 7:379–404.

BARTON, L. V. 1939a. A further report on the storage of vegetable seeds. *Contrib. Boyce Thompson Inst.* 10:205–20.

BARTON, L. V. 1939b. Storage of elm seeds. *Contrib. Boyce Thompson Inst.* 10:221–33.

BARTON, L. V. 1939c. Storage of some flower seeds. *Contrib. Boyce Thompson Inst.* 10:399–427.

BARTON, L. V. 1941. Relation of certain air temperatures and humidities to viability of seeds. *Contrib. Boyce Thompson Inst.* 12:85–102.

BARTON, L. V. 1953. Seed storage and viability. *Contrib. Boyce Thompson Inst.* 17:87–103.

BARTON, L. V. 1960. Storage of seeds of *Lobelia cardinalis* L. *Contrib. Boyce Thompson Inst.* 20:395–401.

BARTON, L. V. 1961. *Seed Preservation and Longevity.* New York: Interscience.

BARTON, L. V. 1965. Dormancy in seeds imposed by the seed coat. In W. Ruhland, ed., *Encyclopedia of Plant Physiology [Handbuch der Pflanzenphysiologie]* 15 (2): 727–45. New York: Springer.

BARTON, L. V. 1966a. The effect of storage conditions on the viability of bean seeds. *Contrib. Boyce Thompson Inst.* 23:281–84.

BARTON, L. V. 1966b. Effects of temperature and moisture on viability of stored lettuce, onion, and tomato seeds. *Contrib. Boyce Thompson Inst.* 23:285–90.

BARTON, L. V., and H. R. GARMAN. 1946. Effect of age and storage condition of seeds on the yields of certain plants. *Contrib. Boyce Thompson Inst.* 14:243–55.

BARTON-WRIGHT, E. C., R. G. BOOTH, and W. J. S. PRINGLE. 1944. Analysis of barley from King Tutankhamen's tomb. *Nature* (Lond.) 153:288.

BARTOSZ, G. 1981. Non-specific reactions: molecular basis for ageing. *J. Theor. Biol.* 91:233–35.

BASKIN, J. M., and C. C. BASKIN 1977. Dormancy and germination in seeds of common ragweed with reference to Beal's buried seed experiment. *Am. J. Bot.* 64:1174–76.

BASKIN, J. M., and C. C. BASKIN 1978. The seed bank in a population of an endemic plant species and its ecological significance. *Biol. Conserv.* 14:125–30.

BASS, L. N. 1965. Effect of maturity, drying rate, and storage conditions on longevity of Kentucky bluegrass seed. *Proc. Assoc. Off. Seed Anal.* 55:43–46.

BASS, L. N. 1970. Prevention of physiological necrosis (red cotyledons) in lettuce seeds (*Lactuca sativa* L.). *J. Am. Soc. Hortic. Sci.* 95:550–53.

BASS, L. N. 1973a. Controlled atmosphere and seed storage. *Seed Sci. Technol.* 1:463–92.

BASS, L. N. 1973b. Response of seeds of 27 *Cucumis melo* cultivars to three storage conditions. *Proc. Assoc. Off. Seed Anal.* 63:83–87.

BASS, L. N. 1979. Physiological and other aspects of seed preservation. In I. Rubenstein, R. L. Phillips, C. E. Green, and B. G. Gengenbach, eds., *The Plant Seed: Development, Preservation, and Germination*, pp. 145–70. New York: Academic Press.

BASS, L. N. 1980a. Flower seed storage. *Seed Sci. Technol.* 8:591–99.

BASS, L. N. 1980b. Seed viability during long-term storage. *Hortic. Rev.* 2:117–41.

BASS, L. N., D. C. CLARK, and E. JAMES. 1962. Vacuum and inert-gas storage of lettuce seed. *Proc. Assoc. Off. Seed Anal.* 52:116–22.

BASS, L. N., D. C. CLARK, and E. JAMES. 1963a. Vacuum and inert-gas storage of safflower and sesame seeds. *Crop Sci.* 3:237–40.

BASS, L. N., D. C. CLARK, and E. JAMES. 1963b. Vacuum and inert-gas storage of crimson clover and sorghum seeds. *Crop Sci.* 3:425–28.

BASS, L. N., and P. C. STANWOOD. 1978. Long-term perservation of sorghum seed as affected by seed moisture, temperature, and atmospheric environment. *Crop Sci.* 18:575–77.

BASU, R. N. 1976. Physico-chemical control of seed deterioration. *Seed Res.* (New Delhi) 4:15–23.

BASU, R. N. 1977. Seed treatment for vigour, viability, and productivity. *Indian Farming* 27 (1):27–28.

BASU, R. N., T. K. BOSE, K. CHATTOPADHYAY, M. DASGUPTA, N. DHAR, C. KUNDU, R. MITRA, P. PAL, and G. PATHAK. 1975. Seed treatment for the maintenance of vigour and viability. *Indian Agric.* 19:91–96.

BASU, R. N., K. CHATTOPADHYAY, P. K. BANDOPADHYAY, and S. L. BASAK. 1978. Maintenance of vigour and viability of stored jute seeds. *Seed Res.* (New Delhi) 6:1–13.

BASU, R. N., K. CHATTOPADHYAY, and P. PAL. 1974. Maintenance of seed viability in rice (*Oryza sativa* L.) and jute (*Corchorus capsularis* L. and *C. olitorius* L.). *Indian Agric.* 18:75–79.

BASU, R. N., and M. DAS GUPTA. 1974. Control of seed deterioration in wheat (*Triticum aestivum* L.). *Indian Agric.* 18:285–88.

BASU, R. N., and M. DASGUPTA. 1978. Control of seed deterioration by free radical controlling agents. *Indian J. Exp. Biol.* 16:1070–73.

BASU, R. N., and G. DEY. 1983. Soaking and drying of stored sunflower seeds for maintaining viability, vigour of seedlings, and yield potential. *Indian J. Agric. Sci.* 53:563–69.

BASU, R. N., and N. DHAR. 1979. Seed treatment for maintaining vigour, viability, and productivity of sugar beet (*Beta vulgaris*). *Seed Sci. Technol.* 7:225–33.

BASU, R. N., and P. PAL. 1978. Seed treatment to maintain viability, vigor, and yield potential of stored rice seed. *Int. Rice Res. Newsl.* 3 (2):5.

BASU, R. N., and P. PAL. 1979. Physicochemical control of seed deterioration in rice. *Indian J. Agric. Sci.* 49:1–6.

BASU, R. N., and P. PAL. 1980. Control of rice seed deterioration by hydration-dehydration pretreatments. *Seed Sci. Technol.* 8:151–60.

Basu, R. N., D. Pan, and B. Punjabi. 1979. Control of lettuce seed deterioration. *Indian J. Plant Physiol.* 22:247–54.

Basu, R. N., and A. B. Rudrapal. 1979. Iodine treatment of seed for the maintenance of vigour and viability. *Seed Res.* (New Delhi) 7:80–82.

Basu, R. N., and A. B. Rudrapal. 1980. Iodination of mustard seed for the maintenance of vigour and viability. *Indian J. Exp. Biol.* 18:492–94.

Battle, W. R. 1948. Effect of scarification on longevity of alfalfa seed. *Agron. J.* 40:758–59.

Baxter, R. W. 1964. Paleozoic starch in fossil seeds from Kansas coal balls. *Trans. Kans. Acad. Sci.* 67:418–22.

Baxter, W. H. 1857. Sixteenth and final report of a committee, consisting of Professor Daubeny, Professor Henslow, and Professor Lindley, appointed to continue their experiments on the growth and vitality of seeds. *Br. Assoc. Adv. Sci. Rep., Rep. Res. Sci.* 27:43–56.

Beal, W. J. 1879. Experiments and other work of the Horticultural Department. *Annu. Rep. Mich. State Board Agric.* 18:188–202.

Beal, W. J. 1885. The vitality of seeds buried in the soil. *Proc. Soc. Prom. Agric. Sci.* 6:14–15.

Beal, W. J. 1889. Vitality and growth of seeds buried in soil. *Proc. Soc. Prom. Agric. Sci.* 10:15–16.

Beal, W. J. 1894. The vitality of seeds buried in the soil. *Proc. Soc. Prom. Agric. Sci.* 15:283–84.

Beal, W. J. 1899. The vitality of seeds twenty years in the soil. *Proc. Soc. Prom. Agric. Sci.* 20:86–87.

Beal, W. J. 1905. The vitality of seeds. *Bot. Gaz.* 40:140–43.

Beal, W. J. 1910. The vitality of seeds buried in the soil. *Proc. Soc. Prom. Agric. Sci.* 31:21–23.

Beattie, J. H., and V. R. Boswell. 1939. Longevity of onion seed in relation to storage conditions. *U.S. Dep. Agric. Circ.* 512:1–22.

Beattie, J. H., A. M. Jackson, and R. E. Currin. 1932. Effect of cold storage and age of seed on germination and yield of peanuts. *U.S. Dep. Agric. Circ.* 233:1–12.

Becquerel, P. 1906. Sur la longévité des graines. *C. R. Hebd. Seances Acad. Sci.* (Paris) 142:1549–51.

Becquerel, P. 1907. Recherches sur la vie latente des graines. *Ann. Sci. Nat. Bot.*,ser. 9, 5:193–311.

Becquerel, P. 1934. La longévité des graines macrobiotiques. *C. R. Hebd. Seances Acad. Sci.* (Paris) 199:1662–64.

Becquerel, P. 1950. La vie latente des graines aux confins du zéro absolu. *C. R. Hebd. Seances Acad. Sci.* (Paris) 231:1274–77.

Bejnar, W. 1958. Wpływ wieku nasion buraków pastewnych na ich wartość jako materiału siewnego. [The influence of the age of mangel seeds on their seeding value.] *Rocz. Nauk Roln., Ser. A Prod. Rosl.* 77:477–87.

Beker, M. E. 1977. Biomembrany mikroorganizmov pri obezvozhivanii, regidratatsii i reaktivatsii. [Biomembranes of microorganisms during drying, rehydration, and reactivation.] In *Biomembrany: Struktura, funktsii, metody issledovaniya*, pp. 216–35. Riga: Zinatne.

Bennici, A., M. B. Bitonti, C. Floris, D. Gennai, and A. M. Innocenti. 1984. Ageing in *Triticum durum* wheat seeds: early storage in carbon dioxide prolongs longevity. *Environ. Exp. Bot.* 24:159–65.

Berg, C. C., R. A. Nilan, and C. F. Konzak. 1965. The effect of pressure and seed

water content on the mutagenic action of oxygen in barley seeds. *Mutat. Res.* 2:263–73.
BERINGER, H., and F. NORTHDURFT. 1979. Plastid development and tocochromanol accumulation in oil seeds. In L. A. Appelqvist and C. Liljenberg, eds., *Advances in the Biochemistry and Physiology of Plant Lipids*, pp. 133–37. New York: Elsevier/North Holland.
BERJAK, P. 1978. Viability extension and improvement of stored seeds. *S. Afr. J. Sci. [S. Afr. Tydskr. Wet.]* 74:365–68.
BERJAK, P., M. DINI, and H. O. GEVERS. In press. Deteriorative changes in embryos of long-stored, uninfected maize caryopses. *S. Afr. J. Bot.*
BERJAK, P., and T. A. VILLIERS. 1970. Ageing in plant embryos. I. The establishment of the sequence of development and senescence in the root cap during germination. *New Phytol.* 69:929–38.
BERJAK, P., and T. A. VILLIERS. 1972a. Ageing in plant embryos. II. Age-induced damage and its repair during early germination. *New Phytol.* 71:135–44.
BERJAK, P., and T. A. VILLIERS. 1972b. Ageing in plant embryos. IV. Loss of regulatory control in aged embryos. *New Phytol.* 71:1069–74.
BERJAK, P., and T. A. VILLIERS. 1972c. Ageing in plant embryos. V. Lysis of the cytoplasm in non-viable embryos. *New Phytol.* 71:1075–79.
BERNARD, C. 1878. *Leçons sur les phénomènes de la vie communs aux animaux et aux végétaux*. Paris: Baillière. [Lectures on the Phenomena of Life Common to Plants and Animals. Vol. 1. Trans. H. E. Hoff, R. Guillemin, and L. Guilleman. Springfield, Ill.: Thomas, 1974.]
BERNHARDT, D., K.-H. KÖHLER, and M. HECKER. 1981. Makromolekülsynthesen und Processing als Indikatoren der Vitalität von Samen: Modelluntersuchungen an *Agrostemma githago*. In W. Lampeter, ed., *Saatgutvitalität und Pflanzenertrag*, pp. 604–35. Halle-Wittenberg: Martin-Luther-Universität.
BERRIE, A. M. M., R. DON, D. BULLER, M. ALLAM, and W. PARKER. 1975. The occurrence and function of short chain length fatty acids in plants. *Plant Sci. Lett.* 6:163–73.
BEWLEY, J. D., and M. BLACK. 1982. *Physiology and Biochemistry of Seeds in Relation to Germination*. Vol. 2. New York: Springer.
BHATTACHARYYA, B., and K. GUPTA. 1983. Sterols in relation to ageing of seeds of *Helianthus annuus* and *Cicer arietinum*. *Phytochemistry* (Oxf.) 22:1913–16.
BHAUMIK, M., and S. MUKHERJI. 1981. Accumulation of growth inhibitors in relation to viability of stored jute seeds. *Sci. Cult.* 47:63–64.
BIASUTTI OWEN, E. 1956. The storage of seeds for maintenance of viability. *Bull. Commonw. Bur. Pastures Field Crops* 43:1–81.
BIBBEY, R. O. 1948. Physiological studies of weed seed germination. *Plant Physiol.* 23:467–84.
BIER, A. 1925a. Über Keimverzug und seine Bedeutung nach Versuchen an Samen der gelben Lupine. *Angew. Bot.* 7:335–56.
BIER, A. 1925b. Keimverzug. *Mitt. Dtsch. Dendrol. Ges.* 35:187–91.
BISWAS, M., and S. M. SIRCAR. 1976. Note on the effect of storage conditions on the viability of seeds of rice. *Indian J. Agric. Sci.* 46:442–44.
BLACKMAN, F. F. 1909. The longevity and vitality of seeds. *New Phytol.* 8:31–36.
BLASDALE, W. C. 1899. A description of some Chinese vegetable food materials and their nutritive and economic value. *U.S. Dep. Agric. Office Exp. Stn. Bull.* 68:1–48.
BLISS, C. I. 1970. *Statistics in Biology*. Vol. 2. New York: McGraw-Hill.
BLISS, R. D., K. A. PLATT-ALOIA, and W. W. THOMSON. 1984. Changes in plasmalemma

organization in cowpea radicle during imbibition in water and NaCl solutions. *Plant Cell Environ.* 7:601–76.

BLUMER, J. C. 1910. The vitality of pine seed in serotinous cones. *Torreya* 10:108–11.

BOAKYE-BOATENG, K. B., and D. J. HUME. 1975. Effects of storage conditions on germination of soybean (*Glycine max* L. Merr.) seed. *Ghana J. Agric. Sci.* 8:109–14.

BOCKHOLT, A. J., J. S. ROGERS, and T. R. RICHMOND. 1969. Effects of various storage conditions on longevity of cotton, corn, and sorghum seeds. *Crop Sci.* 9:151–53.

BOGDÁN, I., Z. PAPP, and M. SZABÓ. 1963. Kétszázéves gabonaszemek az országos levéltárban. [Two-hundred-year-old cereal grains in the National Archives.] *Agrartort. Szemle* 5:50–66.

BOMME, U., H. FUCHS, and H. HECHT. 1982. Einfluss von Lagerdauer, Aufbewahrungstemperatur, Blaugel und Vakuum auf die Keimfähigkeit von Angelika (*Angelica archangelica* L.)-Samen. *Gartenbauwissenschaft* 47:110–13.

BONASTRE. 1828. Sur quelques substances végétales trouvées dans l'intérieur des cercueils des momies égyptiennes. *J. Pharm. Chimie* (Paris) 14:430–36.

BONDIE, J., T. LUNPERIS, and W. C. STEERE. 1979. The automatic seed analyzer. *Proc. Beltwide Cotton Prod. Mech. Conf.*, pp. 72–75.

BOOS, G. V. 1966. Povyshenie zhiznennosti semyan ogurtsov i tomatov v zakrytom grunte. [Increasing seed viability in cucumbers and tomatoes under glasshouse conditions.] *Tr. Prikl. Bot. Genet. Sel.* 38 (1):178–90.

BOSHNAKOV, P., and K. IVANOV. 1980. Aktivnost na nyakoi enzimi pri s''khranyavaneto na zelenchukovi semena. [Enzyme activity during the storage of vegetable seeds.] *Nauchni Tr. Vissh. Selskostop. Inst. Plovdiv* 25 (2):91–98.

BOURLAND, F. M., and A. A. L. IBRAHIM. 1982. Effects of accelerated aging treatments on six cotton cultivars. *Crop Sci.* 22:637–40.

BOVERIS, A., S. A. PUNTARULO, A. H. ROY, and R. A. SÁNCHEZ. 1984. Spontaneous chemiluminescence of soybean embryonic axes during imbibition. *Plant Physiol.* 76:447–51.

BOVERIS, A., R. A. SANCHEZ, A. I. VARSAVSKY, and E. CADENAS. 1980. Spontaneous chemiluminescence of soybean seeds. *FEBS [Fed. Eur. Biochem. Soc.] Lett.* 113:29–32.

BOVERIS, A., A. I. VARSAVSKY, S. GONÇALVES DA SILVA, and R. A. SÁNCHEZ. 1983. Chemiluminescence of soybean seeds: spectral analysis, temperature dependence, and effect of inhibitors. *Photochem. Photobiol.* 38:99–104.

BOWEN, H. C., and P. D. WOOD. 1968. Experimental storage of corn underground and its implications for Iron Age settlements. *Bull. Inst. Archaeol. Univ. Lond.* 7:1–14.

BRAHM, C., and J. BUCHWALD. 1904. Botanische und chemische Untersuchungen an prähistorischen Getreidekörnern aus alten Gräberfunden. I. Zur Kenntnis der Kleberzellen und der Kleberverteilung in den Getreidekörnern. *Z. Unters. Nahr. Genussm.* 7:12–19.

BRAUN, A. 1878. Ueber den Samen. *Samml. Gemeinver. Wiss. Vortr.* (Berlin) 13:393–424.

BRAY, C. M. 1979. Nucleic acid and protein synthesis in the embryo of germinating cereals. In D. L. Laidman and R. G. Wyn Jones, eds., *Recent Advances in the Biochemistry of Cereals*, pp. 147–73. New York: Academic Press.

BRAY, C. M., and T.-Y. CHOW. 1976a. Lesions in post-ribosomal supernatant fractions associated with loss of viability in pea (*Pisum arvense*) seed. *Biochim. Biophys. Acta* 442:1–13.

BRAY, C. M., and T.-Y. CHOW. 1976b. Lesions in ribosomes of non-viable pea (*Pisum arvense*) embryonic axis tissue. *Biochim. Biophys. Acta* 442:14–23.

BRAY, C. M., and J. DASGUPTA. 1976. Ribonucleic acid synthesis and loss of viability in pea seed. *Planta* (Berl.) 132:103–8.
BRENCHLEY, W. E. 1918. Buried weed seeds. *J. Agric. Sci.* (Camb.) 9:1–31.
BRETT, C. C. 1952. Factors affecting the viability of grass and legume seed in storage and during shipment. *Proc. Int. Grassl. Congr.* 6 (1):878–84.
BRISON, F. R. 1941a. The influence of storage conditions upon the germination of onion seed. *Proc. Am. Soc. Hortic. Sci.* 40:501–3.
BRISON, F. R. 1941b. Influence of storage conditions upon the germination of onion seed. *Proc. Trans. Tex. Acad. Sci.* 25:69–71.
BROCKLEHURST, P. A., and J. DEARMAN. 1983. Interactions between seed priming treatments and nine seed lots of carrot, celery, and onion. I. Laboratory germination. *Ann. Appl. Biol.* 102:577–84.
BROCKLEHURST, P. A., and R. S. S. FRASER. 1980. Ribosomal RNA integrity and rate of seed germination. *Planta* (Berl.) 148:417–21.
BROCKMANN, R., and L. ACKER. 1977. Verhalten der Lipoxygenase in wasserarmen Milieu. I. Einfluss der Wasseraktivität auf die enzymatische Lipidoxidation. *Lebensm-Wiss. Technol.* 10:24–27.
BROCQ-ROUSSEU and E. GAIN. 1908. Sur la durée des peroxydiastases des graines. *C. R. Hebd. Seances Acad. Sci.* (Paris) 146:545–48. [*Ann. Serv. Antiquites* (Cairo) 11 (1911): 40–43.]
BROWN, C. M. 1962. Effect of age of seed on performance of several oat varieties. *Agron. J.* 54:519–20.
BROWN, E. O., and R. H. PORTER. 1942. The viability and germination of seeds of *Convolvulus arvensis* L. and other perennial weeds. *Iowa Agric. Exp. Stn. Res. Bull.* 294:473–504.
BRUCH, E. C. 1961. California Department of Agriculture buried seed project, 1932–1960. *Calif. Dep. Agric. Bull.* 50:29–30.
BRUCHER, H. 1948. Eine Schnellmethode zur Bestimmung der Keimfähigkeit von Samen. *Physiol. Plant.* 1:343–58.
BRYANT, T. R. 1972. Gas exchange in dry seeds: circadian rhythmicity in the absence of DNA replication, transcription, and translation. *Science* (Wash., D.C.) 178:634–36.
BUCH, T. G. 1960. Voprosy khraneniya semyan iv i topolei. [Problems concerning the storage of willow and poplar seeds.] *Tr. Gl. Bot. Sada* (Mosc.) 7:219–39.
BUCHVAROV, P. Z., and N. D. ALEKHINA. 1984. ATP content in soya seeds after natural and accelerated ageing. *Dokl. Bolg. Akad. Nauk [C. R. Acad. Bulg. Sci.]* 37:919–22.
BUCHVAROV, P. [Z.], and T. S. GANTCHEFF. 1984. Influence of accelerated and natural aging on free radical levels in soybean seeds. *Physiol. Plant.* 60:53–56.
BUCHVAROV, P. Z., T. G. KUDREV, S. S. BACHEV, and E. D. MOLLÉ. 1983. Chemiluminiscence and leakage of electrolytes from soya seeds after accelerated and natural ageing. *Dokl. Bolg. Akad. Nauk [C. R. Acad. Bulg. Sci.]* 36:1089–91.
BUCHVAROV, P. Z., and G. M. NIKOLAEV. 1984. Investigating the state of water in seeds of soya after natural and accelerated aging by the method of the spin echo of the NMR. *Dokl. Bolg. Akad. Nauk [C. R. Acad. Bulg. Sci.]* 37:915–18.
BUCHWALD, N. F., and H. A. JENSEN. 1974. Examination of the germination capacity of seeds from an approximately 60 years old seed collection. *K. Vet. Landbohoejsk. Arsskr.* (Denm.), pp. 118–24.
BUKHTOYAROVA, Z. T., N. P. KRASNOOK, and I. A. VISHNYAKOVA. 1979a. Izmenenie aktivnosti fosfataz v semenakh risa s razlichnoi vskhozhest'yu. [Change in the phos-

phatase activity of rice seeds of different germinability.] *Prikl. Biokhim. Mikrobiol.* 15:494–98.

BUKHTOYAROVA, Z. T., N. P. KRASNOOK, and I. A. VISHNYAKOVA. 1979b. Izmenenie dykhatel'nogo metabolizma v semenakh risa razlichnoi zhiznesposobnosti. [Alteration of respiratory metabolism in rice seeds of varying viability.] *Prikl. Biokhim. Mikrobiol.* 15:930–36.

BUKHTOYAROVA, Z. T., E. A. MORGUNOVA, N. P. KRASNOOK, and I. A. VISHNYAKOVA. 1982. Aktivnost' i izofermentnyi sostav kisloi fosfatazy semyan risa s razlichnoi zhiznesposobnost'yu. *Fiziol. Rast.* (Mosc.) 29:93–97. [Activity and isoenzyme composition of acid phosphatase in rice seeds with different viability. *Sov. Plant Physiol.* 29:78–82.]

BULAT, H. 1963. Das allmähliche, durch ungünstige Lagerungsbedingungen beschleunigte Absterben der Samen bzw. Rückgang der Keimfähigkeit im Bilde des topografischen Tetrazoliumverfahrens. *Proc. Int. Seed Test. Assoc.* 28:713–51.

BULLER, D. C., W. PARKER, and J. S. G. REID. 1976. Short-chain fatty acids as inhibitors of gibberellin-induced amylolysis in barley endosperm. *Nature* (Lond.) 260:169–70.

BURGASS, R. W., and A. A. POWELL. 1984. Evidence for repair processes in the invigoration of seeds by hydration. *Ann. Bot.* (Lond.) 53:753–57.

BURGERSTEIN, A. 1895. Beobachtungen über die Keimkraftdauer von ein- bis zehnjährigen Getreidesamen. *Verh. Zool. Bot. Ges. Wien.* 45:414–21.

BURGESS, J. L. 1938. Report on project to determine the percentage and duration of viability of different varieties of soybeans grown in North Carolina. *Proc. Assoc. Off. Seed Anal.* 23:69.

BURLISON, W. L., C. A. VAN DOREN, and J. C. HACKLEMAN. 1940. Eleven years of soybean investigations: varieties, seeding, storage. *Ill. Agric. Exp. Stn. Bull.* 462:123–67.

BURNS, R. E., J. L. WEIMER, and P. R. HENSON. 1958. Factors affecting the longevity of blue lupine seeds. *U.S. Dep. Agric., Agric. Res. Serv.,* ser. ARS-34, 5:1–5.

BURNSIDE, O. C., C. R. FENSTER, L. L. EVETTS, and R. F. MUMM. 1981. Germination of exhumed weed seed in Nebraska. *Weed Sci.* 29:577–86.

BURRIS, J. S. 1980. Maintenance of soybean seed quality in storage as influenced by moisture, temperature, and genotype. *Iowa State J. Res.* 54:377–89.

BURTON, G. W., and J. B. POWELL. 1965. Six chlorophyll-deficient seedlings in pearl millet, *Pennisetum typhoides*, and a suggested system for their nomenclature. *Crop Sci.* 5:1–3.

BURTON, W. G. 1982. *Post-harvest Physiology of Food Crops.* New York: Longman.

BUSSARD, L. 1935. Contribution a l'étude des variations de la faculté germinative des semences au cours de leur conservation. *Ann. Agron.* (Paris) 5:249–77.

BUSSE, J. 1935. Samenaufbewahrung im Vakuum. *Z. Forst. Jagdwes.* 67:321–26.

BUTLER, W. L., and J. E. BAKER. 1965. A haemoprotein from the lipid of peanuts. *Nature* (Lond.) 205:1319–21.

BUTTROSE, M. S. 1973. Rapid water uptake and structural changes in imbibing seed tissues. *Protoplasma* 77:111–22.

BUTTROSE, M. S., and A. SOEFFKY. 1973. Ultrastructure of lipid deposits and other contents in freeze-etched coleoptile cells of ungerminated rice grains. *Aust. J. Biol. Sci.* 26:357–64.

BYRD, H. W., and J. C. DELOUCHE. 1971. Deterioration of soybean seed in storage. *Proc. Assoc. Off. Seed Anal.* 61:41–57.

CADENAS, E., and H. SIES. 1984. Low-level chemiluminescence as an indicator of singlet molecular oxygen in biological systems. *Methods Enzymol.* 105:221–31.

Bibliography

CALLAGHAN, P. T., K. W. JOLLEY, and J. LELIEVRE. 1979. Diffusion of water in the endosperm tissue of wheat grains as studied by pulsed field gradient nuclear magnetic resonance. *Biophys. J.* 28:133–42.

CANDOLLE, A.[-L.-P. P.] DE. 1846. Sur la durée relative de la faculté de germer dans des graines appartenant a diverses familles. *Ann. Sci. Nat. Bot.*, ser. 3, 6:373–82.

CANDOLLE, A.[-L.-P. P. DE]. 1870. Recherches nouvelles sur les Alpes proposées au Club Alpin Suisse. *Echo des Alpes*, April pp. 66–70.

CANDOLLE, A.-L.-P. P. DE. 1883. *Origine des plantes cultivées*. Paris: Baillière. [*Origin of Cultivated Plants*. New York: Appleton, 1885.]

CANDOLLE, A. P. DE. 1832. *Physiologie végétale*. Paris: Béchet Jeune.

CANDOLLE, C. DE. 1895. Sur la vie latente des graines. *Arch. Sci. Phys. Nat.* 33:497–512.

CANODE, C. L. 1965. Germination of normal and hulled grass seed stored under three conditions. *Crop Sci.* 5:409–11.

CANODE, C. L. 1972. Germination of grass seed as influenced by storage conditions. *Crop Sci.* 12:79–80.

CARPENTER, J. A. 1969. The interrelation of age, permeability, and viability of subterranean clover seeds. *Aust. J. Exp. Agric. Anim. Husb.* 9:513–16.

CARRUTHERS, W. 1895. The germination of mummy wheat. *Nature Notes* (Lond.) 6:1–3.

CARRUTHERS, W. 1911. On the vitality of farm seeds. *J. R. Agric. Soc. Engl.* 72:168–83.

CARTLEDGE, J. L., L. V. BARTON, and A. F. BLAKESLEE. 1936. Heat and moisture as factors in the increased mutation rate from *Datura* seeds. *Proc. Am. Philos. Soc.* 76:663–85.

CARTLEDGE, J. L., and A. F. BLAKESLEE. 1933. Mutation rate increased by aging seeds as shown by pollen abortion. *Science* (Wash., D.C.) 78:523.

CARTLEDGE, J. L., and A. F. BLAKESLEE. 1934. Mutation rate increased by aging seeds as shown by pollen abortion. *Proc. Natl. Acad. Sci. U.S.A.* 20:103–10.

CARTLEDGE, J. L., and A. F. BLAKESLEE. 1935. Mutation rate from old *Datura* seeds. *Science* (Wash., D.C.) 81:492–93.

CARTLEDGE, J. L., M. J. MURRAY, and A. F. BLAKESLEE. 1935. Increased mutation rate from aged *Datura* pollen. *Proc. Natl. Acad. Sci. U.S.A.* 21:597–600.

CARUGNO, N., G. GIOVANNOZZI-SERMANNI, and G. WITTMER. 1961. Analisi cromatografiche in fase vapore di olii di tabacco e osservazioni preliminari sull'attività citologica di alcuni loro componenti. *Tabacco* (Rome) 65:219–28.

CARVALHO, N. M. DE, I. A. T. SPINA, and A. M. F. X. DE CAMARGO. 1980. Relações entre o tamanho e o potencial de armazenamento das sementes em duas cultivares de soja. *Rev. Bras. Semen.* 2:35–44.

CATON-THOMPSON, G., and E. W. GARDNER. 1934. *The Desert Fayum*. London: Royal Anthropological Institute.

CAZZUOLA, F. 1877. Alcuni esperimenti fatti sulla durata della facoltà germinativa dei semi. *Boll. R. Soc. Toscan. Ortic.* 2:270–74.

CHABOT, J. F., and A. C. LEOPOLD. 1982. Ultrastructural changes of membranes with hydration in soybean seeds. *Am. J. Bot.* 69:623–33.

CHAKRAVERTY, R. K. 1975. Germination viability and seedling growth in two species of *Corchorus*. *Sci. Cult.* 41:393–95.

CHAKRAVERTY, R. K. 1976. Viability and growth inhibitors in *Corchorus* seeds under storage. *Sci. Cult.* 42:435–37.

CHANEY, R. W. 1951. How old are the Manchurian lotus seeds? *Gard. J. N.Y. Bot. Gard.* 1:137–39.

CHAPMAN, G. W., and J. A. ROBERTSON. 1977. Changes in phospholipid levels during high moisture storage of soybeans. *J. Am. Oil Chem. Soc.* 54:195–98.

CHATTERJEE, A., P. K. SAHA, P. DAS GUPTA, S. N. GANGULY, and S. M. SIRCAR. 1976. Chemical examination of viable and non-viable rice seeds. *Physiol. Plant.* 38:307–8.

CHAUHIN, K. P. S., and S. K. BANERJEE. 1983a. Studies on the induction of variability in quantitative characters of soybean and barley after accelerated ageing of seeds. *Seed Res.* (New Delhi) 11:129–41.

CHAUHIN, K. P. S., and S. K. BANERJEE. 1983b. Correlative studies in accelerated aged seed of soybean and barley. *Seed Res.* (New Delhi) 11:191–203.

CHAUHIN, K. P. S., and M. S. SWAMINATHAN. 1984. Cytogenetical effects of ageing in seeds. *Genetica* (The Hague) 64:69–76.

CHAUVIN, H. 1878. Germination des vieilles graines. *Bull. Soc. Hortic. Cote-d'Or*, ser. 3, 5:112.

CHEAH, K. S. E., and D. J. OSBORNE. 1978. DNA lesions occur with loss of viability in embryos of ageing rye seed. *Nature* (Lond.) 272:593–99.

CHEN, C. C., C. H. ANDREWS, C. C. BASKIN, and J. C. DELOUCHE. 1972. Influence of quality of seed on growth, development, and productivity of some horticultural crops. *Proc. Int. Seed Test. Assoc.* 37:923–39.

CHEN, S. S. C. 1972. Metabolic activities of dormant seeds during dry storage. *Naturwissenschaften* 59:123–24.

CHEN, S. S. C. 1978. Application of a volatile radioactive tracer to detect metabolic activities in dry wild oat seeds. In J. H. Crowe and J. S. Clegg, eds., *Dry Biological Systems*, pp. 175–84. New York: Academic Press.

CHEPIL, W. S. 1946. Germination of weed seeds: II. The influence of tillage treatments on germination. *Sci. Agric.* (Ottawa) 26:347–57.

CHERRY, J. P. 1983. Protein degradation during seed deterioration. *Phytopathology* 73:317–21.

CHILDE, V. G. 1936. *Man Makes Himself.* London: Watts.

CHIN, H. F. and E. H. ROBERTS. 1980. *Recalcitrant Crop Seeds.* Kuala Lumpur: Tropical Press.

CHING, T. M. 1958. Some histological and physiological studies on abnormal crimson clover seedlings. *Proc. Assoc. Off. Seed Anal.* 48:96–99.

CHING, T. M. 1961. Respiration of forage seed in hermetically sealed cans. *Agron. J.* 53:6–8.

CHING, T. M. 1972. Aging stresses on physiological and biochemical activities of crimson clover (*Trifolium incarnatum* L. var. Dixie) seeds. *Crop Sci.* 12:415–18.

CHING, T. M. 1973. Adenosine triphosphate content and seed vigor. *Plant Physiol.* 51:400–402.

CHING, T. M. 1982. Adenosine triphosphate and seed vigor. In A. A. Khan, ed., *The Physiology and Biochemistry of Seed Development, Dormancy, and Germination*, pp. 487–505. New York: Elsevier.

CHING, T. M., and W. CALHOUN. 1968. Productivity of 10-year-old canned forage seeds. *Agron. J.* 60:393–94.

CHING, T. M., and R. DANIELSON. 1972. Seedling vigor and adenosine triphosphate level of lettuce seeds. *Proc. Assoc. Off. Seed Anal.* 62:116–24.

CHING, T. M., and I. SCHOOLCRAFT. 1968. Physiological and chemical differences in aged seeds. *Crop Sci.* 8:407–9.

Bibliography

CHIPPINDALE, H. G., and W. E. J. MILTON. 1934. On the viable seeds present in the soil beneath pastures. *J. Ecol.* 22:508–31.

CHIRKOVSKII, V. I. 1953. Vliyanie stareniya semyan na razvitie rastenii u tabaka. [The effect of seed aging on the growth of tobacco plants.] *Dokl. Akad. Nauk SSSR* 92:439–42.

CHIRKOVSKII, V. I. 1960. Vliyanie ekologicheskikh faktorov pri formirovanii semyan tabaka na protsess ikh stareniya. [The effect of ecological factors during development of tobacco seeds on the aging process.] *Sb. Nauchno-Issled. Rab. Vses. Nauchno-Issled. Inst. Tab. Makhorki* (Krasnodar) 151:65–76.

CHMELAR, F. 1946. Klicenítezce bobtnajících (tyrdých) zrn jetele cerveného (*Trifolium pratense*) vlozenýchdo licidla pred 26 lety. [Germination of imbibition-resistant (hard) seeds of red clover (*Trifolium pratense*) placed in a germinator 26 years previously.] *Sb. Cesk. Akad. Zemed.* 19:222–33.

CHRISTENSEN, C. M. 1972. Microflora and seed deterioration. In E. H. Roberts, ed., *Viability of Seeds*, pp. 59–93. Syracuse, N.Y.: Syracuse University Press.

CHRISTENSEN, C. M., and H. H. KAUFMANN. 1969. *Grain Storage*. Minneapolis, Minn.: University of Minnesota Press.

CHRISTIANSEN, M. N., R. P. MOORE, and C. L. RHYNE. 1960. Cotton seed quality preservation by a hard seed coat characteristic which restricts internal water uptake. *Agron. J.* 52:81–84.

CHRISTIDIS, B. G. 1954. Seed vitality and other cotton characters as affected by the age of seed. *Plant Physiol.* 29:124–31.

CHUDOBA, Z. 1972. Laboratoryjne próby oznaczenia straty suchej masy nasion podczas ich przechowywania. [Laboratory experiments to determine losses of dry matter in seeds during storage.] *Biul. Inst. Hodowli Aklim. Rosl.*, no. 1/2, pp. 11–14.

CHUNG, D. S., and H. B. PFOST. 1967a. Adsorption and desorption of water vapor by cereal grains and their products. Pt. I: Heat and free energy changes of adsorption and desorption. *Trans. Am. Soc. Agric. Eng.* 10:549–51, 555.

CHUNG, D. S., and H. B. PFOST. 1967b. Adsorption and desorption of water vapor by cereal grains and their products. Pt. III: A hypothesis for explaining the hysteresis effect. *Trans. Am. Soc. Agric. Eng.* 10:556–57.

CIESLAR, A. 1904. Einiges über die Rolle des Lichtes im Walde. *Mitt. Forstl. Versuchswes. Osterr.* 30:1–105.

CIFFERI, R. 1942. Una soperchiera inglese ai danni di Cosimo Ridolfi. *Georgofili* 8:70–73.

CLAPHAM, A. R., T. G. TUTIN, and E. F. WARBURG. 1962. *Flora of the British Isles*. 2d ed. Cambridge: Cambridge University Press.

CLARK, L. E. 1963. Effect of various bags on viability and vigor of seed and subsequent effect of vigor on yield. *Annu. Rep. Int. Crop Improv. Assoc.* 45:133–36.

CLARK, V. A. 1904. Seed selection according to specific gravity. *N.Y. Agric. Exp. Stn. Geneva Bull.* 256:367–425.

COHN, M. A., and R. L. OBENDORF. 1978. Occurrence of a stelar lesion during imbibitional chilling of *Zea mays* L. *Am. J. Bot* 65:50–56.

COKER, W. C. 1909. Vitality of pine seeds and the delayed opening of cones. *Am. Nat.* 43:677–81.

COLE, E. W., and M. MILNER. 1953. Colorimetric and fluorimetric properties of wheat in relation to germ damage. *Cereal Chem.* 30:378–91.

COLES, J. 1973. *Archaeology by Experiment*. London: Hutchinson.

COLNAGO, L. A., and P. R. SEIDL. 1983. Application of carbon-13 nuclear magnetic resonance to the germination of soybean seeds *in vivo*. *J. Agric. Food Chem.* 31:459–61.

COLUMELLA, L. J. M. 1941. *De Re Rustica*. Vol. 1. Ed. H. B. Ash. Cambridge: Harvard University Press.

CÔME, D. 1968. Relations entre l'oxygène et les phénomènes de dormance embryonnaire et d'inhibition tégumentaire. *Bull. Soc. Fr. Physiol. Veg.* 14:31–45.

CÔME, D. 1971. Dégazage des enveloppes séminales lors de leur imbibition. I. Cas général. *Physiol. Veg.* 9:439–46.

COMPTON, R. H. 1911. The anatomy of the mummy pea. *New Phytol.* 10:249–55.

CONDUCTOR. [Loudon, J. C.] 1836. Vitality of seeds. *Gard. Mag.* (Lond.) 12:695.

CONGER, A. D., and M. L. RANDOLPH. 1968. Is age-dependent genetic damage in seeds caused by free radicals? *Radiat. Bot.* 8:193–96.

CONGER, B. V., R. A. NILAN, and C. F. KONZAK. 1968. Post-irradiation oxygen sensitivity of barley seeds varying slightly in water content. *Radiat. Bot.* 8:31–36.

CONGER, B. V., R. A. NILAN, and C. F. KONZAK. 1969. The role of water content in the decay of radiation-induced oxygen-sensitive sites in barley seeds during post-irradiation hydration. *Radiat. Res.* 39:45–56.

COOK, R. 1980. The biology of seeds in the soil. In O. T. Solbrig, ed., *Demography and Evolution in Plant Populations*, pp. 107–29. Oxford: Blackwell.

COOLBEAR, P., A. FRANCIS, and D. GRIERSON. 1984. The effect of low temperature pre-sowing treatment on the germination performance and membrane integrity of artificially aged tomato seeds. *J. Exp. Bot.* 35: 1609–17.

CORNER, E. J. H. 1976. *The Seeds of Dicotyledons*. Vol. 1. New York: Cambridge University Press.

CORSI, G., and S. AVANZI. 1969. Embryo and endosperm response to ageing in *Triticum durum* seeds as revealed by chromosomal damage in the root meristem. *Mutat. Res.* 7:349–55.

COSTA, C. L. V., and N. M. DE CARVALHO. 1983. Efeito do tamanho sobre o comportamento de sementes de milho submetidas ao envelhecimento artificial. *Rev. Bras. Semen.* 5:23–27.

COWAN, D. A., T. G. A. GREEN, and A. T. WILSON. 1979. Lichen metabolism. 1. The use of tritium labeled water in studies of anhydrobiotic metabolism in *Ramalina celastri* and *Peltigera polydactyla*. *New Phytol.* 82:489–503.

CRÈVECOEUR, M., R. DELTOUR, and R. BRONCHART. 1982. Quantitative freeze-fracture study of plasmalemma and nuclear envelope of *Zea mays* root cells during early germination. *J. Ultrastruct. Res.* 80:1–11.

CROCIONI, A. 1934. Influenza dell'età del seme sullo sviluppo della pianta. *Nuovi Ann. Agric.* 14:277–90.

CROCKER, W. 1938. Life-span of seeds. *Bot. Rev.* 4:235–74.

CROCKER, W. 1948. *Growth of Plants*. New York: Reinhold.

CROCKER, W., and J. F. GROVES. 1915. A method of prophesying the life duration of seeds. *Proc. Natl. Acad. Sci. U.S.A.* 1:152–55.

CROMARTY, A. 1984. Techniques for drying seeds. In J. B. Dickie, S. Linington, and J. T. Williams, eds., *Seed Management Techniques for Genebanks*, pp. 88–125. Rome: International Board for Plant Genetic Resources.

CROSIER, W., and S. PATRICK. 1952. Longevity of hard seeds in winter vetch. *Proc. Assoc. Off. Seed Anal.* 42:75–80.

CROSSLEY, D. I. 1955. Viability of the seed of lodgepole pine after 20 years in artificial storage. *For. Chron.* 31:250–53.

CROWE, J. H. 1971. Anhydrobiosis: an unsolved problem. *Am. Nat.* 105:563–74.

CROWE, J. H., and J. S. CLEGG. 1973. *Anhydrobiosis.* Stroudsburg, Pa.: Dowden, Hutchinson & Ross.
CROWE, J. H., and J. S. CLEGG. 1978. *Dry Biological Systems.* New York: Academic Press.
CROWE, J. H., and L. M. CROWE. 1982. Induction of anhydrobiosis: membrane changes during drying. *Cryobiology* 19:312–28.
CROWE, J. H., L. M. CROWE, and R. MOURADIAN. 1983. Stabilization of biological membranes at low water activities. *Cryobiology* 20:346–56.
CROWE, J. H., M. A. WHITTAM, D. CHAPMAN, and L. M. CROWE. 1984. Interactions of phospholipid monolayers with carbohydrates. *Biochim. Biophys. Acta* 769:151–59.
CSERESNYES, Z. 1971. Vigoarea germenilor de grîu în funcţie de vîrsta seminţei. [The vigor of wheat seeds as a function of age.] *An. Inst. Cercet. Cereale Plante Teh. Fundulea,* ser. C. *Amelior. Genet. Fiziol. Tehnol. Agric.* (Bucharest) 37:77–85.
CUGINI, G. 1880. Intorno ad un mezzo atto a riconoscere se i semi oleiferi siano ancora capaci di germinare. *Nuovo G. Bot. Ital.* 12:250–53.
CULLIS, P. R., and B. DE KRUIJFF. 1979. Lipid polymorphism and the functional roles of lipids in biological membranes. *Biochim. Biophys. Acta* 559:399–420.
CUMING, A. C., and D. J. OSBORNE. 1978a. Membrane turnover in imbibed and dormant embryos of the wild oat (*Avena fatua* L.). I. Protein turnover and membrane replacement. *Planta* (Berl.) 139:209–17.
CUMING, A. C., and D. J. OSBORNE. 1978b. Membrane turnover in imbibed and dormant embryos of the wild oat (*Avena fatua* L.). II. Phospholipid turnover and membrane replacement. *Planta* (Berl.) 139:219–26.
CUNNINGHAM, T. M., and K. W. CREMER. 1965. Control of the understorey in wet eucalypt forests. *Aust. For.* 29:4–14.
ČURIOVÁ, S. 1984. Zjišťování skladovatelnosti osiv metodou urychleného stárnutí. [Determination of seed storability by the accelerated aging method.] *Rostl. Vyroba* 30:9–18.
ČURIOVÁ, S., and M. VLASÁK. 1984. Ordůdová variabilita při urychlenem stárnutí semen. [Cultivar variability during accelerated aging of seeds.] *Sb. UVTIZ [Ustav. Vedeckotech. Inf. Zemed.] Genet. Slechteni* 20:127–31.
CZYŻEWSKI, W. 1958. Badanie przyczyn wywołujących utratę zdolności kiełkowania nasion marchwi zwyczajnej *Daucus carota* L. [Studies on the causes inducing the loss of germinability of carrot seed *Daucus carota* L.] *Zesz. Nauk. Wyzsz. Szk. Roln. Wroclawiu, Roln.* 5:69–94.
CZYŻEWSKI, W. 1963. Procesy inaktywacji biologicznej w nasionach cebuli *Allium cepa* L. w czasie ich przechowywania. [The process of biological inactivation in onion seeds *Allium cepa* L. during storage.] *Zesz. Nauk. Wyzsz. Szk. Roln. Wroclawiu, Roln.* 18:11–49.
D'AFRICA, G. 1940. Sulla perdita della facoltà germinativa nei semi delle Gramineae. *Lav. Ist. Bot. Giard. Colon. Palermo* 11:168–73.
DAFTARY, R. D., and Y. POMERANZ. 1965. Storage effects in wheat. Changes in lipid composition in wheat during storage deterioration. *J. Agric. Food Chem.* 13:442–46.
DAHLGREN, K. V. O. 1923. *Geranium bohemicum* L. x *G. bohemicum* *deprehensum Erik Almq., ein grun-weiss-marmorierter Bastard. *Hereditas* 4:239–50.
D'AMATO, F. 1951. Mutazioni cromosomiche spontanee in plantule di *Pisum sativum* L. *Caryologia* 3:285–93.
D'AMATO, F., and O. HOFFMANN-OSTENHOF. 1956. Metabolism and spontaneous mutation in plants. *Adv. Genet.* 8:1–28.
DANDY, J. E. 1958. *The Sloane Herbarium.* London: British Museum.

D'ARBAUMONT, J. 1878. De la faculté germinative des graines de melon. *Bull. Soc. Hortic. Cote-d'Or*, ser. 3, 5:80–94.

DARLINGTON, H. T. 1915. Dr. Beal's seed vitality experiments. *Mich. Acad. Sci. Rep.* 17:164–66.

DARLINGTON, H. T. 1922. Dr. W. J. Beal's seed-viability experiment. *Am. J. Bot.* 9:266–69.

DARLINGTON, H. T. 1931. The fifty-year period for Dr. Beal's seed viability experiment. *Am. J. Bot.* 18:262–65.

DARLINGTON, H. T. 1941. The sixty-year period for Dr. Beal's seed viability experiment. *Am. J. Bot.* 28:271–73.

DARLINGTON, H. T. 1951. The seventy-year period for Dr. Beal's seed viability experiment. *Am. J. Bot.* 38:379–81.

DARLINGTON, H. T., and G. P. STEINBAUER. 1961. The eighty-year period for Dr. Beal's seed viability experiment. *Am. J. Bot.* 48:321–25.

DARWIN, C. R. 1855a. Vitality of seeds. *Gard. Chron.*, p. 758. [In P. H. Barrett, ed., *The Collected Papers of Charles Darwin*, 1:260–61. Chicago: University of Chicago Press, 1977.]

DARWIN, C. R. 1855b. Effect of salt-water on the germination of seeds. *Gard. Chron.*, p. 773. [In P. H. Barrett, ed., *The Collected Papers of Charles Darwin*, 1:262–63. Chicago: University of Chicago Press, 1977.]

DARWIN, C. R. 1855c. Longevity of seeds. *Gard. Chron.*, p. 854. [In P. H. Barrett, ed., *The Collected Papers of Charles Darwin*, 1:263. Chicago: University of Chicago Press, 1977.]

DARWIN, C. R. 1881. *The Formation of Vegetable Mould through the Action of Worms, with Observations on their Habits.* London: Murray.

DASGUPTA, M., P. BASU, and R. N. BASU. 1976. Seed treatment for vigour, viability, and productivity of wheat (*Triticum aestivum* L.). *Indian Agric.* 20:265–73.

DASGUPTA, M., K. CHATTOPADHYAY, S. L. BASAK, and R. N. BASU. 1977. Radioprotective action of seed invigoration treatments. *Seed Res.* (New Delhi) 5:104–18.

DAVIES, W. E. 1977. The establishment, screening, and maintenance of a gene pool of white clover (*Trifolium repens* L.). *Proc. Int. Grassl. Congr.* 13:277–80.

DAVIS, W. C. 1931. Phenolase activity in relation to seed viability. *Plant Physiol.* 6:127–38.

DAVIS, W. E. 1926. The use of catalase as a means of determining the viability of seeds. *Proc. Assoc. Off. Seed Anal.* 18:33–39.

DEGEN, A. VON. 1925. The longevity of seeds. In *Report of the Fourth International Seed Testing Congress, Cambridge, 1924*, pp. 139–43. London: HMSO.

DE KRUIJFF, B., P. R. CULLIS, and A. J. VERKLEIJ. 1980. Non-bilayer lipid structures in model and biological membranes. *Trends Biochem. Sci.* 5:79–81.

DE LEO, P., A. DELL'AQUILA, and A. MENSI. 1973. Andamento dell'attività ribonucleasica in semi di pisello durante la conservazione. *Ann. Fac. Agrar. Univ. Bari* 26:449–59.

DELL'AQUILA, A., and P. DE LEO. 1974. Variazioni metaboliche durante l'invecchiamento in semi di pisello: meccanismo di attivazione della ribonucleasi. *Boll. Soc. Ital. Biol. Sper.* 50:1329–34.

DELL'AQUILA, A., P. DE LEO, E. CALDIROLI, and G. ZOCCHI. 1978. Damages at translational level in aged wheat embryos. *Plant Sci. Lett.* 12:217–26.

DELL'AQUILA, A., L. LIOI, and I. SCARASCIA. 1980. Deoxyribonucleic acid synthesis and deoxyribonucleic acid-polymerase activity during early germination of wheat embryos at high and low viability. *Biol. Plant.* (Prague) 22:287–93.

Dell'Aquila, A., D. Pignone, and G. Carella. 1984. Polyethylene glycol 6000 priming effect on germination of aged wheat lots. *Biol. Plant.* (Prague) 26:166–73.
Dell'Aquila, A., and P. Uggenti. 1974. Effetto di alcuni trattamenti ormonali sulla germinazione di semi di pisello in diverse fasi di invecchiamento. *Inf. Bot. Ital.* 6:286–87.
Dell'Aquila, A., G. Zocchi, G. A. Lanzani, and P. De Leo. 1976. Different forms of EF1 and viability in wheat embryos. *Phytochemistry* (Oxf.) 15:1607–10.
Delouche, J. C. 1980. Environmental effects on seed development and seed quality. *HortScience* 15:775–80.
Delouche, J. C. 1981. Harvest and post-harvest factors affecting the quality of cotton planting seed and seed quality evaluation. *Proc. Beltwide Cotton Prod. Res. Conf.*, pp. 289–305.
Delouche, J. C., and C. C. Baskin. 1973. Accelerated aging techniques for predicting the relative storability of seed lots. *Seed Sci. Technol.* 1:427–52.
Delouche, J. C., T. W. Still, M. Raspet, and M. Lienhard. 1962. The tetrazolium test for seed viability. *Miss. Agric. For. Exp. Stn. Tech. Bull.* 51:1–64.
Dent, T. V. 1942. Some records of extreme longevity of seeds of Indian forest plants. *Indian For.* 68:617–31.
De Palozzo, A., and W. G. Jaffe. 1965. Reacciones inmunológicas de extractos de semillas de habichuelas recuperadas de tumbas prehistóricas en el Peru. *Bol. Sci. Quim. Peru* 31:1–5.
Derbyshire, E., N. Harris, D. Boulter, and E. M. Jope. 1977. The extraction, composition, and intra-cellular distribution of protein in early maize grains from an archaeological site in N.E. Arizona. *New Phytol.* 78:499–504.
Des Moulins, C. 1835. Notice sur des graines trouvées dans des tombeaux Romains, et qui ont conservé leur faculté germinative. *Actes Soc. Linn. Bordeaux* 7:65–80.
Des Moulins, C. 1846. *Documents relatifs a la faculté germinative conservée par quelques graines antiques.* Bordeaux: Lafargue.
De Vries, H. 1891. Sur la durée de la vie de quelques graines. *Arch. Neerl. Sci. Exactes Nat.* 24:271–77.
De Vries, H. 1899. *Zaaien en Planten.* Haarlem: Willink & Zoon.
De Vries, H. 1901. *Die Mutationstheorie.* Vol. 1. Leipzig: Von Veit. [*The Mutation Theory,* trans. J. B. Farmer and H. D. Darbishire. Chicago: Open Court, 1909.]
Dey, B., P. K. Sircar, and S. M. Sircar. 1968. Phenolics in relation to nonviability of rice seeds. In S. M. Sircar, ed., *Proceedings of the International Symposium: Plant Growth Substances, Calcutta, 1967,* pp. 57–64. Calcutta: Calcutta University.
Dey, B., and S. M. Sircar. 1968a. Viability and germination of rice seeds (*Oryza sativa* L.): role of endogenous auxin levels. *Indian J. Agric. Sci.* 38:477–85.
Dey, B., and S. M. Sircar. 1968b. The presence of an abscisic acid–like factor in nonviable rice seeds. *Physiol. Plant.* 21:1054–59.
Dey, G., and R. N. Basu. 1982. Studies on the maintenance of seed viability of sunflower (*Helianthus annuus*) by physicochemical treatments. *Indian J. Plant Physiol.* 25:87–97.
Dharmalingam, C., and R. N. Basu. 1978. Control of seed deterioration in cotton (*Gossypium hirsutum* L.). *Curr. Sci.* (Bangalore) 47:484–87.
Dharmalingam, C., V. Ramakrishnan, and K. R. Ramaswamy. 1976. Viability and vigour of stored seeds of blackgram in India (*Vigna mungo* [L.] Hepper). *Seed Res.* (New Delhi) 4:40–50.
Dhindsa, R. S. 1982. Inhibition of protein synthesis by products of lipid peroxidation. *Phytochemistry* (Oxf.) 21:309–13.

DICKIE, J. B., S. MCGRATH, and S. H. LININGTON. 1985. Estimation of provisional seed viability constants for *Lupinus polyphyllus* Lindley. *Ann. Bot.* 55:147–51.

DILLMAN, A. C., and E. H. TOOLE. 1937. Effect of age, condition, and temperature on the germination of flaxseed. *Agron. J.* 29:23–29.

DIMITRIEVICZ, N. 1875. Wie lange bewahren die Samen unserer Culturpflanzen ihre Keimfähigkeit? In F. Haberlandt, ed., *Wissenschaftlich-praktische Untersuchungen auf dem Gebiete des Pflanzenbaues*, 1:98–104. Vienna: Gerold's Sohn.

DIXON, M., and E. C. WEBB. 1979. *Enzymes*. 3d ed. New York: Academic Press.

DJU, D.-C., and C. MCCAY. 1949. Vigor and viability in soybeans. *Soybean Dig.* 9 (5):22–24.

DOBRETSOV, G. E., T. A. BORSCHEVSKAYA, V. A. PETROV, and YU. A. VLADIMIROV. 1977. The increase of phospholipid bilayer rigidity after lipid peroxidation. *FEBS [Fed. Eur. Biochem. Soc.] Lett.* 84:125–28.

DORE, J. 1955. Dormancy and viability of padi seeds. *Malay. Agric. J.* 38:163–73.

DOROSHENKO, A. V. 1937. Plazmoliticheskii metod opredeleniya vskhozhesti semyan. [Plasmolytic method of determining seed germinability.] *Tr. Prikl. Bot. Sel.*, ser. 4, no. 2, pp. 113–19.

DORPH-PETERSEN, K. 1911. Kurze Mitteilungen über Keimuntersuchungen mit Samen verschiedener wildwachsendenen Pflanzen, ausgeführt in der "Dansk Frøkontrol," 1896–1909. *Jahresber. Ver. Angew. Bot.* 8:239–47.

DORPH-PETERSEN, K. 1924. How long do the various seed species retain their germination power?: Investigations made at the Danish State Seed Testing Station in 1891–1903–1920. *Int. Rev. Sci. Pract. Agric.* 2:283–301.

DORPH-PETERSEN, K. 1928. Combien de temps les semences de *Tussilago farfara* gardent-elles leur faculté germinative sous de différentes conditions de température? *Proc. Int. Seed Test. Assoc.* 1 (3–4):72–76.

DORPH-PETERSEN, K. 1934/35. Anlaeg af forsøg over, hvor laenge frø, der henligger i jorden, bevarer spireevnen. [A plan of research on the question of how long seeds retain germinability when buried in the soil.] *Tiddskr. Planteavl* 40:456–58.

DORWORTH, C. E., and C. M. CHRISTENSEN. 1968. Influence of moisture content, temperature, and storage time upon changes in fungus flora, germinability, and fat acidity values in soybeans. *Phytopathology* 58:1457–59.

DOURADO, A. M., and E. H. ROBERTS. 1984a. Chromosome aberrations induced during storage in barley and pea seeds. *Ann. Bot.* (Lond.) 54:767–79.

DOURADO, A. M., and E. H. ROBERTS. 1984b. Phenotypic mutations induced during storage in barley and pea seeds. *Ann. Bot.* (Lond.) 54:781–90.

DUBININ, N. P., and L. G. DUBININA. 1968. Problema potentsial'nykh izmenenii v khromosomakh pri khranenii sukhikh semyan *Crepis capillaris. Genetika* 4 (9):5–23. [The problem of potential changes in the chromosomes during storage of dry *Crepis capillaris* seeds. *Sov. Genet.* 4:1139–52.]

DUBININ, N. P., V. K. SHCHERBAKOV, and G. N. KESLER. 1965. Spektr mutatsii khromosom pri raznykh urovnyakh estestvennogo mutirovaniya kletok. *Dokl. Akad. Nauk SSSR* 161:1434–36. [Spectra of chromosome mutations at various natural mutation levels of cells. *Dokl. Akad. Nauk SSSR, Biol. Sci. Sect.* 161:212–14.]

DUCHARTRE, P. 1867. *Éléments de Botanique*. Paris: Baillière.

DUCKWORTH, R. B. 1962. Diffusion of solutes in dehydrated vegetables. In J. Hawthorn and J. M. Leitch, eds., *Recent Advances in Food Science*, 2:46–49. London: Butterworths.

DUHAMEL DU MONCEAU, H. L. 1754. *Traité de la conservation des graines, et en particulier du froment*. Paris: Guerin-Delatour.

DUHAMEL DU MONCEAU, H. L. 1780. *Des semis et plantations des arbres, et de leur culture*. Paris: Desaint.

DUKE, S. H., and G. KAKEFUDA. 1981. Role of the testa in preventing cellular rupture during imbibition of legume seeds. *Plant Physiol.* 67:449–56.

DUNGAN, G. H., and B. KOEHLER. 1944. Age of seed corn in relation to seed infection and yielding capacity. *Agron. J.* 36:436–43.

DUPART, M.-A., and Y. LE DEUNFF. 1983. Vieillissement des graines de lupin blanc, variété Kalina, et son suivi par conductimétrie. *C. R. Hebd. Seances Acad. Sci.* (Paris), ser. 3, *Sci. Vie*, 296:195–98.

DUREAU DE LA MALLE. 1825. Mémoire sur l'alternance ou sur ce problème: la succession alternative dans la reproduction des espèces végétales vivant en société, est-elle une loi générale de la nature? *Ann. Sci. Nat.*, ser. 1, 5:353–81.

DUVEL, J. W. T. 1904. The vitality and germination of seeds. *U.S. Dept. Agric. Bur. Plant Ind. Bull.* 58:1–96.

DUVEL, J. W. T. 1905. The vitality of buried seeds. *U.S. Dept. Agric. Bur. Plant Ind. Bull.* 83:1–20.

EARLE, F. R., and Q. JONES. 1962. Analysis of seed samples from 113 plant families. *Econ. Bot.* 16:221–50.

EARNSHAW, M. J., B. TRUELOVE, and R. D. BUTLER. 1970. Swelling of *Phaseolus* mitochondria in relation to free fatty acid levels. *Plant Physiol.* 45:318–21.

ECHIGO, T. 1965. Chozōka daizu tanpakushitsu no henshitsu. (1) Tanpakushitsu no yōkaido henka. [Studies of changes that occur in soybean proteins as a result of storage. (1) Changes in solubility of the proteins.] *Tamagawa Daigaku Nogakubu Kenkyu Hokoku [Bull. Fac. Agric. Tamagawa Univ.]* 6:95–102.

EDJE, O.T., and J. S. BURRIS. 1970. Physiological and biological changes in deteriorating soybean seeds. *Proc. Assoc. Off. Seed Anal.* 60:158–66.

EDJE, O. T., and J. S. BURRIS. 1971. Effects of soybean seed vigor on field performance. *Agron. J.* 63:536–38.

EDWARDS, M. 1976. Metabolism as a function of water potential in air-dry seeds of charlock (*Sinapis arvensis* L.). *Plant Physiol.* 58:237–39.

EFEIKIN, A. K. 1961. Vliyanie krupnosti semyan na sokhranenie vskhozhesti u myagkoi pshenitsy. [The effect of seed size on longevity in soft wheat.] *Tr. Chuv. S'kh. Inst.* 5 (1):160–62.

EFFMAN, H., and G. SPECHT. 1967. Bestimmung der Lebensfähigkeit der Samen von Gramineen mit der Säurefuchsinmethode unter Anwendung der Sequenzanalyse. *Proc. Int. Seed Test. Assoc.* 32:27–47.

EGLEY, G. H., and J. M. CHANDLER. 1983. Longevity of weed seeds after 5.5 years in the Stoneville 50-year buried-seed study. *Weed Sci.* 31:264–70.

EGUCHI, T., Y. OSHIKA, and H. YAMADA. 1958. Sosai-shushi no jukudo to jumyō ni kansuru kenkyū. [Studies on the effect of maturity on longevity of vegetable seeds.] *Nogyo Gijutsu Kenkyusho*, Hokoku E, *Engei [Bull. Nat. Inst. Agric. Sci.*, ser. E, *Hortic.]* 7:145–65.

EHRENBERG, L., J. MOUTSCHEN-DAHMEN, and M. MOUTSCHEN-DAHMEN. 1957. Aberrations chromosomiques produites dans des graines par de hautes pressions d'oxygène. *Acta Chem. Scand.* 11:1428–30.

ELDAN, M., A. M. MAYER, A. POLJAKOFF-MAYBER, and M. ELDAN. 1974. Difference in subcellular localization of isocitrate lyase in lettuce seeds of different ages. *Plant Cell Physiol.* 15:169–73.

ELLERTON, D. R., and D. A. PERRY. 1983. The influence of anoxia and ethanol on barley seed death. *Ann. Appl. Biol.* 102:193–202.

ELLIS, R. H. 1984. The meaning of viability. In J. B. Dickie, S. Linington, and J. T. Williams, eds., *Seed Management Techniques for Genebanks*, pp. 146–78. Rome: International Board for Plant Genetic Resources.

ELLIS, R. H., K. OSEI-BONSU, and E. H. ROBERTS. 1982. The influence of genotype, temperature, and moisture on seed longevity in chickpea, cowpea, and soya bean. *Ann. Bot.* (Lond.) 50:69–82.

ELLIS, R. H., and E. H. ROBERTS. 1980. Improved equations for the prediction of seed longevity. *Ann. Bot.* (Lond.) 45:13–30.

ELLIS, R. H., and E. H. ROBERTS. 1981a. The quantification of ageing and survival in orthodox seeds. *Seed Sci. Technol.* 9:373–409.

ELLIS, R. H., and E. H. ROBERTS. 1981b. An investigation into the possible effects of ripeness and repeated threshing on barley seed longevity under six different storage environments. *Ann. Bot.* (Lond.) 48:93–96.

ERDTMAN, G. 1954. *An Introduction to Pollen Analysis*. Waltham, Mass.: Chronica Botanica.

ERNST, A. 1876. Botanische Miscellaneen. *Bot. Ztg.* 34:33–41.

ESBO, H. 1954. Livskraftens bibehållande hos oskalat och skalat frö av timotej vid långtidslagring under ordinära betingelser på frömagasin. [Preservation of viability of unhulled and hulled timothy seeds during prolonged storage under normal conditions in a seed repository.] *K. Skogs. Lantbruksakad. Tidskr.* (Swed.) 93:123–48.

ESBO, H. 1960. Vitality of timothy seed during long storage. *Proc. Int. Seed Test. Assoc.* 25:580–89.

ESITASVILI, G. 1956. Sazamtrosa da nesvis teslis xnovaneba da tesviacin gasroba. [Age and pre-sowing drying of seeds of watermelon and melon.] *Tr. Inst. Polevod. Akad. Nauk. Gruz. SSR* 9:243–48.

EVANS, G. 1957. Red clover seed storage for 23 years. *J. Br. Grassl. Soc.* 12:171–77.

EWART, A. J. 1908. On the longevity of seeds. *Proc. R. Soc. Victoria* 21:1–210.

EXELL, A. W. 1931. The longevity of seeds. *Gard. Chron.*, ser. 3, 89:283.

FEDOSENKO, V. A. 1976. Primenenie metoda sverkhnizkotemperaturnoi konservatsii dlya khraneniya semyan zernovykh bobovykh kultur. [Application of the method of ultra-low conservation temperature for storage of legume seeds.] *Byull. Vses. Nauchno-Issled. Inst. Rastenievod. im. N. I. Vavilova* (Leningr.) 62:77–79.

FEDOSENKO, V. A. 1981. Khranenie semyan ovoshchnykh kul'tur v zhidkom azote. [Storage of seeds of vegetable crops in liquid nitrogen.] *Byull. Vses. Nauchno-Issled. Inst. Rastenievod. im. N. I. Vavilova* (Leningr.) 109:79.

FEENEY, R.E., and J. R. WHITAKER. 1982. The Maillard reaction and its prevention. In J. P. Cherry, ed., *Food Protein Deterioration*, pp. 201–29. Washington, D.C.: American Chemical Society.

FERNÁNDEZ GARCÍA DE CASTRO, M., and C.J. MARTÍNEZ-HONDUVILLA. 1982. Biochemical changes in *Pinus pinea* seeds during storing. *Rev. Esp. Fisiol.* 38:13–20.

FERNÁNDEZ GARCÍA DE CASTRO, M., and C. J. MARTÍNEZ-HONDUVILLA. 1984. Ultrastructural changes in naturally aged *Pinus pinea* seeds. *Physiol. Plant.* 62:581–88.

FIELDING, J. L., and A. GOLDSWORTHY. 1980. Tocopherol levels and ageing in wheat grains. *Ann. Bot.* (Lond.) 46:453–56.

FIELDING, J. L., and A. GOLDSWORTHY. 1982. The evolution of volatiles in relation to ageing in dry wheat seed. *Seed Sci. Technol.* 10:277–82.

FILIMONOV, M. A. 1952. Osobennosti nabukhaniya zhivykh i mertvykh semyan bobovykh trav. [Swelling properties of living and dead legume seeds.] *Dokl. Akad. Nauk SSSR* 84:377–79.

FILIMONOV, M. A. 1958. Nabukhanie i zhiznesposobnost' semyan. [Swelling of seeds and their viability.] *Byull. Mosk. Ova Ispyt. Prir. Otd. Biol.* 63 (5):41–51.

FILIPENKO, G. I. 1983. Osobennosti fiziologo-biokhimicheskikh izmenenii v semenakh pshenitsy i yachmenya v protsesse uskorennogo stareniya. [Characteristics of physiological and biochemical changes in wheat and barley seeds in the accelerated aging procedure.] *Byull. Vses. Nauchno-Issled. Inst. Rastenievod. im. N. I. Vavilova* (Leningr.) 128:19–21.

FILTER, P. 1932. Untersuchungen über die Lebensdauer von Handels- und anderen Saaten, mit besonderer Berücksichtigung der harten Samen und der Altersverfärbung bei den Leguminosen. *Landwirtsch. Vers. Stn.* 114:149–70.

FILUTOWICZ, A., and W. BEJNAR. 1954. Wpływ wieku nasion buraków cukrowych na ich wartość siewną. [The influence of the age of sugar beet seeds on their seeding value.] *Rocz. Nauk Roln.*, ser. A, *Prod. Rosl.*, 69:323–40.

FINEAN, J. B. 1969. Biophysical contributions to membrane structure. *Q. Rev. Biophys.* 2:1–23.

FINNEY, D. J. 1971. *Probit Analysis*. 3d ed. New York: Cambridge University Press.

FIRSOVA, M. K. 1959. *Metody opredeleniya kachestva semyan*. Moscow: Gosudarstvennoe Izdatel'stvo Sel'skokhozyaistvennoi Literatury.

FISHER, R. A., and F. YATES. 1963. *Statistical Tables for Biological, Agricultural, and Medical Research*. 6th ed. New York: Hafner.

FIVAZ, A. E. 1931. Longevity and germination of seeds of *Ribes*, particularly *R. rotundifolium*, under laboratory and natural conditions. *U.S. Dept. Agric. Tech. Bull.* 261:1–40.

FLICHE, P. 1905. Deux observations relatives à la flore des jeunes taillis. *C. R. Hebd. Seances Acad. Sci.* (Paris) 140:1129–32.

FLOOD, R. G. 1978. Contribution of impermeable seed to longevity in *Trifolium subterraneum* (subterranean clover). *Seed Sci. Technol.* 6:647–54.

FLOOD, R. G., and A. SINCLAIR. 1981. Fatty acid analysis of aged permeable and impermeable seeds of *Trifolium subterraneum* (subterranean clover). *Seed Sci. Technol.* 9:475–77.

FLORES, F. B. 1938. Viability of seeds of cotton as affected by moisture and age under different methods of storing. *Philipp. J. Agric.* 9:347–56.

FLORIS, C. 1966. The possible role of the endosperm in the ageing of the embryo in the wheat seed. *G. Bot. Ital.* 73:349–50.

FLORIS, C. 1967. Effetto della bassa temperatura sulla germinazione di semi vecchi di *Triticum durum* Desf. *Genet. Agrar.* 21:231–37.

FLORIS, C. 1970. Ageing in *Triticum durum* seeds; behavior of embryos and endosperms from aged seeds as revealed by the embryo-transplantation technique. *J. Exp. Bot.* 21:462–68.

FLORIS, C., and M. C. ANGUILLESI. 1974. Ageing of isolated embryos and endosperms of durum wheat: an analysis of chromosome damage. *Mutat. Res.* 22:133–38.

FLORIS, C., and P. MELETTI. 1972. Survival and chlorophyll mutation in *Triticum durum* plants raised from aged seeds. *Mutat. Res.* 14:118–22.

FLORIS, C., P. MELETTI, and A. ONNIS. 1972. La conservazione dei semi. *Inf. Bot. Ital.* 4:150–55.

FLOYD, A. G. 1966. Effect of fire upon weed seeds in the wet sclerophyll forests of northern New South Wales. *Aust. J. Bot.* 14:243–56.

FOARD, D. E., and A. H. HABER. 1966. Mitoses in thermodormant lettuce seeds with reference to histological location, localized expansion, and seed storage. *Planta* (Berl.) 71:160–70.

FOUNTAIN, D. W., and J. D. BEWLEY. 1973. Polyribosome formation and protein synthesis in imbibed but dormant lettuce seeds. *Plant Physiol.* 52:604–7.

FRANCIS, A., and P. COOLBEAR. 1984. Changes in the membrane phospholipid composition of tomato seeds accompanying loss of germination capacity caused by controlled deterioration. *J. Exp. Bot.* 35:1764–70.

FRANKEL, E. N. 1980. Lipid oxidation. *Prog. Lipid Res.* 19:1–22.

FRASER, M. T. 1916. Parallel tests of seeds by germination and by electrical response. *Ann. Bot.* (Lond.) 30:181–89.

FRASER, R. S. S. 1980. Why bank plant genes? *Trends Biochem. Sci.* 5 (10):1–2.

FRIEDLANDER, A., and S. NAVARRO. 1972. The role of phenolic acids in the browning, spontaneous heating and deterioration of stored soybeans. *Experientia* (Basel) 28:761–63.

FRÖHLICH, H., and A. HENKEL. 1964. Der Einfluss des Alters des Gurkensaatgutes auf Triebkraft, Wachstum und Ertrag. *Arch. Gartenbau* 12:287–94.

FU, J.-R., Z.-J. LI, and D.-Y. CAI. 1983. Huasheng zhongzi lie bian zhong chao wei jiegoude yanjiu. [Ultrastructural studies on the deterioration of peanut seeds.] *Zhiwu Shengli Xuebao [Chih Wu Sheng Li Hseuh Pao: Plant. Physiol. J.]* 9:93–101.

FUJIMOTO, K., W. E. NEFF, and E. N. FRANKEL. 1984. The reaction of DNA with lipid oxidation products, metals, and reducing agents. *Biochim. Biophys. Acta* 795:100–107.

FUNK, C. R., J. C. ANDERSON, M. W. JOHNSON, and R. W. ATKINSON. 1962. Effect of seed source and seed age on field and laboratory performance of field corn. *Crop Sci.* 2:318–20.

GAIN, E. 1900. Sur les embryons du blé et de l'orge pharaoniques. *C. R. Hebd. Seances Acad. Sci.* (Paris) 130:1643–46.

GAIN, E. 1901. Sur le vieillisement de l'embryon des Graminées. *C. R. Hebd. Seances Acad. Sci.* (Paris) 133:1248–50.

GALINOPOULOU, D., W. P. WILLIAMS, and P. J. QUINN. 1982. Structural studies of plant membrane lipid dispersions subjected to oxidation in the presence of decomposing peroxychromate. *Biochim. Biophys. Acta* 713:315–22.

GALLESCHI, L., and C. FLORIS. 1978. Metabolism of ageing seed: glutamic acid decarboxylase and succinic semialdehyde dehydrogenase activity of aged wheat embryos. *Biochem. Physiol. Pflanz. [BPP]* 173:160–66.

GALLESCHI, L., I. GRILLI, and C. FLORIS. 1977. Glutamic acid decarboxylase in aged wheat seeds. *G. Bot. Ital.* 111:287–88.

GALLIARD, T. 1980. Degradation of acyl lipids: hydrolytic and oxidative enzymes. In P. K. Stumpf and E. E. Conn, eds., *The Biochemistry of Plants*, 4:85–116. New York: Academic Press.

GALLIARD, T., and H. W. CHAN. 1980. Lipoxygenases. In P. K. Stumpf and E. E. Conn, eds., *The Biochemistry of Plants*, 4:131–61. New York: Academic Press.

GANE, R. 1948a. The effect of temperature, water content, and composition of the atmosphere on the viability of carrot, onion, and parsnip seeds in storage. *J. Agric. Sci.* (Camb.) 38:84–89.

GANE, R. 1948b. The effect of temperature, humidity, and atmosphere on the viability of Chewing's fescue grass seed in storage. *J. Agric. Sci.* (Camb.) 38:90–92.

GARBOE, A. 1951. Lidt om "pludseligt" fremkommende planter og om varigheden af frøs spireevne: et biologisk-historisk problem. [A short note on "spontaneously" appearing plants and on the persistence of seed viability: a biological and historical problem.] *Nat. Verden* 7/8:193–200.

GARDNER, H. W. 1979. Lipid hydroperoxide reactivity with proteins and amino acids: a review. *J. Agric. Food Chem.* 27:220–29.

GARNER, F. H., and H. G. SANDERS. 1935. Investigations in crop husbandry. II. On the age of seed beans. *J. Agric. Sci.* 25:361–68.
GARRARD, A. 1955. The germination and longevity of seeds in an equatorial climate. *Gard. Bull.* (Singapore) 14:534–45.
GÄRTNER, C. F. VON. 1849. *Versuche und Beobachtungen über die Bastarderzeugung im Pflanzenreich.* Stuttgart: Hering.
GÁSPÁR, S., A. BUS, and J. BÁNYAI. 1981. Relationship between 1000-seed weight and germination capacity and seed longevity in small seeded Fabaceae. *Seed Sci. Technol.* 9:457–67.
GAUTIER, A. 1896. Remarques de M. Armand Gautier à propos de la note de M. V. Jodin, "Sur l'état dit de *vie latente.*" *C. R. Hebd. Seances Acad. Sci.* (Paris) 122:1351–52.
GELMOND, H., I. LURIA, L. W. WOODSTOCK, and M. PERL. 1978. The effect of accelerated aging of sorghum seeds on seedling vigour. *J. Exp. Bot.* 29:489–95.
GÉRARDIN, S. 1810. *Essai de physiologie végétale.* Paris: Schoell.
GERASIMOVA, E. N. 1935a. Priroda i prichiny mutatsii. II. Nasledovanie mutatsii, proiskhodyashchikh pri starenii semyan: poyavlenie "gomozigotnykh" dslokantov v potomstve rastenii, vyrashchennykh iz starykh semyan. [The nature and causes of mutations. II. The inheritance of mutations caused by aging of seeds: occurrence of "homozygous" dislocants in the progeny of plants grown from old seeds.] *Biol. Zh.* 4:635–42.
GERASSIMOVA, H. [GERASIMOVA, E. N.] 1935b. The nature and causes of mutations. II. Transmission of mutations arising in aged seed: occurrence of "homozygous dislocants" among progeny of plants raised from aged seeds. *Cytologia* (Tokyo) 6:431–37.
GERM, H. 1956. Über den örtlichen Verlauf des Absterbens von Samen. *Bodenkultur, Sonderheft* 7:39–42.
GESTO, M. D. V., and A. VAZQUEZ. 1976. The effect of ageing and soaking on the phenolic content and germination of *Phaseolus* kidneybean seeds. *An. Edafol. Agrobiol.* 35:1067–78.
GHOSAL, K. K., and J. L. MONDAL. 1978. Nature and consequence of chromosomal damage in aged seeds of tetraploid and hexaploid wheat. *Seed Res.* (New Delhi) 6:129–34.
GHOSH, B., J. ADHIKARY, and N. C. BANERJEE. 1981. Changes of some metabolites in rice seeds during ageing. *Seed Sci. Technol.* 9:469–73.
GHOSH, B., and M. M. CHAUDHURI. 1984. Ribonucleic acid breakdown and loss of protein synthetic capacity with loss of viability of rice embryos (*Oryza sativa*). *Seed Sci. Technol.* 12:669–77.
GHOSH, B., T. SENGUPTA, and S. M. SIRCAR. 1978. Physiological changes of rice seeds during storage. *Indian J. Exp. Biol.* 16:411–13.
GHOSH, T., and M. BASAK. 1958. Method of storing jute seed and effect of age on yield of fibre. *Indian J. Agric. Sci.* 28:235–42.
GILMOUR, J., and M. WALTERS. 1973. *Wild Flowers.* 5th ed. London: Collins.
GIOVANNOZZI-SERMANNI, G., and M. E. SCARASCIA-VENEZIAN. 1956. Ricerche analitiche in *Nicotiana rustica* con cromotografia da carta: modificazioni di alcune sostanze di riserva in semi di diversa età. *Tabacco* (Rome) 60:71–77.
GIRARDIN, J. 1848. Moyens frauduleux de déguiser l'altération de la graine de vesce et l'ancienneté de la graine de trèfle incarnat. *J. Pharm. Chimie* (Paris), ser. 3, 14:414–17. [In J. Girardin, *Mélanges d'agriculture, d'économie rurale et publique et de sciences physiques appliquées,* vol. 1, pp. 358–63. Paris: Masson-Dusacq, 1852.]
GISQUET, P., H. HITIER, C. IZARD, and A. MOUNAT. 1951. Mutations naturelles observées

chez *N. tabacum* L. et mutations expérimentales provoquées par l'extrait à froid de graines vieillies prématurement. *Ann. Inst. Exp. Tabac Bergerac* 1 (2):5–35.

GLASS, R. L., J. G. PONTE, C. M. CHRISTENSEN, and W. F. GEDDES. 1959. Grain storage studies. XXVIII. The influence of temperature and moisture level on the behavior of wheat stored in air or nitrogen. *Cereal Chem.* 36:341–56.

GŌ, T., I. ISHIKAWA, and H. IKEDA. 1979. *Acacia mearnsii* De Wild. no shushi no jumyō. (I) Hashu-go 17 nen made. [Longevity of *Acacia mearnsii* De Wild. seeds. (I) Results of 17 years' storage.] *Nippon Ringaku Kaishi [J. Jpn. For. Soc.]* 61:53–57.

GODWIN, H. 1968. Evidence for longevity of seeds. *Nature* (Lond.) 220:708.

GODWIN, H., and E. H. WILLIS. 1964a. Cambridge University natural radiocarbon measurements VI. *Radiocarbon* 6:116–37.

GODWIN, H., and E. H. WILLIS. 1964b. The viability of lotus seeds (*Nelumbium nucifera*, Gaertn.). *New Phytol.* 63:410–12.

GOFF, E. S. 1890. The comparative vitality of hulled and unhulled timothy seed. *Wisc. Agric. Exp. Stn. Rep.* 7:202–4.

GOIRAN, A. 1893. A proposito di una singolare stazione di *Hieracium staticaefolium* Vill. *Boll. Soc. Bot. Ital.* 93–97.

GÖKSIN, A. 1942. Altersermittlung beim Saatgut der Fichte und Kiefer. *Forstwiss. Zentralbl.* (Hamb.) 86:111–17.

GOLA, G. 1906. Sull'attività respiratoria di alcuni semi durante il periodo della quiescenza. *Atti R. Accad. Sci. Torino, Cl. Sci. Fis. Mat. Nat.* 41:330–37.

GOLDSWORTHY, A., J. L. FIELDING, and M. B. J. DOVER. 1982. "Flash imbibition": a method for the re-invigoration of aged wheat seed. *Seed Sci. Technol.* 10:55–65.

GONZÁLEZ, B., C. J. MARTÍNEZ-HONDUVILLA, and M. FERNÁNDEZ GARCÍA DE CASTRO. 1982. Cambios en proteínas y proteasas durante el envejecimiento artificial en semillas de *Pinus pinea*. *Rev. Esp. Fisiol.*, supl., 38:307–14.

GONZÁLEZ-JULIÁ, B., M. FERNÁNDEZ GARCÍA DE CASTRO, and C. J. MARTÍNEZ-HONDUVILLA. 1983. Cambios en las fracciones de albúminas y globulinas en semillas de *Pinus pinea* durante el envejecimiento artificial. *Rev. Esp. Fisiol.* 39:409–16.

GOODSELL, S. F., G. HUEY, and R. ROYCE. 1955. The effect of moisture and temperature during storage on cold test reaction of *Zea mays* stored in air, carbon dioxide, or nitrogen. *Agron. J.* 47:61–64.

GORDON-KAMM, W. J., and P. L. STEPONKUS. 1984. Lamellar-to-hexagonal II phase transitions in the plasma membrane of isolated protoplasts after freeze-induced dehydration. *Proc. Natl. Acad. Sci. U.S.A.* 81:6373–77.

GÓRECKI, R. J. 1982. Viability and vigour of ageing pea seeds with various densities. *Acta Soc. Bot. Pol.* 51:481–88.

GÓRECKI, R. J., G. E. HARMAN, and L. R. MATTICK. 1985. The volatile exudates from germinating pea seeds of different viability and vigor. *Can. J. Bot.* 63:1035–39.

GÓRECKI, R. J., and S. JAGIELSKI. 1982. Wartość przechowalnicza nasion grochu, bobiku i łubinu żółtego o różnym ciężarze właściwym. [Storage quality of pea, field bean, and yellow lupin seeds of different specific gravity.] *Zesz. Nauk. Akad. Roln. Tech. Olsztynie, Roln.* 32:57–67.

GORI, C. 1956. Aberrazioni e ritardo di sviluppo in plantule e piante di seme invecchiato di *Vicia faba* L. *Caryologia* 9:76–81.

GOSS, W. L. 1924. The vitality of buried seeds. *J. Agric. Res.* (Wash., D.C.) 29:349–62.

GOSS, W. L. 1937. Germination of flower seeds stored for ten years in the California State Seed Laboratory. *Calif. Dep. Agric. Bull.* 26:326–33.

Goss, W. L. 1939. Germination of buried weed seeds. *Calif. Dep. Agric. Bull.* 28:132–35.

Grabe, D. F. 1966. Significance of seedling vigor in corn. *Proc. Hybrid Corn Res. Conf.* 21:39–44.

Grabe, D. F. 1967. Low seed vigor: hidden threat to crop yields. *Crops Soils Mag.* (Madison) 19 (6):11–13.

Graber, L. F. 1922. Scarification as it affects longevity of alfalfa seed. *Agron. J.* 14:298–302.

Grabiec, S., K. Bogdański, and A. Marczukajtis. 1968. Variability of chemiluminescence intensity of grain tissues in rye of various periods of storage durations. *Bull. Acad. Pol. Sci.*, Ser. *Sci. Biol.* 26:761–64.

Gračanin, M. 1927. Über das Verhältnis zwischen der Katalaseaktivität und der Samenvitalität. *Biochem. Z.* 180:205–10.

Grange, A. 1980. Vieillissement des graines de *Phaseolus vulgaris* (L.) var. Contender. I. Effets sur la germination, la vigueur, la teneur en eau et la variation des formes d'azote. *Physiol. Vég.* 18:579–86.

Grange, A., M.-N. Miège, J.-F. Manen, and J. Miège. 1980. Vieillissement des graines de *Phaseolus vulgaris* (L.) var. Contender. II. Effets sur les BAPAses et les inhibiteurs de trypsin. *Physiol. Vég.* 18:587–96.

Grange, A., M. N. Miège, and J. Miège. 1973. Evolution des protéines de graines de *Phaseolus vulgaris* var. Contender pendant la phase quiescente en relation avec le degré de maturation des graines et les modalités de leur conservation. *Bull. Soc. Bot. Suisse [Ber. Schweiz. Bot. Ges.]* 83:42–53.

Gratkowski, H. 1973. Pregermination treatments for redstem ceanothus seeds. *U.S. For. Serv. Res. Pap.*, ser. PNW, 156:1–11.

Gray, J. 1978. Measurement of lipid oxidation: a review. *J. Am. Oil Chem. Soc.* 55:539–46.

Griffiths, D. J., and R. A. D. Pegler. 1964. The effects of long-term storage on the viability of S23 perennial ryegrass seed and on subsequent plant development. *J. Br. Grassl. Soc.* 19:183–90.

Griffon, E. 1901. Revue des travaux de physiologie et de chimie végétales parus de 1893 a 1900. *Rev. Gen. Bot.* 13:276–84.

Grilli, I., M. C. Anguillesi, P. Meletti, C. Floris, and L. Galleschi. 1982. Influence of age and storage temperature on RNA metabolism in durum wheat seeds. *Z. Pflanzenphysiol.* 107:211–21.

Grime, J. P. 1979. *Plant Strategies and Vegetation.* New York: Wiley.

Gross, E. 1917. Veränderungen der Getreidesamen bei 10jähriger Lagerung. *Z. Landwirtsch. Versuchswes. Oesterr.* 20:471–87.

Gruner, S. M., K. J. Rothschild, and N. A. Clark. 1982. X-ray diffraction and electron microscope study of phase separation in rod outer segment photoreceptor membrane multilayers. *Biophys. J.* 39:241–51.

Grzesiuk, S., and K. Kulka. 1971a. Nucleic acids and nucleases in cereal seeds of various age. *Bull. Acad. Pol. Sci.*, ser. *Sci. Biol.* 19:363–66.

Grzesiuk, S., and K. Kulka. 1971b. Proteins in ageing oats seeds. *Bull. Acad. Pol. Sci.*, ser. *Sci. Biol.* 19:435–40.

Grzesiuk, S., and J. Łuczyńska. 1972. Nucleic acid (NA) and proteins in artificially ageing soyabean seeds (*Glycine max* Merr.). *Bull. Acad. Pol. Sci.*, ser. *Sci. Biol.* 20:891–96.

Grzesiuk, S., and J. Tłuczkiewicz. 1982. Viability and vigour of ageing winter wheat grains. *Acta Soc. Bot. Pol.* 51:251–262.

GUÉRIN-MÉNEVILLE, F.-E. 1857. Note sur le blé Drouillard, variété de froment provenant d'un tombeau d'Égypte. *C. R. Hebd. Seances Acad. Sci.* (Paris) 44:473–74.

GUILLAUMIN, A. 1928. Le maintien des graines dans un milieu privé d'oxygène comme moyen de prolonger leur faculté germinative. *C. R. Hebd. Seances Acad. Sci.* (Paris) 187:571–72.

GUILLAUMIN, A. 1937. Prolongation de la faculté germinative des graines. *C. R. Hebd. Seances Acad. Agric. Fr.* 23:803–4.

GUILLEMET, R. 1931. Sur le pouvoir lipolytique de différentes variétés de graines de ricin; facteurs susceptibles de le modifier. *C. R. Seances Soc. Biol. Fil.* 108:779–81.

GUMIŃSKA, Z. 1958. Utlenione związki próchniczne jako stymulatory kiełkowania nasion. [Oxidized humic compounds as stimulants of seed germination.] *Acta Soc. Bot. Pol.* 27:501–22.

GUMIŃSKA, Z., and J. SULEJ. 1964. Wpływ humianu sodowego i wersenianu sodowego na kiełkowanie nasion. [The effect of sodium humate and sodium versanate on seed germination.] *Biul. Inst. Hodowli Aklim. Rosl.*, no. 3, pp. 29–31.

GUNTHARDT, H., L. SMITH, M. E. HAFERKAMP, and R. A. NILAN. 1953. Studies on aged seeds. II. Relation of age of seeds to cytogenetic effects. *Agron. J.* 45:438–41.

GUPTA, P. C. 1976. Viability of stored soybean seeds in India. *Seed Res.* (New Delhi) 4:32–39.

GUSTAFSSON, A. 1937. The different storability of chromosomes and the nature of mitosis. *Hereditas* 22:281–335.

GUY, R. 1982. Influence du stockage sur la durée de germination des semences. *Rev. Suisse Vitic. Arboric. Hortic.* 14:99–101.

GUYOT, L. 1960. Sur la présence dans les terres cultivées et incultes de semences dormantes des espèces adventices. *Bull. Serv. Carte Phytogeog.*, ser. B, 5:197–254.

GVOZDEVA, Z. V. 1965. Nadezhnyi sposob dlitel'nogo khraneniya semyan kollektsionnykh obraztsov. [Safe means for long-term storage of seed collection samples.] *Agrobiologiya* 572–76.

GVOZDEVA, Z. V. 1966. O dlitel'nom khranenii semyan. [On the long-term storage of seeds.] *Tr. Prikl. Bot. Genet. Sel.* 38 (1):133–45.

GVOZDEVA, Z. V. 1970. Prodolzhitel'nost' zhizni semyan maslichnykh i tekhnicheskikh kultur pri razlichnykh usloviyakh khraneniya. [Seed longevity of oil-bearing and industrial crops under various storage conditions.] *Tr. Prikl. Bot. Genet. Sel.* 42 (1):204–16.

GVOZDEVA, Z. V. 1971. Prodolzhitel'nost' zhizni semyan nekotorykh ovoshchnykh kul'tur pri razlichnykh sposobakh khraneniya. [Longevity of some vegetable crop seeds under various storage conditions.] *Tr. Prikl. Bot. Genet. Sel.* 44 (3):187–210.

GVOZDEVA, Z. V., and T. A. YARCHUK. 1969. Prodolzhitel'nost' zhizni semyan kukuruzy v zavisimosti ot uslovii khraneniya. [Longevity of maize seed in relation to the conditions of storage.] *Tr. Prikl. Bot. Genet. Sel.* 41 (2):27–42.

GVOZDEVA, Z. V., and N. V. ZHUKOVA. 1971. Vliyanie uslovii khraneniya na prodolzhitel'nost' zhizni semyan fasoli, nuta i soi. [Influence of storage conditions on longevity of seeds of bean, chickpea, and soybean.] *Tr. Prikl. Bot. Genet. Sel.* 45 (3):161–68.

HABER, A. H., and M. L. RANDOLPH. 1967. Gamma-ray induced ESR signals in lettuce: evidence for seed-hydration-resistant and -sensitive free radicals. *Radiat. Bot.* 7:17–28.

HABER, E. S. 1950. Longevity of the seed of sweet corn inbreds and hybrids. *Proc. Am. Soc. Hortic. Sci.* 55:410–12.

HABERLANDT, F. 1873. Die Keimfähigkeit unserer Getreidekörner, ihre Dauer und die Mittel ihrer Erhaltung. *Wien. Landwirtsch. Ztg.* 126.

HAFENRICHTER, A. L., R. B. FOSTER, and J. L. SCHWENDIMAN. 1965. Effect of storage at four locations in the West on longevity of forage seeds. *Agron. J.* 57:143–47.

HAFERKAMP, M. E., L. SMITH, and R. A. NILAN. 1953. Studies on aged seeds. I. Relation of age of seed to germination and longevity. *Agron. J.* 45:434–37.

HAHN, E. L. 1950. Spin echoes. *Phys. Rev.* 80:580–94.

HALDER, S., and K. GUPTA. 1980. Effect of storage of sunflower seeds in high and low relative humidity on solute leaching and internal biochemical changes. *Seed Sci. Technol.* 8:317–21.

HALDER, S., and K. GUPTA. 1981. Effect of RH on sunflower seed viability with special reference to membrane permeability and biochemical changes. *Geobios* (Jodhpur) 8:6–9.

HALDER, S., S. KOLE, and K. GUPTA. 1983. On the mechanism of sunflower seed deterioration under two different types of accelerated ageing. *Seed Sci. Technol.* 11:331–39.

HALLAM, N. D. 1972. Embryogenesis and germination in rye (*Secale cereale* L.). 1. Fine structure of the developing embryo. *Planta* (Berl.) 104:157–66.

HALLAM, N. D. 1973. Fine structure of viable and non-viable rye and other embryos. In W. Heydecker, ed., *Seed Ecology*, pp. 115–44. State College, Pa.: Pennsylvania State University Press.

HALLAM, N. D., and D. J. OSBORNE. 1974. Fine structure and histochemistry of ancient viable and non-viable seeds. In J. V. Sanders and D. J. Goodchild, eds., *Proceedings of the Eighth International Congress on Electron Microscopy, Canberra, 1974*, 2:598–99. Canberra: Australian Academy of Sciences.

HALLAM, N. D., B. E. ROBERTS, and D. J. OSBORNE. 1973. Embryogenesis and germination in rye (*Secale cereale* L.). III. Fine structure and biochemistry of the non-viable embryo. *Planta* (Berl.) 110:279–90.

HALLOIN, J. M. 1975. Solute loss from deteriorated cottonseed: relationship between deterioration, seed moisture, and solute loss. *Crop Sci.* 15:11–15.

HALMER, P. and J. D. BEWLEY. 1984. A physiological perspective on seed vigour testing. *Seed Sci. Technol.* 12:561–75.

HAMMERSCHMIDT, P. A., and D. E. PRATT. 1978. Phenolic antioxidants of dried soybeans. *J. Food Sci.* 43:556–59.

HAMMOND, N. G. L., and H. H. SCULLARD. 1970. *The Oxford Classical Dictionary*. 2d ed. Oxford: Clarendon.

HANELT, P. 1977. Ökologische und systematische Aspekte der Lebensdauer von Samen. *Biol. Rundsch.* 15:81–91.

HARGIN, K. D., and W. R. MORRISON. 1980. The distribution of acyl lipids in the germ, aleurone, starch, and non-starch endosperm of four wheat varieties. *J. Sci. Food Agric.* 31:877–88.

HARMAN, G. E., and A. L. GRANETT. 1972. Deterioration of stored pea seed: changes in germination, membrane permeability, and ultrastructure resulting from infection by *Aspergillus ruber* and from aging. *Physiol. Plant Pathol.* 2:271–78.

HARMAN, G. E., A. A. KHAN, and K. L. TAO. 1976. Physiological changes in the early stages of germination of pea seeds induced by aging and by infection by a storage fungus, *Aspergillus ruber*. *Can. J. Bot.* 54:39–44.

HARMAN, G. E., and L. R. MATTICK. 1976. Association of lipid oxidation with seed ageing and death. *Nature* (Lond.) 260:323–24.

HARMAN, G. E., B. L. NEDROW, B. E. CLARKE, and L. R. MATTICK. 1982. Association

of volatile aldehyde production during germination with poor soybean and pea seed quality. *Crop Sci.* 22:712–16.
HARMAN, G. E., B. NEDROW, and G. NASH. 1978. Stimulation of fungal spore germination by volatiles from aged seeds. *Can J. Bot.* 56:2124–27.
HARPER, J. L. 1977. *Population Biology of Plants.* New York: Academic Press.
HARRINGTON, G. T. 1916. Agricultural value of impermeable seeds. *J. Agric. Res.* (Wash., D.C.) 6:761–96.
HARRINGTON, J. F. 1960. Germination of seeds from carrot, lettuce, and pepper plants grown under severe nutrient deficiencies. *Hilgardia* 30:219–35.
HARRINGTON, J. F. 1970. Seed and pollen storage for conservation of plant gene resources. In O. H. Frankel and E. Bennett, eds., *Genetic Resources in Plants: Their Exploration and Conservation*, pp. 501–21. Oxford: Blackwell.
HARRINGTON, J. F. 1972. Seed storage and longevity. In T. T. Kozlowski, ed., *Seed Biology*, 3:145–245. New York: Academic Press.
HARRINGTON, J. F. 1973. Biochemical basis of seed longevity. *Seed Sci. Technol.* 1:453–61.
HARRIS, R. H., and H. L. WALSTER. 1953. Observations on a sixty-four-year-old wheat sample. *Cereal Chem.* 30:58–62.
HARRISON, B. J. 1966. Seed deterioration in relation to storage conditions and its influence upon germination, chromosomal damage, and plant performance. *J. Natl. Inst. Agric. Bot.* 10:644–63.
HARRISON, B. J., and J. MCLEISH. 1954. Abnormalities of stored seed. *Nature* (Lond.) 173:593–94.
HARRISON, J. G. 1977. The effect of seed deterioration on the growth of barley. *Ann. Appl. Biol.* 87:485–94.
HARRISON, J. G., and D. A. PERRY. 1976. Studies on the mechanism of barley seed deterioration. *Ann. Appl. Biol.* 84:57–62.
HARRISON, J. S., and W. E. TREVELYAN. 1963. Phospholipid breakdown in baker's yeast during drying. *Nature* (Lond.) 200:1189–90.
HARTMANN, W. 1970. Der Einfluss Samenalters auf die Vernaliserbarkeit von *Oenothera biennis* L. *Biochem. Physiol. Pflanz. [BPP]* 161:469–71.
HARTY, R. L., J. M. HOPKINSON, B. H. ENGLISH, and J. ALDER. 1983. Germination, dormancy, and longevity in stored seed of *Panicum maximum*. *Seed Sci. Technol.* 11:341–51.
HAUSER, H. 1975. Lipids. In F. Franks, ed., *Water: A Comprehensive Treatise*, 4:209–303. New York: Plenum.
HAUSSKNECHT, C. 1892. Pflanzengeschichtliche, systematische und floristische Besprechungen und Beitrage. *Mitt. Thuring. Bot. Ver.* 2:45–67.
HAYDAR, M., and D. HADZIYEV. 1974. Pea mitochondrial lipids and their oxidation during mitochondria swelling. *J. Sci. Food Agric.* 25:1285–1305.
HECKER, M. 1974. Untersuchungen über die Protein-, RNA-, und DNA-Synthese in gealterten Samen von *Agrostemma githago* L. unterschiedlicher Keimungsbereitschaft. *Biol. Rundsch.* 12:277–79.
HECKER, M., and D. BERNHARDT. 1976. Proteinbiosynthesen in dormanten und nachgereiften Embryonen und Samen von *Agrostemma githago*. *Phytochemistry* (Oxf.) 15:1105–9.
HEDGES, R. E. M., and J. A. J. GOWLETT. 1984. Accelerating carbon dating. *Nature* (Lond.) 308:403–4.
HEGARTY, T. W. 1978. The physiology of seed hydration and dehydration, and the

relation between water stress and the control of germination: a review. *Plant Cell Environ.* 1:101–19.
HEINRICH, M. 1913. Der Einfluss der Luftfeuchtigkeit, der Wärme und des Sauerstoffs der Luft auf lagerndes Saatgut. *Landwirtsch. Vers. Stn.* 81:289–376.
HEIT, C. E. 1967a. Propagation from seed. Pt. 6: Hardseededness—a critical factor. *Am. Nurseryman* 125 (10):10–12, 88–96.
HEIT, C. E. 1967b. Propagation from seed. Pt. 10. Storage methods for conifer seeds. *Am. Nurseryman* 126 (8):14–15, 38–54.
HELBAEK, H. 1958. Grauballemandens sidste måltid. [Grauballe man's last meal.] *Kuml*, pp. 83–116.
HELBAEK, H. 1963. Palaeo-ethnobotany. In D. Brothwell and E. Higgs, eds., *Science in Archaeology*, pp. 177–85. London: Thames & Hudson.
HELDREICH, T. VON. 1873. *Glaucium serpieri* Heldr. *Gartenflora* 22:323–24.
HELDREICH, T. DE. 1876. Sertulum plantarum novarum vel minus cognitarum Florae Hellenicae. In *Atti del Congresso Internazionale Botanico, Florence, 1874*, pp. 136–37. Florence: Reale Società Toscana di Orticultura.
HENREY, B. 1975. *British Botanical and Horticultural Literature before 1800*. Vol. 2. New York: Oxford University Press.
HENSLOW, [J. S.]. 1860. On the supposed germination of mummy wheat. *Br. Assoc. Adv. Sci. Rep., Trans. Sect.* 30:110–11.
HERNØ, A. 1944. Undersøgelser over spireevne, vandindhold m.m. i frø af en raekke vigtige kulturplanter, opbevaret gennem en aarraekke paa et almindeligt frølager. [Examination of germinating capacity, moisture content, etc. in seeds of a number of important crop plants, stored for several years in an ordinary warehouse.] *Tidsskr. Planteavl.* 48:551–602.
HERRERA, T., W. H. PETERSON, E. J. COOPER, and H. J. PEPPLER. 1956. Loss of cell constituents on reconstitution of active dry yeast. *Arch. Biochem. Biophys.* 63:131–43.
HEYDECKER, W. 1972. Vigour. In E. H. Roberts, ed., *Viability of Seeds*, pp. 209–52. Syracuse, N.Y.: Syracuse University Press.
HEYDECKER, W. 1974. Small-scale seed storage. *J. R. Hortic. Soc.* 99:216–20.
HEYDECKER, W., J. HIGGINS, and Y. J. TURNER. 1975. Invigoration of seeds? *Seed Sci. Technol.* 3:881–88.
HIBBARD, R. P. and E. V. MILLER. 1929. Biochemical studies on seed viability. I. Measurements of conductance and reduction. *Plant Physiol.* 3:335–52.
H[ILL], A. W. 1917. The flora of the Somme battlefield. *Kew Bull.*, pp. 297–300.
HILL, M. O., and P. A. STEVENS. 1981. The density of viable seed in soils of forest plantations in upland Britain. *J. Ecol.* 69:693–709.
HILL, R. A., J. LACEY, and P. J. REYNOLDS. 1983. Storage of barley grain in Iron Age type underground pits. *J. Stored Prod. Res.* 19:163–71.
HILLMAN, G. C., G. V. ROBINS, D. ODUWOLE, K. D. SALES, and D. A. C. MCNEIL. 1983. Determination of thermal histories of archeological cereal grains with electron spin resonance spectroscopy. *Science* (Wash., D.C.) 222:1235–36.
HILLMAN, J. R. 1978. *Isolation of Plant Growth Substances*. New York: Cambridge University Press.
HINTON, H. E. 1968. Reversible suspension of metabolism and the origin of life. *Proc. R. Soc. Lond. B, Biol. Sci.* 171:43–56.
HOEKSTRA, F. A. 1984. Imbibitional chilling injury in pollen. Involvement of the respiratory chain. *Plant Physiol.* 74:815–21.
HOEKSTRA, F. A., and T. VAN ROEKEL. 1983. Isolation-inflicted injury to mitochondria

from fresh pollen gradually overcome by an active strengthening during germination. *Plant Physiol.* 73:995–1001.

HOFFPAUIR, C. L., S. E. POE, L. U. WILES, and M. HICKS. 1950. Germination and free fatty acids in seed stock lots of cottonseed. *J. Am. Oil Chem. Soc.* 27:347–48.

HOLMES, G. D., and G. BUSZEWICZ. 1958. The storage of seed of temperate forest tree species. *For. Abstr.* 19:313–22, 455–76.

HOME, F. 1757. *The Principles of Agriculture and Vegetation.* Edinburgh: Hamilton & Balfour.

HOOKER, J. D. 1855. Longevity of seeds. *Gard. Chron.*, pp. 805.

HOPKINS, E. F., J. F. RAMÍREZ SILVA, V. PAGAN, and A. G. VILLAFAÑE. 1947. Investigations on the storage and preservation of seed in Puerto Rico. *P. R. Agric. Exp. Stn. Bull.* 72:1–47.

HOSHINO, M., F. MATSUMOTO, and J. IKEDA. 1960. Mame-ka-shushi no chozō ni kansuru 2,3 no jikken. [Experiments on the storage of legume seeds.] *Nippon Sakumotsu Gakkai Kiji [Jpn. J. Crop Sci.]* 29:177–78.

HOULBERG U., and E. MADSEN. 1985. Chlorophyll and isocitrate lyase content of germinating seeds as indices of vigour. *Seed Sci. Technol.* 13:143–48.

HOVE, E. L., and P. L. HARRIS. 1951. Note on the linoleic acid-tocopherol relationship in fats and oils. *J. Am. Oil Chem. Soc.* 28:405.

HUANG, A. H. C., and R. A. MOREAU. 1978. Lipases in the storage tissues of peanut and other oil seeds during germination. *Planta* (Berl.) 141:111–16.

HUBER, T. A., and M. B. MCDONALD. 1982. Gibberellic acid influence on aged and unaged barley seed germination and vigor. *Agron. J.* 74:386–89.

HUDSON, D. 1949. *Martin Tupper: His Rise and Fall.* London: Constable.

HUGHES, P. A., and R. F. SANDSTED. 1975. Effect of temperature, relative humidity, and light on the color of "California Light Red Kidney" bean seed during storage. *HortScience* 10:421–23.

HULL, A. C. 1973. Germination of range plant seeds after long periods of uncontrolled storage. *J. Range Manage.* 26:198–200.

HUMMEL, B. C. W., L. S. CUENDET, C. M. CHRISTENSEN, and W. F. GEDDES. 1954. Grain storage studies. XIII. Comparative changes in respiration, viability, and chemical composition of mold-free and mold-contaminated wheat upon storage. *Cereal Chem.* 31:143–50.

HUNTER, I. R., D. F. HOUSTON, and E. B. KESTER. 1951. Development of free fatty acids during storage of brown (husked) rice. *Cereal Chem.* 28:232–39.

HUNTER, R. E., and J. T. PRESLEY. 1963. Morphology and histology of pinched root tips of *Gossypium hirsutum* L. seedlings grown from deteriorated seeds. *Can. J. Plant Sci.* 43:146–50.

HURD, R. G. 1969. Effect of tomato seed age on germination and seedling growth. *Annu. Rep. Glasshouse Crops Res. Inst.* (Littlehampton), pp. 151–54.

HUSS, E. 1956. Om barrskogsfröets kvalitet och andra på såddresultatet inverkande faktorer. [On the quality of forest tree seed and other factors affecting sowing.] *Medd. Statens Skogsforskningsinst.* (Stockholm) 46 (9):1–59.

IBN AL-'AWWĀM. 1866. *Le Livre de l'agriculture d'Ibn-al-Awam (Kitab al-Felahah).* Vol. 2. Trans. J.-J. Clement-Mullet. Paris: Franck.

IBRAHIM, A. E., and E. H. ROBERTS. 1983. Viability of lettuce seeds. I. Survival in hermetic storage. *J. Exp. Bot.* 34:620–30.

IBRAHIM, A. E., E. H. ROBERTS, and A. J. MURDOCH. 1983. Viability of lettuce seeds. II. Survival and oxygen uptake in osmotically controlled storage. *J. Exp. Bot.* 34:631–40.

IGLESIAS, H. A., and J. CHIRIFE. 1982. *Handbook of Food Isotherms: Water Sorption Parameters for Food and Food Components.* New York: Academic Press.

IKEDA, N., S. UDO, I. SAISHO, and S. MINAMIKATA. 1960. Hakka-shushi no chozō ni kansuru kenkyū. [Studies on the storage of mint seed.] *Okayama Daigaku Nogakubu Gakujutsu Hokoku [Sci. Rep. Fac. Agric. Okayama Univ.]* 16:1–5.

INNOCENTI, A. M., and S. AVANZI. 1971. Seed aging and chromosome breakage in *Triticum durum* Desf. *Mutat. Res.* 13:225–31.

INNOCENTI, A. M., and B. BITONTI. 1977. Età del seme e variazioni nel rapporto istoni/DNA in meristemi quiescenti di *Triticum durum* cv. Capelli. *Atti Soc. Toscana Sci. Nat. Pisa Mem.*, ser. B, 84:1–6.

INNOCENTI, A. M., and M. B. BITONTI. 1979. Histones/DNA ratio in young and old root meristems of *Triticum durum* caryopses. *Carylogia* 32:441–48.

INNOCENTI, A. M., and M. B. BITONTI. 1980. Differente invecchiamento nelle cariossidi "chiare" e "scure" di *Haynaldia villosa* Schur: uno studio citofotometrico nei meristemi radicali quiescenti. *G. Bot. Ital.* 114:29–35.

INNOCENTI, A. M., and M. B. BITONTI. 1981. Changes in histones/DNA ratio in scutellum nuclei during ageing of *Triticum durum* caryopses. *Caryologia* 34:179–86.

INNOCENTI, A. M., and M. B. BITONTI. 1984. Variation of histone/DNA ratio in the embryonic areas of *Triticum durum* aged seeds. *Caryologia* 37:387–91.

INNOCENTI, A. M., M. B. BITONTI, and A. BENNICI. 1983. Cytophotometric analyses and *in vitro* culture test in the embryo first node of old *Triticum durum* caryopses. *Caryologia* 36:83–87.

INNOCENTI, A. M., and C. FLORIS. 1979. Metabolic activity distribution in the embryo of naturally aged seeds of durum wheat. *Biochem. Physiol. Pflanz. [BPP]* 174:404–10.

IONESOVA, A. S. 1970. *Fiziologiya semyan dikorastushchikh rastenii pustyni.* Tashkent: FAN.

ISELY, D., and L. N. BASS. 1959. Seeds and packaging materials. *Proc. Hybrid Corn Res. Conf.* 14:101–10.

IVANOV, S. L., and E. I. BERDICHEVSKII. 1933. Vliyanie razlichnoi vlazhnosti na khranenie maslichnykh semyan i sostav masla. [The effect of varying water content on storage of oilseeds and on oil composition.] *Bot. Zh.* 18:321–30.

JACKSON, W. D., and H. N. BARBER. 1958. Patterns of chromosome breakage after irradiation and ageing. *Heredity.* 12:1–25.

JAIN, N. K., and J. R. SAHA. 1971. Effect of storage length on seed germination in jute (*Corchorus* spp.). *Agron J.* 63:636–38.

JAKOVUK, A. S., M. G. ROMASHKIN, and N. I. YATSUN. 1970. Izmenenie biologicheskikh svoistv semyan tabaka pod vozdeistviem ul'trazvuka. [Changes in the biological properties of tobacco seeds following ultrasonic treatment.] *Biol. Zh. Arm.* 23 (1):67–74.

JAMES, E. 1961. An annotated bibliography on seed storage and deterioration: a review of 20th-century literature in the English language. *U.S. Dep. Agric., Agric. Res. Serv.* (ser. ARS-34) 15 (1):1–81.

JAMES, E. 1963. An annotated bibliography on seed storage and deterioration: a review of 20th-century literature reported in foreign languages. *U.S. Dep. Agric., Agric. Res. Serv.* (ser. ARS-34) 15 (2):1–30.

JAMES, E. 1967. Preservation of seed stocks. *Adv. Agron.* 19:87–106.

JAMES, E. 1968. Limitations of glutamic acid decarboxylase activity for estimating viability in beans (*Phaseolus vulgaris* L.). *Crop Sci.* 8:403–4.

JAMES, E., L. N. BASS, and D. C. CLARK. 1964. Longevity of vegetable seed stored 15 to 30 years at Cheyenne, Wyoming. *Proc. Am. Soc. Hortic. Sci.* 84:527–34.

JAMES, E., L. N. BASS, and D. C. CLARK. 1967. Varietal differences in longevity of vegetable seeds and their response to various storage conditions. *Proc. Am. Soc. Hortic. Sci.* 91:521–28.

JAMIESON, G. S. 1943. *Vegetable Fats and Oils.* 2d ed. New York: Reinhold.

JENSEN, C. 1941. *Is It Possible that Seeds through Treatment with Light May Keep Their Germinating Power through a Longer Span of Years than Normal?* Copenhagen: Qvist.

JENSEN, C. 1942. Über die Möglichkeit, mit Hilfe von Lichtbehandlung die Keimfähigkeit von Samen zu verlängern. *Z. Bot.* 37:487–99.

JENSEN, C. O., W. SACKS, and F. A. BALDAUSKI. 1951. The reduction of triphenyltetrazolium chloride by dehydrogenases of corn embryo. *Science* (Wash., D.C.) 113:65–66.

JENSEN, H. A. 1971. Undersøgelser of spireevnen hos frø fra en ca. 50 år gammel frøsamling. [Investigations of the germinability of seeds from an approximately 50-year-old seed collection.] *Statsfrokontrollens Beret.* (Den.) 100:71–76.

JENSSEN, C. 1879. Untersuchungen über den Kulturwerth der Handels-Saaten unserer gewöhnlichsten Klee- und Grasarten. *Landwirtsch. Jahrb.* 8:133–331.

JOHNSON, T. 1907. Elektrische Samenprüfung. *Angew. Bot.* 6:102–12.

JONES, D. B., J. P. DIVINE, and C. E. F. GERSDORFF. 1942. The effect of storage of corn on the chemical properties of its proteins and on its growth-promoting value. *Cereal Chem.* 19:819–30.

JONES, D. B., and C. E. F. GERSDORFF. 1941. The effect of storage on the protein of wheat, white flour, and whole wheat flour. *Cereal Chem.* 18:417–34.

JONES, J. A. 1928. Overcoming delayed germination of *Nelumbo lutea. Bot. Gaz.* 85:341–43.

JONES, J. W. 1926. Germination of rice seed as affected by temperature, fungicides, and age. *Agron. J.* 18:576–92.

JONES, Q., and F. R. EARLE. 1966. Chemical analyses of seeds. II. Oil and protein content of 759 species. *Econ. Bot.* 20:127–55.

JOSEPH, H. C. 1929. Germination and keeping quality of parsnip seeds under various conditions. *Bot. Gaz.* 87:195–210.

JOUANNET, F. 1835. Notice statistique sur la commune de La Monzie–St. Martin: histoire, antiquités, moeurs. *Calend. Admin. Dordogne*, pp. 183–93.

JUEL, I. 1941/42. Der Auxingehalt in Samen verschiedenen Alters, sowie einige Untersuchungen betreffend die Haltbarkeit der Auxine. *Planta* (Berl.) 32:227–33.

JUST, L. 1877. Ueber die Einwirkung höherer Temperaturen auf die Erhaltung der Keimfähigkeit der Samen. *Beitr. Biol. Pflanz.* 2:311–48.

JUSTICE, O. L., and L. N. BASS. 1978. Principles and practices of seed storage. *U.S. Dep. Agric. Handb.* 506:1–289.

KADOTA, T. 1937. Nasu F1 shushi no nenrei to zasshu-kyōsei. [Vigor of F1 plants as related to age of seeds in eggplants.] *Engei Gakkai Zasshi [J. Jpn. Soc. Hortic. Sci.]* 8:351–55.

KAGRAMANYAN, R. S. 1971. Izuchenie estestvennogo mutatsionnogo protsessa v trekh posledovatel'nykh mitoticheskikh tsiklakh v starykh semenakh *Crepis capillaris. Genetika* 7 (9):51–59. [Study of the natural mutation process in three successive mitotic cycles in old *Crepis capillaris* seeds. *Sov. Genet.* 7:1141–47.]

KALASHNIK, M. F., and A. I. NAUMENKO. 1979. Posevnye i urozhainye kachestva semyan sorgo v zavisimosti ot dlitel'nosti khraneniya. [Sowing and yield characteristics of sorghum seed as a function of length of storage.] *Kukuruza*, no. 3, p. 27.

KALOYEREAS, S. A. 1958. Rancidity as a factor in the loss of viability of pine and other seeds. *J. Am. Oil Chem. Soc.* 35:176–79.

KALOYEREAS, S. A., W. MANN, and J. C. MILLER. 1961. Experiments in preserving and revitalizing pine, onion, and okra seeds. *Econ. Bot.* 15:213–17.

KAMRA, S. K. 1967. Studies on storage of mechanically damaged seed of Scots pine (*Pinus silvestris* L.). *Stud. For. Suec.* 42:1–19.

KAPLAN, L., and S. L. MAINA. 1977. Archeological botany of the Apple Creek site, Illinois. *J. Seed Technol.* 2 (2):40–53.

KAREL, M. 1975. Free radicals in low moisture systems. In R. B. Duckworth, ed., *Water Relations of Foods*, pp. 435–53. New York: Academic Press.

KAREL, M., and S. YONG. 1981. Autoxidation-initiated reactions in foods. In L. B. Rockland and G. F. Stewart, eds., *Water Activity: Influences on Food Quality*, pp. 511–29. New York: Academic Press.

KARPOV, B. A. 1980. Iskhodnye faktory dolgovechnosti semyan zernovykh kultur. [Initial influences on the longevity of seeds of grain crops.] *S'kh. Biol.* 15:42–44.

KARSSEN, C. M. 1980/81. Environmental conditions and endogenous mechanisms involved in secondary dormancy of seeds. *Isr. J. Bot.* 29:45–64.

KARSSEN, C. M. 1982. Seasonal patterns of dormancy in weed seeds. In A. A. Khan, ed., *The Physiology and Biochemistry of Seed Development, Dormancy, and Germination*, pp. 243–70. New York: Elsevier.

KARYAKIN, A. V., D. N. LAZAREV, and G. A. BARINOVA. 1956. Lyuminestsentnyi metod opredeleniya zhiznesposobnosti semyan sel'skokhozyaistvennykh kultur. [A luminescence method for determining viability of crop seeds.] *Dokl. Akad. Nauk SSSR* 106:739–42.

KASAHARA, Y., K. NISHI, and Y. UEYAMA. 1965. 50 yonenkan maido shita igusa-shushi oyobi suiden zassō-shushi no hatsuga to seiiku. [Studies on the germination and growth of seeds of rush and other weeds buried for more than fifty years.] *Nogaku Kenkyu [Agric. Res.]* 51:75–101.

KATO, A., K. FUKUEI, and S. TANIFUJI. 1982. Histone synthesis during the early stages of germination in *Vicia faba* embryonic axes. *Plant Cell Physiol.* 23:967–76.

KATO, Y. 1954. Descriptive and experimental cytology in *Allium*. II. Chromosome breakage in the seedling of *Allium*. *Bot. Mag.* (Tokyo) 67:122–28.

KAUL, B. L. 1969. Aging in relation to seed viability, nuclear damage, and sensitivity to mutagens. *Caryologia* 22:25–33.

KAUR, J., and A. K. SRIVASTAVA. 1982. Effect of different storage conditions on some biochemical parameters associated with the loss of seed viability in soybean. *J. Res. Punjab Agric. Univ.* 19:368–73.

KAZAKOV, E. D., and A. N. BOLKOVA. 1950. Izmenenie zhira pri mnogoletnem khranenii zerna. [Change in lipid during several years of grain storage.] *Dokl. Akad. Nauk. SSSR* 72:559–60.

KECK, K., and O. HOFFMANN-OSTENHOF. 1952. Chromosome fragmentation in *Allium cepa* induced by seed extracts of *Phaseolus vulgaris*. *Caryologia* 4:289–94.

KEEPAX, C. 1977. Contamination of archaeological deposits by seeds of modern origin with particular reference to the use of flotation machines. *J. Archaeol. Sci.* 4:221–29.

KEMP, W. 1844. An account of some seeds buried in a sand-pit which germinated. *Ann. Mag. Nat. Hist.* 13:89–91.

KERNER, A. 1890. *Pflanzenleben*. Vol. 1. Leipzig: Bibliographisches Institut.

KHAN, A. A., J. W. BRAUN, K.-L. TAO, W. F. MILLIER, and R. F. BENSIN. 1976. New methods for maintaining seed vigor and improving performance. *J. Seed Technol.* 1 (2):33–57.

KHARLUKHI, L., and P. K. AGRAWAL. 1984. Evidence for participation of pentose phosphate pathway during seed deterioration on storage. *Indian J. Exp. Biol.* 22:612–14.

KHEKKER, M., D. BERNGARDT, and K.-KH. KELER. [HECKER, M., D. BERNHARDT, and K.-H. KÖHLER.] 1977. Vliyanie temperatury khraneniya semyan na ikh prorastanie i biosintez belkov. [Influence of storage temperature of seeds on their germination and protein biosynthesis.] *Fiziol. Biokhim. Kul't. Rast.* 9:457–60.

KHOROSHAILOV, N. G. 1973. Otvetnye reaktsii raznokachestvennykh semyan razlichnykh sel'skokhozyaistvennykh kultur na usloviya khraneniya. [The response of different crop seeds of varying quality to the conditions of storage.] In F. E. Reimers, ed., *Fiziologobiokhimicheskie problemy semenovedeniya i semenovodstva*, 1:93–99. Irkutsk: Akademiya Nauk SSSR Sibirskoe Otdelenie.

KHOROSHAILOV, N. G., and N. V. ZHUKOVA. 1971. Opyt dlitel'nogo khraneniya semyan. [Experience of long-term storage of seeds.] *Tr. Prikl. Bot. Genet. Sel.* 44 (3):175–86.

KHOROSHAILOV, N. G., and N. V. ZHUKOVA. 1973. Dlitel'noe khranenie kollektsionnykh obraztsov semyan. [Prolonged storage of collected seed samples.] *Tr. Prikl. Bot. Genet. Sel.* 49 (3):269–79.

KHOROSHAILOV, N. G., and N. V. ZHUKOVA. 1978. Dlitel'noe khranenie semyan mirovoi kollektsii VIR. [Prolonged storage of seeds in the world collection of the VIR.] *Byull. Vses. Nauchno-Issled. Inst. Rastenievod.* 77:9–19.

KHRISTOV, K. 1979. Uskorenoto stareene na semenata: metod za poluchavane na mutatsii pri tsarevitsata. [Accelerated aging of seeds: a method for producing mutations in maize.] *Genet. Sel.* 12:26–35.

KHRISTOV, K., and P. KHRISTOVA. 1978. V"rkhy nyakoi biologichni i mutatsionni izmeneniya pri stareene na semena ot tsarevitsa. [On some biological and mutagenic changes during aging of maize seeds.] *Rasteniev"d. Nauki* 15 (8):13–23.

KHRISTOV, K., and P. KHRISTOVA. 1980. Genetichen analiz na mutatantni linii tsarevitsa, indutsirani s uskoreno stareene na semena. [Genetic analysis of mutated maize lines induced by accelerated aging of seeds.] *Genet. Sel.* 13:420–28.

KIERMEIER, F., and E. CODURO. 1955. Der Einfluss des Wassergehaltes auf Enzymreaktionen in wasserarmen Lebensmitteln. III. Miteilung. Uber Enzymreaktionen in Mehl, Teig und Brot. *Z. Lebensm. Unters. Forsch.* 102:7–12.

KIETREIBER, M. 1975. Vitalitätsprüfung der Aleuronzellen des Endosperms von *Zea mays* (Aleuron-Tetrazoliumtest). *Seed Sci. Technol.* 3:803–9.

KIEWNICK, L. 1964. Untersuchungen über den Einfluss der Samen- und Bodenmikroflora auf die Lebensdauer der Spelzfrüchte des Flughafers (*Avena fatua* L.). II. Zum Einfluss der Mikroflora auf die Lebensdauer der Samen im Boden. *Weed Res.* 4:31–43.

KIKUCHI, T. 1954. Hai-ishoku-hō ni yoru shushi no rōka ni kansuru kenkyū. Dai 1 pō. Komugi no ko-shushi ni tsuite. [Studies on the deterioration of seeds by the embryo transplantation method. I. Old wheat seeds.] *Utsunimaya Daigaku Nogakubu Gakujutsu Hokoku [Bull. Coll. Agric. Utsunimaya Univ.]* 2:277–92.

KING, M. W., and E. H. ROBERTS. 1979. *The Storage of Recalcitrant Seeds: Achievements and Possible Approaches*. Rome: International Board for Plant Genetic Resources.

KING-PARKS, H. 1885. On the supposed germinating powers of mummy wheat. *J. Sci.* (Lond.) 22:604–10.

KIVILAAN, A., and R. S. BANDURSKI. 1973. The ninety-year period for Dr. Beal's seed viability experiment. *Am. J. Bot.* 60:140–45.

KIVILAAN, A., and R. S. BANDURSKI. 1981. The one-hundred-year period for Dr. Beal's seed viability experiment. *Am. J. Bot.* 68:1290–92.

KIYASHKO, YU. G. 1981. Fiziologo-biokhimicheskie izmeneniya semyan soi protsesse stareniya. [Physiological-biochemical changes in soybean seeds during the aging process.] *Byull. Vses. Nauchno-Issled. Inst. Rastenievod. im. N. I. Vavilova* (Leningr.) 114:38–40.

KJAER, A. 1940. Germination of buried and dry stored seeds. I. 1934–1939. *Proc. Int. Seed Test. Assoc.* 12:167–90.
KJAER, A. 1948. Germination of buried and dry stored seeds. II. 1934–1944. *Proc. Int. Seed Test. Assoc.* 14:19–26.
KJØLLER, A., and S. ØDUM. 1971. Evidence for longevity of seeds and microorganisms in permafrost. *Arctic* 24:230–33.
KLING, M. 1931. Ein Beitrag zur Keimfähigkeit der Unkrautsamen. Ein 21jähriger Versuch. *Fortschr. Landwirtsch.* (Vienna) 6:577–78.
KLING, M. 1942/43. Ein Beitrag zur Keimfähigkeit der Unkrautsamen. Ein 32jähriger Versuch. *Prakt. Bl. Pflanzenbau Pflanzenschutz* 42:43–47.
KLISURSKA, D., and A. DENCHEVA. 1984. Fiziologo-biokhimichni osnovi na zhiznenostta na semenata. III. Vliyanie na usloviyata na s''khranenie na semenata v''rkhy aktivnostta i izoenzimniya spekt''r na peroksidazata, izolirana ot endospermata na semena ot tsarevitsa. [Physiological and biochemical basis of seed viability. III. Effect of seed storage conditions on the activity and isozyme spectrum of peroxidase isolated from the endosperm of maize seed.] *Fiziol. Rast.* (Sofia) 10 (3):19–27.
KNECHT, H. 1931. Über die Beziehungen zwischen Katalaseaktivität und Vitalität im ruhenden Samen. *Bot. Zbl., Beih.* 48 (1):229–313.
KOCH, W. 1968. Zur Lebensdauer von Unkrautsamen. *Saatgutwirtschaft* 20:251–53.
KOCH, W. 1969. Einfluss von Umweltfaktoren auf die Samenphase annueller Unkräuter insbesondere unter dem Gesichtspunkt der Unkrautbekämpfung. *Arb. Univ. Hohenheim (Landwirtsch. Hochsch.)* 50:1–204.
KOLE, S. N., and K. GUPTA. 1982. Biochemical changes in safflower (*Carthamus tinctorius*) seeds under accelerated ageing. *Seed Sci. Technol.* 10:47–54.
KOLESNIKOV, G. P., T. P. ZHURBA, and A. I. APROD. 1980. Fiziologo-biofizicheskie pokazateli snizheniya vskhozhesti semyan v protsesse ikh stareniya. [Physiological and biophysical indices of decreased seed germinability as a consequence of aging.] *Byull. Nauchno-Tekh. Inf. Vses. Nauchno-Issled. Inst. Risa* (Krasnodar) 27:26–29.
KOLK, H. 1962. Viability and dormancy of dry stored weed species. *Vaxtodling* 18:1–192.
KONDŌ, M. 1926. Über die Dauer der Erhaltung der Keimkraft bei verschiedenen Samenarten in Japan. *Ber. Ohara Inst. Landwirtsch. Biol. Okayama Univ.* 3:127–33.
KONDŌ, M. 1936. Untersuchungen über harte Samen von *Astragalus sinicus*, die in 17 bzw. 18 Jahre lang unter Wasser nicht gequollen hatten, besonders über die Vererbungverhältnisse dieser Hartschaligkeit. *Ber. Ohara Inst. Landwirtsch. Biol. Okayama Univ.* 7:321–27.
KONDŌ, M., and T. OKAMURA. 1932/33. Storage of rice. VII. On the influence of varying moisture content and germinating power upon the preservation of vitamin-B in hulled rice. *Ber. Ohara Inst. Landwirtsch. Biol. Okayama Univ.* 5:407–12.
KONDŌ, M., and T. OKAMURA. 1934a. Storage of rice. IX. Relation between varying moisture content and change in quality of hulled rice stored in containers air-tight as well as with carbon-dioxide. *Ber. Ohara Inst. Landwirtsch. Biol. Okayama Univ.* 6:149–74.
KONDŌ, M., and T. OKAMURA. 1934b. Storage of rice. X. Studies on four lots of unhulled rice stored forty-six to eighty-four years in granaries. *Ber. Ohara Inst. Landwirtsch. Biol. Okayama Univ.* 6:175–85.
KONDŌ, M., and T. OKAMURA. 1938. Storage of rice. XX. Studies on unhulled rice stored about one hundred years in a granary. *Ber. Ohara. Inst. Landwirtsch. Biol. Okayama Univ.* 8:47–52.

KOOSTRA, P. [T.] 1973. Changes in seed ultrastructure during senescence. *Seed Sci. Technol.* 1:417–25.
KOOSTRA, P. T., and J. F. HARRINGTON. 1969. Biochemical effects of age on membranal lipids of *Cucumis sativus* L. seed. *Proc. Int. Seed Test. Assoc.* 34:329–40.
KORSMO, E. 1930. *Unkräuter im Ackerbau der Neuzeit*. Berlin: Springer.
KOSAR, W. F., and R. C. THOMPSON. 1957. Influence of storage humidity on dormancy and longevity of lettuce seed. *Proc. Am. Soc. Hortic. Sci.* 70:273–76.
KOVAL'CHUK, P. P. 1973a. Izuchenie vliyaniya razlichnykh vozdeistvii i uslovii vyrashchivaniya na dlitel'nost' sokhraneniya posevnykh kachestv i zhiznesposobnost' semyan razlichnykh kul'tur i sortov. [An investigation of the effect of different influences and cultivation conditions on the length of preservation of planting quality and viability of seeds of different crops and cultivars.] In F. E. Reimers, ed., *Fiziologo-biokhimicheskie problemy semenovedeniya i semenovodstva*, 1:111–20. Irkutsk: Akademiya Nauk SSSR Sibirskoe Otdelenie.
KOVAL'CHUK, P. P. 1973b. Zavisimost' posevnykh i urozhainykh kachestv semyan pshenitsy i yachmenya ot dlitel'nosti ikh khraneniya. [Dependence of sowing qualities and yielding ability of wheat and barley seeds upon length of storage.] *Sel. Semenovod. Resp. Mezhved. Temat. Nauch. Sb.* 24:93–99.
KÖVES, E., M. VARGA, and H. ÁCS. 1965. Investigation of the content of the growth-regulating substances of rice grains with reduced viability. *Bot. Kozl.* 52:185–92.
KOZ'MINA, N. P., A. T. NAUMOVA, and N. V. SEREBRYAKOVA. 1964. Materialy k obosnovaniyu rezhimov khraneniya semyan bobovykh kul'tur. [Concerning the basis of storage regimes for legume seeds.] In A. L. Kursanov, ed., *Biologicheskie osnovy povysheniya kachestva semyan sel'skokhozyaistvennykh rastenii*, pp. 245–53. Moscow: Nauka.
KRASNOOK, N. P., E. A. MORGUNOVA, E. T. BUKHTOYAROVA, and I. A. VISHNYAKOVA. 1979. Izmenenie aktivnosti i izozimnogo sostava tsitoplazmaticheskikh i mitokhondrial'nykh degidrogenaz i tsitokhromoksidazy semyan risa s razlichnoi vskhozhest'yu. *Fiziol. Rast.* (Mosc.) 26:64–70. [Changes in activity and isoenzyme composition of cytoplasmic and mitochondrial dehydrogenases and cytochrome oxidase in rice seeds with different germinating capacities. *Sov. Plant Physiol.* 26:45–50.]
KRASNOOK, N. P., E. A. MORGUNOVA, I. A. VISHNYAKOVA, and R. I. POVAROVA. 1976. Aktivnost' degidrogenaz i oksidaz semyan risa razlichnoi zhiznesposobnosti. *Fiziol. Rast.* (Mosc.) 23:156–62. [Activities of dehydrogenases and oxidases in rice seeds of different viabilities. *Sov. Plant Physiol.* 23:130–34.]
KREYGER, J. 1963. General considerations concerning the storage of seeds. *Proc. Int. Seed Test. Assoc.* 28:827–36.
KRISHTOFOVICH, E. N. 1974. Issledovanie elektroforeticheskikh spektrov al'buminov zarodysha i endosperma semyan pshenitsy s razlichnoi vskhozhest'yu. [An investigation of the electrophoretic profiles of embryonic and endosperm albumins from wheat seeds of differing germinability.] *Byull. Vses. Nauchno-Issled. Inst. Rastenievod. im. N. I. Vavilova* (Leningr.) 38:31–33.
KRISHTOFOVICH, E. N. 1975. Dykhanie i fermentativnaya aktivnost' semyan pshenitsy s raznoi vskhozhest'yu. [Respiration and enzymic activity of wheat seeds with different germinability.] *Byull. Vses. Nauchno-Issled. Inst. Rastenievod. im. N. I. Vavilova* (Leningr.) 47:18–19.
KRISHTOFOVICH, E. N. 1980. Fiziologo-biokhimicheskie pokazateli semyan pshenitsy pri dlitel'nom khranenii. [Physiological and biochemical characteristics of wheat seeds in long-term storage.] *Tr. Prikl. Bot. Genet. Sel.* 66 (3):83–89.
KRISHTOFOVICH, E. N., and N. F. POKROVSKAYA. 1972. Vliyanie uslovii i prodolzhi-

tel'nosti khraneniya semyan zernovykh kultur na ikh khimicheskii sostav i posevnye kachestva. [The effect of conditions and duration of storage on the chemical composition and sowing qualities of seeds of cereal crops.] *Byull. Vses. Nauchno-Issled. Inst. Rastenievod. im. N. I. Vavilova* (Leningr.) 27:25–29.

KROGMAN, D. W., and A. T. JAGENDORF. 1959. Inhibition of the Hill reaction by fatty acids and metal chelating agents. *Arch. Biochem. Biophys.* 80:421–30.

KRONSTAD, W. E., R. A. NILAN, and C. F. KONZAK. 1959. Mutagenic effects of oxygen on barley seeds. *Science* (Wash., D.C.) 129:1618.

KUENEMAN, E. A. 1983. Genetic control of seed longevity in soybeans. *Crop Sci.* 23:5–8.

KUGLER, I. 1952. Keimfähigkeit und Abgabe fluoresszierender Stoffe bei Samen. *Naturwissenschaften* 39:213.

KULESHOV, N. N. 1963. *Agronomicheskoe semenovedenie.* Moscow: Izdatelstvo Sel'skokhozyaistvennoi Lituraturi, Zhurnalov i Plakatov.

KULKA, K. 1971a. Biochemiczne aspekty starzenia się ziarna owsa i jęczmienia. [Biochemical aspects of aging in oat and barley seeds.] *Zesz. Nauk. Wyzsz. Szk. Roln. Olsztynie,* ser. A., suppl. 6, pp. 3–90.

KULKA, K. 1971b. Oddychanie i sucha masa kiełkującego ziarna owsa i jeczmienia o różnym wieku. [The respiration and dry weight of germinating oat and barley grains of different ages.] *Zesz. Probl. Postepow Nauk. Roln.* 113:133–40.

KULKA, K. 1973. Fizjologiczne i biochemiczne mechanizmy starzenia się nasion. [Physiological and biochemical mechanisms of seed aging.] *Biul. Inst. Hodowli Aklim. Rosl.,* no. 5/6, pp. 37–44.

KULKA, K., and T. KONOPKA. 1978. Biosynteza i zmiany zawartości kwasów nukleinowych oraz aktywność enzymów w osiach zarodkowych starzejących się nasion grochu. [Biosynthesis and changes in nucleic acids and activity of enzymes in embryonic axes of aging pea seeds.] *Zesz. Probl. Postepow Nauk Roln.* 202:67–82.

KUN, E., and L. G. ABOOD. 1949. Colorimetric estimation of succinic dehydrogenase by triphenyltetrazolium chloride. *Science* (Wash., D.C.) 109:144–46.

KUNDU, C., and R. N. BASU. 1981. Hydration-dehydration treatment of stored carrot seeds for the maintenance of vigour, viability, and productivity. *Sci. Hortic.* (Amsterdam) 15:117–25.

KUNERT, G. 1966. Lagerung und Lipolyse Untersuchungen am Endosperm gelagerter *Ricinus*-Samen. *Flora,* ser. A, 157:299–304.

KUNTH. 1826. Recherches sur les plantes trouvées dans les tombeaux égyptiens par M. Passalacqua. *Ann. Sci. Nat.* (Paris) 8:418–23.

KUNTZ, I. D., and W. KAUZMANN. 1974. Hydration of proteins and polypeptides. *Adv. Protein Chem.* 28:239–345.

KUNZÉ, R. E. 1881. *The Germination and Vitality of Seeds.* New York: Torrey Botanical Club.

KUPERMAN, F. M. 1950. Eshche raz o mekhanicheskikh povrezhdeniyakh semyan. [More on the mechanical injury of seeds.] *Sel. Semenovod.* (Mosc.), no. 3, pp. 45–48.

KURDINA, V. N. 1963. Sorbtsionnye svoistva i intensivnost' dykhaniya semyan ovoshchnykh kul'tur pri khranenii. [Sorption affinities and intensity of respiration of vegetable seeds during storage.] *Izv. Timiryazevsk. S'kh. Akad.* (Mosc.), no. 4, pp. 65–75.

KUWADA, H. 1980. Ikusei-fukunibaitai-sakumotsu to sono ryōshin-sakumotsu tono seiriseitai-gakuteki seishitsu no sai ni tsuite. Dai XXV hō. Shushi no hatsugaryoku— ijikikan. [Studies on the differences in physiological and ecological characteristics of an artifically produced amphidiploid compared to those of its parents. XXV. Seed longevity.] *Kagawa Daigaku Nogakubu Gakujutsu Hokoku [Tech. Bull. Fac. Agric. Kagawa Univ.]* 32:9–12.

KUZ'MINA, R. I., L. V. KASHKINA, and V. L. ABRAMOV. 1980. O fraktsiyakh vody v zerne kukuruzy. *Fiziol. Rast.* (Mosc.) 27:266–71. [Water fractions in corn kernels. *Sov. Pl. Physiol.* 27:200–205.]

KWUN, K. C. 1980. Byu no hwa chong ja eu bala e mi chi neun ethylene mit gibberellin eu young hyang. [Effect of ethylene and gibberellin on the germination of aged rice seeds.] *Nongsa Shihon Yon'gu Pogo (Changmul P'yon), Crop [Res. Rep. Off. Rural Dev., Crop]* (Suwon) 22:82–89.

LAFFERTY, H. A. 1931. The loss of vitality in stored farm seeds. *Irel. Dep. Agric. J.* 30:237–45.

LAKON, G. 1942. Topographischer Nachweis der Keimfähigkeit der Getreidefrüchte durch Tetrazoliumsalze. *Ber. Dtsch Bot. Ges.* 60:299–305.

LAKON, G. 1954. Der Keimwert der nackten Karyopsen im Saatgut von Hafer und Timothee. *Saatgutwirtschaft* 6:233–35, 259–62.

LA QUINTINE, J. DE. [LA QUINTINYE, DE.] 1693. *The Compleat Gard'ner.* Vol. 2. Trans. J. Evelyn. London: Gillyflower & Partridge.

LAUGHLAND, J., and D. H. LAUGHLAND. 1939. The effect of age on the vitality of soybean seeds. *Sci. Agric.* (Ottawa) 20:236–37.

LAUREMBERG, P. [LAUREMBERGUS, P.]. 1654. *Horticultura.* Vol. 1. Frankfurt am Main: Meriani.

LAWRENCE, T. 1967. The effect of age on the germination and emergence of Russian wild ryegrass seed harvested by two methods at progressive stages of maturity. *Can. J. Plant Sci.* 47:181–85.

LAZUKOV, M. I. 1969a. Dolgovechnost' semyan ovoshchnykh krestotsvetnykh kul'tur i ikh gistokhimicheskaya kharakteristika. [Longevity of seeds of cultivated cruciferous species and their histochemical characteristics.] *Dokl. Timiryazevsk. S'kh. Akad.* (Mosc.) 148:71–76.

LAZUKOV, M. I. 1969b. Izmenchivost' nekotorykh priznakov u ovoshchykh korneplodnykh rastenii semeistva krestotsvetnykh pri poseve starymi semenami. [Variability of some parameters of cruciferous root crop plants as a consequence of sowing old seeds.] *Dokl. Timiryazevsk. S'kh. Akad.* (Mosc.) 153:119–24.

LAZUKOV, M. I. 1970. Chastota khromosomnykh abberatsii pri starenii semyan ovoshchykh rastenii semeistva Cruciferae. [Frequency of chromosomal abberations during aging of vegetable seeds of the family Cruciferae.] *Dokl. Timiryazevsk. S'kh. Akad.* (Mosc.) 165:109–12.

LEBEDEVA, L. I., YU. YA. KERKIS, and T. T. MYAGKAYA. 1970. Vliyanie stareniya semyan na spontannye i radiatsionnye perestroiki khromosom. *Genetika* 6 (11):145–48. [The influence of seed aging on spontaneous and radiation-induced chromosome rearrangements. *Sov. Genet.* 6:1529–32.]

LECHERT, H. T. 1981. Water binding on starch: NMR studies on native and gelatinized starch. In L. B. Rockland and G. F. Steward, eds., *Water Activity Influences on Food Quality*, pp. 223–45. New York: Academic Press.

LEE, T.-C., W. T. WU, and V. R. WILLIAMS. 1965. The effect of storage time on the compositional patterns of rice fatty acids. *Cereal Chem.* 42:498–505.

LEES, E. 1851. Records of observations on plants appearing upon newly broken ground, raised embankments, deposits of soil, etc. *Phytologist* 4:131–37.

LEIBOVITZ, B. E., and B. V. SIEGEL. 1980. Aspects of free radical reactions in biological systems: aging. *J. Gerontol.* 35:45–56.

LENGLEN. 1949. Le blé d'Égypte. *C. R. Hebd. Seances Acad. Agric. Fr.* 35:335–42.

LEOPOLD, A. C. 1961. Senescence in plant development. *Science* (Wash., D.C.) 134:1727–32.
LEOPOLD, A. C. 1983. Volumetric components of seed imbibition. *Plant Physiol.* 73:677–80.
LEOPOLD, A. C., and M. E. MUSGRAVE. 1980. Respiratory pathways in aged soybean seeds. *Plant Physiol.* 49:49–54.
LERMAN, J. C., and E. M. CIGLIANO. 1971. New carbon-14 evidence for six-hundred-year-old *Canna compacta* seed. *Nature* (Lond.) 232:568–70.
LESAGE, P. 1911. Sur l'emploi des solutions de potasse à la reconnaisance de la faculté germinative de certaines graines. *C. R. Hebd. Seances Acad. Sci.* (Paris) 152:615–17.
LESAGE, P. 1922. Sur la détermination de la faculté germinative autrement que par la germination des graines. *C. R. Hebd. Seances Acad. Sci.* (Paris) 174:766–67.
LEVENGOOD, W. C., J. BONDIE, and C.-L. CHEN. 1975. Seed selection for potential viability. *J. Exp. Bot.* 26:911–19.
LEVINE, Y. K., and M. H. F. WILKINS. 1971. Structure of oriented lipid bilayers. *Nat. New Biol.* 230:69–72.
LEWIS, D. 1955. Seed storage. *John Innes Hortic. Inst. Annu. Rep.* 46:15–16.
LEWIS, J. 1958. Longevity of crop and weed seeds. 1. First interim report. *Proc. Int. Seed Test. Assoc.* 23:340–54.
LEWIS, J. 1973. Longevity of crop and weed seeds: survival after 20 years in soil. *Weed Res.* 13:179–91.
LEWIS, J. In press. Longevity of crop and weed seeds: survival after 30 years in soil. *Seed Sci. Technol.*
LIBBY, W. F. 1951. Radiocarbon dates. II. *Science* (Wash., D.C.) 114:291–96.
LIBBY, W. F. 1955. *Radiocarbon Dating.* 2d ed. Chicago: University of Chicago Press.
LIBERTY HYDE BAILEY HORTORIUM, CORNELL UNIVERSITY. 1976. *Hortus Third.* New York: Macmillan.
LIKHACHEV, B. S. 1977. Izmenenie zhiznesposobnosti semyan kukuruzy pri ekstremal'nykh usloviyakh khraneniya. *Dokl. Vses. Akad. S'kh. Nauk im. V. I. Lenina* (Mosc.), no. 4, pp. 38–40. [Changes in the viability of corn seeds during storage under extreme conditions. *Sov. Agric. Sci.*, no. 4, pp. 44–46.]
LIKHACHEV, B. S. 1980. Nekotorye metodicheskie voprosy izucheniya biologii stareniya semyan. [Some methodological questions concerning the study of the biology of seed aging.] *S'kh. Biol.* 15:842–44.
LIKHACHEV, B. S., L. I. MUSORINA, Z. N. SHEVCHENKO, L. G. VITMER, and G. V. ZELENSKII. 1978. Ispol'zovanie ekstremal'nykh uslovii khraneniya semyan v modelirovanii protsessov ikh stareniya. [Use of extreme seed storage conditions for modeling the aging process.] *Byull. Vses. Nauchno-Issled. Inst. Rastenievod. im. N. I. Vavilova* (Leningr.) 77:57–62.
LIKHATCHEV, B. S., G. V. ZELENSKY, YU. G. KIASHKO, and Z. N. SHEVCHENKO. 1984. [LIKHACHEV, B. S., G. V. ZELENSKII, YU. G. KIYASHKO, and Z. N. SHEVCHENKO.] Modelling of seed ageing. *Seed Sci. Technol.* 12:385–93.
LIKHOLAT, T. V., and N. G. LYUBARSKAYA. 1983. Vliyanie uskorennogo stareniya na prorastanie semyan pshenitsy s razlichnym soderzhaniem belka. *Fiziol. Rast.* (Mosc.) 30:88–94. [Effect of accelerated aging on germination of wheat seed with different protein content. *Sov. Plant Physiol.* 30:64–69.]
LIN, Y.-H., R. A. MOREAU, and A. H. C. HUANG. 1982. Involvement of glyoxysomal lipase in the hydrolyss of storage triacylglycerols in the cotyledons of soybean seedlings. *Plant Physiol.* 70:108–12.

LINDSTROM, E. W. 1942. Inheritance of seed longevity in maize inbreds and hybrids. *Genetics* 27:154.

LINDSTROM, E. W. 1943. Genetic relations of inbred lines of corn. *Iowa Agric. Exp. Stn. Corn Res. Inst. Annu. Rep.* 8 (2):41–44.

LINKO, P., Y.-Y. CHENG, and M. MILNER. 1960. Changes in the soluble carbohydrates during browning of wheat embryos. *Cereal Chem.* 37:548–56.

LINKO, P., and M. MILNER. 1959a. Free amino acid and keto acids of wheat grains and embryos in relation to water content and germination. *Cereal Chem.* 36:280–94.

LINKO, P., and M. MILNER. 1959b. Enzyme activation in wheat grains in relation to water content: Glutamic acid-alanine transaminase and glutamic acid decarboxylase. *Plant Physiol.* 34:392–96.

LIPKIN, B. YA. 1927. K voprosu o prodolzhitel'nost vremeni sokhraneniya semenami vskhozhesti u razlichnykh khvoinykh drevesnykh porod: Resul'taty pyatiletnikh issledovanii. [On the length of time during which seed germinability is preserved in various coniferous species: results of studies over five years.] *Zap. Beloruss. Gos. Akad. S'kh. im. Okt. Rev.* (Gorki) 5:25–33.

LIPMAN, C. B. 1936. Normal viability of seeds and bacterial spores after exposure to temperatures near the absolute zero. *Plant Physiol.* 11:201–5.

LITYŃSKI, M. 1957. Wpływ wilgotności środowiska na żywotność nasion niektórych gatunków roślin warzywnych. *Rocz. Nauk Roln.*, Ser. A. *Prod. Rosl.* 76:217–94. [Effect of environmental moisture on the vitality of seeds of certain species of vegetables. Warsaw: Centralny Instytut Informacji Naukowo-Technicznej i Ekonomicznej, 1964.]

LITYŃSKI, M., and Z. CHUDOBA. 1964. Obserwacje nad długowiecznościa 30 gatunków roślin uprownych. [Observations on the longevity of 30 species of cultivated plants.] *Hodowla Rosl. Aklim. Nasienn.* 8:361–70.

LIVINGSTON, R. B., and M. L. ALLESSIO. 1968. Buried viable seed in successional field and forest stands, Harvard Forest, Massachusetts. *Bull. Torrey Bot. Club* 95:58–69.

LOISELEUR-DESLONGCHAMPS, J. L. A. 1843. *Considérations sur les céréales et principalement sur les froments*. Paris: Bouchard-Huzard.

LOMAGNO CARAMIELLO, R., and F. MONTACCHINI. 1976. L'invecchiamento delle cariossidi di *Triticum vulgare* Vill. e la sua influenza sull'impianto e lo sviluppo di colture in vitro. *Alliona* 21:77–82.

LONC, W. 1980. Wpływ warunków i czasu przechowywania ziarna siewnego pszenicy ozimej i żyta na wysokość plonów. [The influence of length and condition of seed storage on yields of winter wheat and rye.] *Biul. Inst. Hodowli Aklim. Rosl.* 138:69–76.

LONC, W. 1984. Wpływ warunków przechowywania ziarna siewnego pszenicy ozimej, żyta i jęczmienia jarego na zdolność kiełkowania i plony. [The influence of seed storage conditions on germination and yields of winter wheat, rye, and summer barley.] *Biul. Inst. Hodowli Aklim. Rosl.* 153:165–84.

LOOMIS, E. L., and O. E. SMITH. 1980. The effect of artificial aging on the concentration of Ca, Mg, Mn, K, and Cl in imbibing cabbage seed. *J. Am. Soc. Hortic. Sci.* 105:647–50.

LOPINOT, N. H., and D. E. BRUSSELL. 1982. Assessing uncarbonized seeds from open-air sites in mesic environments: an example from southern Illinois. *J. Archaeol. Sci.* 9:95–108.

LOTAN, J. E. 1974. Cone serotiny-fire relationships in lodgepole pine. *Proc. Tall Timbers Fire Ecol. Conf.* 14:267–78.

LOUGHEED, E. C., D. P. MURR, P. M. HARNEY, and J. T. SYKES. 1976. Low-pressure storage of seeds. *Experientia* (Basel) 32:1159–61.
LOVATO, A. 1976. Vitalità e conservazione delle sementi orticole. *Riv. Agron.* 10:129–42.
LOWE, A. E. 1940. Viability of buffalo grass seeds found in the walls of a sod house. *Agron. J.* 32:891–93.
LOWIG, E. 1970. Versuche zur Frage nach der Veränderung von Samenfarben. *Saatgutwirtschaft* 22:177–78, 267–68, 306.
LUCY, J. A. 1978. Structural interactions between vitamin E and polyunsaturated phospholipids. In C. de Duve and O. Hayaishi, eds., *Tocopherol, Oxygen, and Biomembranes*, pp. 109–20. New York: Elsevier–North Holland.
ŁUCZYŃSKA, J. 1973. Activity of some enzymes from soya bean seeds ageing at various air humidities. *Bull. Acad. Pol. Sci., Ser. Sci. Biol.* 21:155–59.
ŁUCZYŃSKA, J. 1976. Biochemiczne aspekty procesu starzenia sie nasion. [Biochemical aspects of the aging process in seeds.] *Wiad. Bot.* 20:169–79.
LUNDBERG, B., E. SVENS, and S. EKMAN. 1978. The hydration of phospholipids and phospholipid-cholesterol complexes. *Chem. Phys. Lipids* 22:285–92.
LUNN, G., and E. MADSEN. 1981. ATP-levels of germinating seeds in relation to vigor. *Physiol. Plant.* 53:164–69.
LUO, G.-H., C.-B. SHAO, A.-K. WANG, and J.-Y. GUO. 1983. Zai bu tong qiti zhucang xia hua sheng zhongzi huolide yan jiu. [Studies on the vigor of peanut seed stored in different gases.] *Zhiwu Xuebao [Chih Wu Hsueh Pao: Acta Bot. Sin.]* 25:444–49.
LUPTON, F. G. H. 1953. Mummy wheats. *Agriculture* (Lond.) 60:286–88.
LUTE, A. M. 1940. Persistence of impermeability in alfalfa seeds. *J. Colo. Wyo. Acad. Sci.* 2:35.
LUTHRA, J. C. 1936. Ancient wheat and its viability. *Curr. Sci.* (Bangalore) 4:489–90.
LUZZATI, V. 1968. X-ray diffraction studies of lipid-water systems. In D. Chapman, ed., *Biological Membranes: Physical Fact and Function*, pp. 71–123. New York: Academic Press.
MCALISTER, D. F. 1943. The effect of maturity on the viability and longevity of the seeds of western range and pasture grasses. *Agron. J.* 35:442–53.
MCCOY, O. D., and J. F. HARRINGTON. 1970. Effect of age, temperature, and kinetin on the germination of Great Lakes lettuce seed. *Proc. Assoc. Off. Seed Anal.* 60:167–72.
MCDONALD, C. E., and M. MILNER. 1954. The browning reaction in wheat germ in relation to "sick" wheat. *Cereal Chem.* 31:279–95.
MCDONALD, M. B., and D. O. WILSON. 1980. ASA-610 ability to detect changes in soybean seed quality. *J. Seed Technol.* 5 (1):56–66.
MCDONNELL, E. M., F. G. PULFORD, R. B. MIRBAHAR, A. D. TOMOS, and D. L. LAIDMAN. 1982. Membrane lipids and phosphatidyl choline turnover in embryos from germinating low and high vigour wheat (*Triticum aestivum*). *J. Exp. Bot.* 33:631–42.
MCDONOUGH, W. T. 1974. Tetrazolium viability, germinability, and seedling growth of old seeds of 36 mountain range plants. *U.S. For. Serv. Res. Note Int.* 185:1–6.
MCGEE, D. C. 1983. Symposium: Deterioration mechanisms in seeds. Introduction. *Phytopathology* 73:314–15.
MACKAY, D. B., and J. H. B. TONKIN. 1967. Investigations in crop seed longevity. I. Analysis of long-term experiments with special reference to the influence of species, cultivar, provenance, and season. *J. Nat. Inst. Agric. Bot.* 11:209–25.
MACKAY, D. B., J. H. B. TONKIN, and R. J. FLOOD. 1970. Experiments in crop seed storage at Cambridge. *Landwirtsch. Forsch., Sonderheft* 24:189–96.

MacKenzie, P. 1841. Duration of vitality in seeds. *Gard. Chron.*, p. 661.
McKersie, B. D., and T. Senaratna. 1983. Membrane structure in germinating seeds. In C. Nozzolillo, P. J. Lea, and F. A. Loewus, eds., *Mobilization of Reserves in Germination*, pp. 29–52. New York: Plenum.
McKersie, B. D., and R. H. Stinson. 1980. Effect of dehydration on leakage and membrane structure in *Lotus corniculatus* L. seeds. *Plant Physiol.* 66:316–20.
McKnight, R. C., and F. E. Hunter. 1966. Mitochondrial membrane ghosts produced by lipid peroxidation induced by ferrous ions. II. Composition and enzymatic activity. *J. Biol. Chem.* 241:2757–65.
McLeish, J., D. Hockaday, B. J. Harrison, and R. Carpenter. 1968. Ageing in seeds. *John Innes Inst. Annu. Rep.* 59:45–47.
MacLeod, A. M. 1952. Enzyme activity in relation to barley viability. *Bot. Soc. Edinb. Trans. Proc.* 36:18–33.
MacLeod, J., G. Staes, and G. van Eeckhoute. 1890. Cultuurproeven met *Matthiola annua* en *Delphinium ajacis*. *Bot. Jaarb.* 2:83–108.
Madhava Rao, S., A. P. Jayalakshmi, and G. Dharmalingam. 1973. A note on storage life of improved seeds and hybrids. *Seed Res.* (New Delhi) 1:101–3.
Madsen, S. B. 1962. Germination of buried and dry stored seed. III. 1934–1960. *Proc. Int. Seed Test. Assoc.* 27:920–28.
Mahama, A., and A. Silvy. 1982. Influence de la teneur en eau sur la radiosensibilité des semences d'*Hibiscus cannabinus* L. I. Rôle des différents états de l'eau. *Environ. Exp. Bot.* 22:233–42.
Makarov, V. V., and A. P. Prokhorova. 1964. Issledovanie izmeneniya semennykh i biokhimicheskikh svoistv zerna zlakovykh kultur pri dlitel'nom khranenii. [An investigation into the modification of seed and biochemical properties of grain of cereal crops during prolonged storage.] In I. G. Strona, ed., *Voprosy semenovodstva, semenovedeniya i kontrol'no-semennogo dela*, 2:198–206. Kiev: Urozhai
Malysheva, A. G. 1964. Izmenenie biokhimicheskikh svoistv i vskhozhesti semyan maslichnykh kul'tur v protsesse ikh khraneniya. [Changes in the biochemical properties and germination of oil crop seeds in the course of their storage.] In A. L. Kursanov, ed., *Biologicheskie osnovy povysheniya kachestva semyan sel'skokhozyaistvennykh rastenii*, pp. 259–66. Moscow: Nauka.
Mamicpic, N. G., and W. P. Caldwell. 1963. Effects of mechanical damage and moisture content upon viability of soybeans in sealed storage. *Proc. Ass. Off. Seed Anal.* 53:215–20.
Mandal, A. K., and R. N. Basu. 1982. Dip-dry treatment for the maintenance of vigour and viability of stored wheat seed. *Indian Agric.* 26:271–78.
Mandal, A. K., and R. N. Basu. 1983. Maintenance of vigour, viability and yield potential of stored wheat seed. *Indian J. Agric. Sci.* 53:905–12.
Mandal, K., and R. N. Basu. 1981. Role of embryo and endosperm in rice seed deterioration. *Proc. Indian Natl. Sci. Acad.*, Pt. B, *Biol. Sci.* 47:109–14.
Mándy, G., and L. Szabó. 1970a. A Phaseoleae-tribusba tartozó kultúrnövény-fajták magvai csírázóképességének változása a sok éves tárolás alatt. [Changes in the germinability of seed of cultivated species within the tribe Phaseoleae over several years of storage.] *Bot. Kozl.* 57:287–90.
Mándy, G., and L. Szabó. 1970b. Bükkönyfajok (*Vicia* sp.) és-fajták csírázóképességének változása huzamosabb tárolás alatt. [Changes in germination capacity of vetch species (*Vicia* sp.) and varieties during prolonged storage.] *Takarmanytermesztesi Kutato Intez. Kozl.* (Iregszemsce) 10 (1):41–46.
Mándy, G., and L. Szabó. 1973. Köles-és moharfajták csírázóképességének vizsgálata.

[Study of the germinating capacity of proso and foxtail millets.] *Takarmanytermesztesi Kutato Intez. Kozl.* (Iregszemsce) 13 (2):95–104.

MARBACH, I., and A. M. MAYER. 1974. Permeability of seed coats to water as related to drying conditions and metabolism of phenolics. *Plant Physiol.* 54:817–20.

MARCUS, A., J. FEELEY, and T. VOLCANI. 1966. Protein synthesis in imbibed seeds. III. Kinetics of amino acid incorporation, ribosome activation, and polysome formation. *Plant Physiol.* 41:1167–72.

MARQUARDT, H. 1949. Mutationsauslösung durch Putrescin-Hydrochloride und Kaltextrakt aus überalterten *Oenothera*samen. *Experientia* (Basel) 5:401–3.

MARTIN, S. C. 1970. Longevity of velvet mesquite seed in the soil. *J. Range Manage.* 23:69–70.

MARTÍNEZ-HONDUVILLA, C. J., and A. SANTOS-RUIZ. 1975. Rapid biochemical test for seed germinability. *Rev. Esp. Fisiol.* 31:289–92.

MARZKE, F. O., S. R. CECIL, A. F. PRESS, and P. K. HAREIN. 1976. Effects of controlled storage atmospheres on the quality, processing, and germination of peanuts. *U.S. Dep. Agric., Agric. Res. Serv.*, ser. ARS-S, 114:1–12.

MATHUR, P. B., M. PRASAD, and K. P. SINGH. 1956. Studies in the cold storage of peanuts. *J. Sci. Food Agric.* 7:354–60.

MATORIN, D. N., T. V. ORTOIDZE, G. M. NIKOLAEV, and P. S. VENEDIKTOV. 1982. Effects of dehydration on electron transport activity in chloroplasts. *Photosynthetica* (Prague) 16:226–33.

MATSUDA, H., and O. HIRAYAMA. 1973. Kome-chozōji ni okeru shishitsu-seibun narabini shishitsu-kasuibunkaikōso no henka. [Changes of lipid components and lipolytic acylhydrolase activities in rice grains during their storage.] *Nippon Nogeikagaku Kaishi [J. Agric. Chem. Soc. Jpn.]* 47:379–84.

MATSUDA, H., and O. HIRAYAMA. 1975. Kome-hainyubu ni sonzaisuru shishitsu-bunkaikōso no seisei narabini seishitsu. [Purification and characterization of lipolytic acylhydrolases from rice endosperm.] *Nippon Nogeikagaku Kaishi [J. Agric. Chem. Soc. Jpn.]* 49:577–83.

MATSUMURA, S., and T. FUJII. 1966. Radiosensitivity of aged wheat seeds. *Wheat Inf. Serv.* 22:21.

MATTHEWS, S., and W. T. BRADNOCK. 1968. Relationship between seed exudation and field emergence in peas and French beans. *Hortic. Res.* 8:89–93.

MAYER, A. M. 1977. Metabolic control of germination. In A. A. Khan, ed., *The Physiology and Biochemistry of Seed Dormancy and Germination*, pp. 357–84. New York: North Holland.

MAYER, A. M., and A. POLJAKOFF-MAYBER. 1982. *The Germination of Seeds.* 3d ed. New York: Pergamon.

MAZOR, L., M. NEGBI, and M. PERL. 1984. The lack of correlation between ATP accumulation in seeds at the early stage of germination and seed quality. *J. Exp. Bot.* 35:1128–35.

MAZZINI, F., E. BONANDIN, G. C. FANTONE, and G. FOSSAT. 1980. Conservazione del riso. II. Relazione tra alcuni parametri fisici e biochimici. *Riso* (Milan) 28:69–78.

MELCHIOR, H. 1964. *A. Engler's Syllabus der Pflanzenfamilien.* Vol. 2. Berlin: Borntraeger.

MELCHIOR, H., and E. WERDERMANN. 1954. *A. Engler's Syllabus der Pflanzenfamilien.* Vol. 1. Berlin: Borntraeger.

MELSHEIMER. 1876. Zur Phanerogamenflora des Kreises Neuwied. *Verh. Naturhist. Ver. Preuss. Rheinlande Westfalens, Korrespondenzbl.* 33:94–95.

MERCADO, A. T., and J. D. HELMER. 1971. Moisture equilibrium and quality evaluation

of five kinds of seeds stored at various relative humidities. *Araneta J. Agric.* 18:69–102.
MERCER, S. P. 1948. *Farm and Garden Seeds.* 2d ed. London: Crosby Lockwood.
MERKENSCHLAGER, F. 1924. Keimungsphysiologische Probleme. *Naturwiss. Landwirtsch.* (Freising-Munich) 1:1–57.
MESHCHERYAKOV, D. P. 1957. O dlitel'nom sokhranenii vskhozhesti semenami bobovykh. [On the prolonged germinability of legume seeds.] *Izv. Timiryazev. S'kh. Akad.* (Mosc.), no. 2, pp. 46–62.
METZER, R. B. 1961. Effects of the pneumatic conveyor on seed viability. *Tex. Agric. Exp. Stn. Misc. Publ.* 508:1–10.
METZGER, J. D., and D. K. SEBESTA. 1982. Role of endogenous growth regulators in seed dormancy of *Avena fatua*. I. Short chain fatty acids. *Plant Physiol.* 70:1480–85.
MIEHE, H. 1923. Über die Lebensdauer der Diastase. *Ber. Dtsch Bot. Ges.* 41:263–68.
MIERZWIŃSKA, T. 1973. Influence of gibberellic acid on hydrolytic enzymes activity in germinating different-aged lupin seeds. *Acta Soc. Bot. Pol.* 42:509–20.
MIERZWIŃSKA, T. 1975. The effect of GA3 on the germination of ageing lupin seeds. *Seed Sci. Technol.* 3:843–49.
MIERZWIŃSKA T. 1977. Effect of gibberellic acid on α-amylase and ribonuclease activities in rye endosperm after various periods of grain storage. *Act. Soc. Bot. Pol.* 46:69–78.
MIKOLAJCZAK, K. L., R. M. FREIDINGER, C. R. SMITH, and I. A. WOLFF. 1968. Oxygenated fatty acids of oil from sunflower seeds after prolonged storage. *Lipids* 3:489–94.
MILNER, M., and W. F. GEDDES. 1946. Grain storage studies. III. The relation between moisture content, mold growth, and respiration of soybeans. *Cereal Chem.* 23:225–47.
MINAMI MANSHŪ TETSUDO KABUSHIKI KAISHA, CHISHITSU KENKYUJO. 1915. *Manshū Kantō-shū Chishitsu Chōsa Hōkoku.* Dairen: Minami Manshū Tetsudo Kabushiki Kaisha.
MINOR, H. C., and E. H. PASCHAL. 1982. Variation in storability of soybeans under simulated tropical conditions. *Seed Sci. Technol.* 10:131–39.
MIRONENKO, A. V., S. F. SHURKAI, T. M. TROITSKAYA, and L. V. IVANTSOV. 1981. Biokhimicheskie izmeneniya belok-lipidnykh kompleksov semyan lyupina pri potere imi zhiznesposobnosti. *Fiziol. Rast.* (Mosc.) 28:349–57. [Biochemical changes of protein-lipid complexes in lupine seeds with loss of their viability. *Sov. Plant Physiol.* 28:249–55.]
MIROV, N. T. 1944. Possible relation of linolenic acid to the longevity and germination of pine seed. *Nature* (Lond.) 154:218–19.
MIROV, N. T. 1946. Viability of pine seed after prolonged cold storage. *J. For.* 44:193–95.
MISHIN, E. N., I. A. TERSKOV, C. P. GABUDA, and V. V. IVANOV. 1973. Izmenenie kharaktera adsorbtsii vody v protesse probuzhdeniya fiziologicheskoi aktivnosti v zerne pshenitsy. [Change in the character of water adsorption in wheat seed during the process of initiation of physiological activity.] *Izv. Sib. Otd. Akad. Nauk SSSR*, no. 5, pp. 31–36.
MITRA, R., and R. N. BASU. 1979. Seed treatment for viability, vigour, and productivity of tomato. *Sci. Hortic.* (Amsterdam) 11:365–69.
MITRA, S., B. GHOSE, and S. M. SIRCAR. 1974. Physiological changes in rice seeds during loss of viability. *Indian J. Agric. Sci.* 44:744–51.
MODENESI, P. 1977. Invecchiamento del seme ed attività alcool deidrogenasica. *Boll. Soc. Ital. Biol. Sper.* 53:645–51.

MODENESI, P. 1978. Ulteriori studi sull'attività alcool deidrogenasica durante l'invecchiamento del seme. *Boll. Soc. Ital. Biol. Sper.* 54:812–18.
MODENESI, P. 1979. Localizzazione istochimica dell'alcool deidrogenasi durante l'invecchiamento del seme. *Boll. Soc. Ital. Biol. Sper.* 55:2395–400.
MONIN, J., and J. GIRARD. 1973. Remarques sur la perte de viabilité des caryopses de blé (*Triticum sativum* var. Vilmorin 53). *C. R. Seances Soc. Biol. Fil.* 167:1409–13.
MOORE, F. D., A. E. MCSAY, and E. E. ROOS. 1983. Probit analysis: a computer program for evaluation of seed germinability and viability loss rate. *Colo. State Univ. Exp. Stn. Tech. Bull.* 147:1–7.
MOORE, F. D., and E. E. ROOS. 1982. Determining differences in viability loss rates during seed storage. *Seed Sci. Technol.* 10:283–300.
MOORE, R. P. 1972. Effects of mechanical injury on viability. In E. H. Roberts, ed., *Viability of Seeds*, pp. 94–113. Syracuse, N.Y.: Syracuse University Press.
MOORJANI, M. N., and D. S. BHATIA. 1954. Storage effects on the proteins of groundnuts. *J. Sci. Ind. Res.*, sect. B, 13:113–14.
MORA-C., M. A., and R. ECHANDI-Z. 1976. Evaluación del efecto de condiciones de almacenamiento sobre la calidad de semillas de arroz (*Oryza sativa* L.) y de maíz (*Zea mays* L.). *Turrialba* 26:413–16.
MORENO-MARTÍNEZ, E., L. MANDUGANO, M. MENDOZA, and G. VALENCIA. 1985. Use of fungicides for corn seed viability preservation. *Seed Sci. Technol.* 13:235–41.
MORENO-MARTÍNEZ, E., R. MORONES-REZA, and R. GUTIÉRREZ-LOMBARDO. 1978. Diferencias entre líneas, cruzas simples y dobles de maíz en su susceptibilidad al daño por condiciones adversas de almacenamiento. *Turrialba* 28:233–37.
MORGUNOVA, E. A., N. P. KRASNOOK, R. I. POVAROVA, and I. A. VISHNYAKOVA. 1975. Degidrogenazy semyan risa s razlichnoi vskhozhest'yu. *Prikl. Biokhim. Mikrobiol.* 11:760–64. [Dehydrogenases of rice seeds with different germinative capacities. *Appl. Biochem. Microbiol.* 11:668–72.]
MORIKAWA, T., and N. INOMATA. 1978. Negi-shushi no chozō to shinseki-kansō-shori ni yori yūki sareta taisaibō-senshokutai-ijō. [Effects of seed storage and presoaking–drying treatment on chromosome aberration in mitosis of *Allium fistulosum* L.] *Ikushugaku Zasshi [Jpn. J. Breed.]* 28:320–28.
MOROHASHI, Y., J. D. BEWLEY, and E. C. YEUNG. 1981. Biogenesis of mitochondria in imbibed peanut cotyledons: influence of the axis. *J. Exp. Bot.* 32:605–13.
MOSTAFA, H. A. M., S. A. MOHAMEDIEN, and S. M. NASSAR. 1982. A study on improving germination of old seeds of sweet pepper (*Capsicum annuum* L.). *Ain Shams Univ. Fac. Agric. Res. Bull.* 1808:1–17.
MUDD, J. B. 1980. Sterol interconversions. In P. K. Stumpf and E. E. Conn, eds., *The Biochemistry of Plants*, 4:509–34. New York: Academic Press.
MUENSCHER, W. C. 1935. *Weeds*. New York: Macmillan. 2d ed. 1955; reissued 1980, Ithaca: Comstock, Cornell University Press.
MUKHOPADHYAY, A., M. M. CHOUDHURI, K. SEN, and B. GHOSH. 1983. Changes in polyamines and related enzymes with loss of viability in rice seeds. *Phytochemistry* (Oxf.) 22:1547–51.
MUMFORD, P. M., and G. PANGGABEAN. 1982. A comparison of the effects of dry storage on seeds of *Citrus* species. *Seed. Sci. Technol.* 10:257–66.
MUNN, M. T. 1948. The 25-year hard seed soaking experiment. *Proc. Assoc. Off. Seed. Anal.* 38:66–68.
MURATA, M., E. E. ROOS, and T. TSUCHIYA. 1980. Mitotic delay in root tips of peas induced by artificial seed aging. *Bot. Gaz.* 141:19–23.
MURATA, M., E. E. ROOS, and T. TSUCHIYA. 1981. Chromosome damage induced by

artifical seed aging in barley. I. Germinability and frequency of aberrant anaphases at first mitosis. *Can. J. Genet. Cytol.* 23:267–80.

MURATA, M., T. TSUCHIYA, and E. E. ROOS. 1982. Chromosome damage induced by artificial seed aging in barley. II. Types of chromosomal aberrations at first mitosis. *Bot. Gaz.* 143:111–16.

MURATA, M., T. TSUCHIYA, and E. E. ROOS. 1984. Chromosome damage induced by artificial seed aging in barley. III. Behavior of chromosomal aberrations during plant growth. *Theor. Appl. Genet.* 67:161–70.

MURATA, M., and B. K. VIG. 1985. Effect of heat-accelerated seed aging on induction of somatic mosaicism in soybean. *Biol. Zentralbl.* 104:35–41.

MURÍN, A. 1961. Starnutie semien ako príčina mitotických a chromozómových proúch. [Aging of seeds as the cause of mitotic and chromosomal aberration.] *Biologia* (Bratislava) 16:173–77.

MURÍN, A. 1972. Einfluss der Samenalterung auf Störungen der Chromosomen in Keimen einer diploiden und einer tetraploiden Art. *Acta. Univ. Prirod. Fak. Bot.* (Bratislava) 20:127–32.

NÁDVORNÍK, J. 1947. Změny v klíčení zeleninových semen za sedmiletého uskladnění. [Changes in germinability of vegetable seeds during seven years of storage.] *Sb. Cesk. Akad. Zemed.* 20:376–84.

NÁDVORNÍK, J. 1949. Změny v klíčivosti květinových semen za osmiletého uskladnění. [Changes in germinability of flower seeds during eight years of storage.] *Sb. Cesk. Akad. Zemed.* 22:569–74.

NAGATA, Y., S. AOKI, M. SHIMADA, and K. HAYASHI. 1964. Komugi-shushi rōka no seirikagaku. IV. Hatsuga-ritsa to ascorbate oxidase, adenosine triphosphatase, phytase kassei ni tsuite. [Physiology of seed aging. IV. Germinability, ascorbate oxidase, adenosine triphosphatase, and phytase activity.] *Gifu Daigaku Nogakubu Kenkyu Hokoku [Res. Bull. Fac. Agric. Gifu Univ.]* 19:83–89.

NAGATA, Y., and K. HAYASHI. 1966. Komugi-shushi rōka no seirikagaku. V. Hatsuga-ritsu to pyruvic acid-, glutamic acid decarboxylase kassei ni tsuite. [Physiology of seed aging. V. Germinability, pyruvic acid, and glutamic acid decarboxylase activity.] *Gifu Daigaku Nogakubu Kenkyu Hokoku [Res. Bull. Fac. Agric. Gifu Univ.]* 22:142–48.

NAGATA, Y., H. OGUCHI, A. KASUGA, and A. HAYASHI. 1962. Komugu-shushi rōka no seirikagaku. II. Hatsuga-ritsu to catalase, succinic dehydrogenase kassei ni tsuite. [Physiology of seed aging. II. Germinability, catalase, and succinate dehydrogenase activity.] *Gifu Nogakubu Kenkyu Hokuku [Res. Bull. Fac. Agric. Gifu Univ.]* 16:137–45.

NAGY, J., and I. NAGY. 1982. Az elektromos vezetőképességi értékek és a magvigortulajdonságok összefüggése borsónál (*Pisum sativum* L.). [Relationship between electrical conductivity values and seed vigor in peas (*Pisum sativum* L.).] *Novenytermeles* 31:193–205.

NAKAJIMA, Y. 1926. Weitere Untersuchungen über die Lebensdauer der Weidensamen. *Sci. Rep. Tohoku Imp. Univ.*, ser. 4, 1:261–75.

NAKAMURA, S. 1958. Sosai-shushi no chozō. [Storage of vegetable seeds.] *Engei Gakkai Zasshi [J. Jpn. Soc. Hortic. Sci.]* 27:32–44.

NAKAMURA, S. 1975. The most appropriate moisture content of seeds for their long life span. *Seed Sci. Technol.* 3:747–59.

NAKAYAMA, F., and E. M. SÍVORI. 1968. Planta de "achira" (*Canna* sp.) obtenida de semilla de 550 años approximadamente. *Rev. Fac. Agron. Univ. Nac. La Plata* 44:73–82.

NAKAYAMA, N., I. SUGIMOTO, and T. ASAHI. 1980. Presence in dry pea cotyledons of soluble succinate dehydrogenase that is assembled into the mitochondrial inner membrane during seed imbibition. *Plant Physiol.* 65:229–33.

NAKAYAMA, R., and K.-I. SAITO. 1980. Ingenmame-shushi no jumyō no daiareru-bunseki. [Diallel analysis of the longevity of kidney bean seeds.] *Hirosaki Daigaku Nogakubu Gakujutsu Hokoku [Bull. Fac. Agric. Hirosaki Univ.]* 34:47–59.

NAKAYAMA, Y., K. SAIO, and M. KITO. 1981. Decomposition of phospholipids in soybeans during storage. *Cereal Chem.* 58:260–64.

NAPP-ZINN, K. 1964. Zur Frage nach der Abhängigkeit der Pflanzenentwicklung vom Samenalter. *Ber. Dtsch. Bot. Ges.* 77:235–42.

NARASIMHAREDDY, S. B., and P. M. SWAMY. 1977. Gibberellins and germination of inhibitors in viable and non-viable seeds of peanut (*Arachis hypogea* L.). *J. Exp. Bot.* 28:215–18.

NASH, M. J. 1978. *Crop Conservation and Storage in Cool Temperate Climates.* Oxford: Pergamon.

NAUMENKO, A. I., and N. A. TKACHEV. 1976. Izmenie kachestva semyan kukuruzy v protsesse khraneniya. [Change in quality of maize seeds during storage.] *Sel. Semenovod.*, no. 6, pp. 51–52.

NAVASHIN, M. [S.] 1933a. Origin of spontaneous mutations. *Nature* (Lond.) 131:436.

NAVASHIN, M. S. 1933b. Novye dannye po voprosu o samoproizvol'nykh mutatsiyakh. [New data concerning spontaneous mutations.] *Biol. Zh.* 2:111–15.

NAVASHIN, M. S. 1934. Starenie zarodysha kak prichina mutatsii. [The aging of the embryo as a cause of mutation.] *Sov. Bot.*, no. 6, pp. 27–44.

NAVASHIN, M. S., and E. N. GERASIMOVA. 1935. Priroda i prichiny mutatsii. I. O prirode i znachenii khromosomnykh mutatsii, voznikayushchikh v kletkakh pokoyashchikhsya rastitel'nykh zarodyshei v rezul'tate stareniya poslednikh. [The nature and causes of mutations. I. On the nature and importance of chromosomal mutations occurring in quiescent plant embryos as a consequence of aging.] *Biol. Zh.* 4:593–634.

NAVASHIN, M. [S.], and P. [K.] SHKVARNIKOV. 1933. Ob uskorenii mutatsionnogo protsessa v pokoyashikhsya semenakh pod vliyaniem povyshennoi temperatury. [On the acceleration of the mutation process in quiescent seeds under the influence of elevated temperature.] *Priroda* (Mosc.), no. 10, pp. 54–55.

NAWASCHIN, M. [NAVASHIN, M. S.] 1933. Altern der Samen als Ursache von Chromosomenmutationen. *Planta* (Berl.) 20:233–43.

NAWASCHIN, M., and H. GERASSIMOWA. [NAVASHIN, M. S., and E. N. GERASIMOVA] 1936. Natur und Ursachen der Mutationen. III. Über die Chromosomenmutationen, die in den Zellen von ruhenden Pflanzenkeimen bei deren Altern auftreten. *Cytologia* 7:437–65.

NDIMANDE, B. N., H. C. WIEN, and E. A. KUENEMAN. 1981. Soybean seed deterioration in the tropics. I. The role of physiological factors and fungal pathogens. *Field Crops. Res.* 4:113–21.

NEAL, N. P., and J. R. DAVIS. 1956. Seed viability of corn inbred lines as influenced by age and conditions of storage. *Agron. J.* 48:383–84.

NECHITAILO, G. S. 1969. Radiochuvstvitel'nost' i starenie semyan *Allium fistulosum* L. *Radiobiologiya* 9:774–76. [Radiosensitivity and aging of *Allium fistulosum* L. seeds. *Radiobiology* 9 (5):179–81.]

NELYUBOV, D. N. 1925. O sposobakh opredelenie vskhozhesti semyan pomimo prorashchivaniya. [On methods of determining seed germinability without germination.] *Zap. Otd. Semen. Gl. Bot. Sada* (Leningr.) 4 (7):14–35.

NEORAL, K. 1923. Vliv stáří semene na vývoj, výnos a jakost. [The influence of seed age on development, yield, and quality.] *Morav. Zemsky Vyzk. Ustav. Zemed. Sekce Slecht. Publ.* 1:1–9.

NESTERENKO, V. G. 1960. O vskhozhesti semyan pri khranenii ikh v laboratornykh usloviyakh. [On the germinability of seeds during storage under laboratory conditions.] *Byull. Gl. Bot. Sada* (Mosc.) 36:99–103.

NEWHALL, A. G., and J. K. HOFF. 1960. Viability and vigor of 22-year-old onion seed. *Seed World* 86 (12):4–5.

NICHOLS, C. 1941. Spontaneous chromosome aberrations in *Allium*. *Genetics* 26:89–100.

NIETHAMMER, A. 1929a. Die Charakteristik der Lebenskraft verschiedenen Samenmaterials auf chemischer, physikalischer und rechnerischer Grundlage. *Gartenbauwissenschaft* 1:593–614.

NIETHAMMER, A. 1929b. Fortlaufende Untersuchungen über den Chemismus der Angiospermensamen und die äusseren natürlichen wie künstlichen Keimungsfaktoren. IV. Mitteilung: Untersuchungen über die Farbstoff- und Salzpermeabilität von Frucht- und Samenschalen. *Biochem. Z.* 209:263–75.

NIETHAMMER, A. 1929c. Vergleichende biochemische Untersuchungen über das Reifen und Altern von Samen und Früchten. *Oesterr. Bot. Z.* 78:264–78.

NIETHAMMER, A. 1942. Plasmolysestudien an gärtnerisch wichtigem Saatgut. *Gartenbauwissenschaft* 17:91–94.

NIKLAS, K. J., B. H. TIFFNEY, and A. C. LEOPOLD. 1982. Preservation of unsaturated fatty acids in Palaeogene angiosperm fruits and seeds. *Nature* (Lond.) 296:63–64.

NIKOLOVA, A., and P. B"CHVAROV. 1982. Izmeneniya v zhiznenostta na semene ot soya sled uskorenoto im stareene. [Changes in soybean seed viability following accelerated aging.] *Fiziol. Rast.* (Sofia) 8 (3):44–53.

NIKOLOVA, A. K., and A. V. DENCHEVA. 1984. Fiziologo-biokhimichni osnovi na zhiznenostta na semenata. I. Mobilizatsiya na pezervnite belt"tsi v semena ot tsarevitsa v"v vr"zka s promenite na tyakhnata zhiznenost pod vliyanie na usloviyata na s"khranenie. [Physiological and biochemical basis of seed viability. I. Mobilization of maize seed reserve proteins in relation to changes in viability induced by storage conditions.] *Fiziol. Rast.* (Sofia) 10 (3):48–57.

NILAN, R. A., and H. M. GUNTHARDT. 1956. Studies on aged seeds. III. Sensitivity of aged wheat seeds to X-radiation. *Caryologia* 8:316–22.

NILSSON, N. H. 1931. Sind die induzierten Mutanten nur selektive Erscheinungen? *Hereditas* 15:320–28.

NISHIYAMA, I. 1977. Decrease in germination activity of rice seeds due to excessive desiccation in storage. *Nippon Sakumotsu Gakkai Kiji [Jpn. J. Crop Sci.]* 46:111–18.

NOBBE, F. 1876. *Handbuch der Samenkunde*. Berlin: Hempel & Parey.

NOBBE, F. 1919. Untersuchungen über den Quellprozess der Samen von *Trifolium pratense* und einiger anderer Schmetterlingsblütler. *Landwirtsch. Vers. Stn.* 94:197–218.

NORDEN, A. J. 1981. Effect of preparation and storage environment on lifespan of shelled peanut seed. *Crop Sci.* 21:263–66.

NORDFELDT, S., N. OLSSON, G. ANSTRAND, and V. HELLSTROM. 1962. Influence of storage upon total tocopherols in wheat germs. Effect of germination upon total tocopherols in wheat. *K. Lantbrukshogsk. Ann.* (Swed.) 28:181–88.

NORTON, J. M., and G. E. HARMAN. 1985. Responses of soil microorganisms to volatile exudates from germinating pea seeds. *Can. J. Bot.* 63:1040–45.

NOWAK, J., and T. MIERZWINSKA. 1978. Activity of proteolytic enzymes in rye seeds of different ages. *Z. Pflanzenphysiol.* 86:15–22.

NOWOSIELSKA, B., and J. SCHNEIDER. 1980. Wartość produkcyjna nasion sałaty (*Lactuca sativa* L.) po wieloletnim przechowywaniu. [The productive value of lettuce (*Lactuca sativa* L.) seed after long-term storage.] *Hodowla Rosl. Aklim. Nasienn.* 24:731-39.

NOZZOLILLO, C., and M. DE BEZADA. 1984. Browning of lentil seeds, concomitant loss of viability, and the possible role of soluble tannins in both phenomena. *Can. J. Plant Sci.* 64:815-24.

NUTILE, G. E. 1964. Effect of desiccation on viability of seeds. *Crop Sci.* 4:325-28.

NUTI RONCHI, V., and G. MARTINI. 1962. Germinabilità, sviluppo delle plantule e frequenza di aberrazioni cromosomiche in rapporto all'età del seme nel frumento. *Caryologia* 15:293-302.

OATHOUT, C. H. 1928. The vitality of soybean seed as affected by storage conditions and mechanical injury. *Agron. J.* 20:837-55.

ØDUM, S. 1965. Germination of ancient seeds. *Dan. Bot. Ark.* 24 (2):1-70.

ØDUM, S. 1969. De la présence de semences viables dans des décombres, des problèmes de datation et d'aspects écologiques. In L. Guyot, ed., *Troisième Colloque sur la biologie des mauvaises herbes, Grignon, 1969*, pp. 47-62. Grignon: École Nationale Supérieure Agronomique.

ØDUM, S. 1974. Seeds in ruderal soils, their longevity and contribution to the flora of disturbed ground in Denmark. *Proc. Br. Weed Control Conf.* 12:1131-44.

ØDUM, S. 1978. *Dormant Seeds in Danish Ruderal Soils.* Horshølm: Royal Veterinary and Agricultural University.

OE, H., and Y. NAGATA. 1969. Komugi-shushi rōka no seirikagaku. VII. Vitamin B6 ganryō ni tsuite. [Physiology of seed aging. VII. On the vitamin B6 content.] *Gifu Daigaku Nogakubu Kenkyu Hokoku [Res. Bull. Fac. Agric. Gifu Univ.]* 28:131-37.

OHGA, I. 1923. On the longevity of seeds of *Nelumbo nucifera. Bot. Mag.* (Tokyo) 37:87-95.

OHGA, I. 1926a. A report on the longevity of the fruit of *Nelumbium. J. Bot.* (Lond.) 64:154-57.

OHGA, I. 1926b. On the structure of some ancient, but still viable fruits of Indian lotus, with special reference to their prolonged dormancy. *Jpn. J. Bot.* 3:1-20.

OHGA, I. 1926c. The germination of century-old and recently harvested Indian lotus fruits, with special reference to the effect of oxygen supply. *Am. J. Bot.* 13:754-59.

OHGA, I. 1926d. A comparison of the life activity of century-old and recently harvested Indian lotus fruits. *Am. J. Bot.* 13:760-65.

OHGA, I. 1927a. On the age of the ancient fruit of the Indian lotus which is kept in the peat bed in South Manchuria. *Bot. Mag.* (Tokyo) 41:1-6.

OHGA, I. 1927b. Supramaximal temperature and life duration of the ancient fruit of Indian lotus. *Bot. Mag.* (Tokyo) 41:161-72.

OHGA, I. 1927c. *A Study of the Ancient but Still Viable Fruit of the Indian Lotus Found in the Peat Bed Near Pulantien, South Manchuria.* Dairen: South Manchuria Railway Co.

OHGA, I. [ŌGA, I.] 1930. Minami-Manshū-san ko-renjitsu no hatsuga ni tsuite. [On the germination of old lotus fruit from South Manchuria.] *Nippon Gakujutsu Kyokai Hokoku [Rep. Jpn. Sci. Assoc.]* 5:151-57.

OHGA, I. [ŌGA, I.] 1935. Minami Manshū Furanten-san korenjitsu no sono go. [Subsequent facts concerning the ancient lotus fruits from Pulantien, South Manchuria.] *Shokubutsu oyobi Dobutsu [Bot. Zool.]* 3:71-80.

OHGA, I. [ŌGA, I.] 1936. Wagakuni ni mirareta seikatsuryoku o yusuru korenjitsu ni tsuite. [On the discovery of viable ancient lotus fruit in Japan.] *Shokubutsu oyobi Dobutsu [Bot. Zool.]* 4:1505-18.

OHGA, I. [ŌGA, I.] 1951. Shushi no jumyō—watakushi no yume—higan. Chiba-ken

Kemigawa sotanchi no korenjitsu no hakkutsu. [The longevity of seeds, my dream, earnest wish. The excavation of ancient lotus fruit from peat at Kemigawa, Chiba Prefecture.] *Saishu to Shiiku [Collect. Breed.]* 13:206–8, 212.

OHGA, I. [ŌGA, I.] 1953. Kohasu no kajitsu no jumyō to rajio-kabon-testo. [The longevity of ancient lotus fruits: radiocarbon testing.] *Bungei Shunju [Year Lett.]* 31 (11):44–47.

OHLROGGE, J. B., and T. P. KERNAN. 1982. Oxygen-dependent aging of seeds. *Plant Physiol.* 70:791–94.

OLCOTT, H. S., and T. D. FONTAINE. 1941. The absence of lipase in cottonseed. *Oil Soap* 18:123–24.

OLSSON, A. 1950. Inverkan av utsädets ålder och mognadsgrad på skördeutbyte och kvalitet hos vissa lantbruksväxter. [Effect of age and degree of maturity of seed of certain agricultural crops on the amount and the quality of the harvest.] *K. Lantbruksakad. Tidskr.* (Swed.) 89:241–61.

OOSTING, H. J., and M. E. HUMPHREYS. 1940. Buried viable seeds in a successional series of old field and forest soils. *Bull. Torrey Bot. Club* 67:253–73.

ÖPIK, H. 1972. Some observations on coleoptile cell ultrastructure in ungerminated grains of rice (*Oryza sativa* L.). *Planta* (Berl.). 102;61–71.

ÖPIK, H. 1980. The ultrastructure of coleoptile cells in dry rice (*Oryza sativa* L.) grains after anhydrous fixation with osmium tetroxide vapour. *New Phytol.* 85:521–29.

ORLOVA, N. N. 1967. Izuchenie mutatsionnogo protsessa v pokoyashchikhsya semenakh luka *Allium fistulosum* L., khranyashchikhsya v usloviyakh povyshennoi temperatury i vlazhnosti. *Genetika* 3 (11):15–25. [A study of the mutational process in dormant seeds of Welsh onions *Allium fistulosum* L. stored under conditions of high temperature and humidity. *Sov. Genet.* 3 (11):7–12.]

ORLOVA, N. N. 1972. Estestvennyi mutatsionnyi protsess v semenakh pri ikh khranenii. [The natural mutational process in stored seeds.] *Usp. Sovrem. Genet.* 4:206–28.

ORLOVA, N. N. 1973. Nekotorye geneticheskie aspekty stareniya semyan. [Some genetic aspects of seed aging.] In F. E. Reimers, ed., *Fiziologo-biokhimicheskie problemy semenovedeniya i semenovodstva*, 1:100–106. Irkutsk: Akademiya Nauk SSSR Sibirskoe Otdelenie.

ORLOVA, N. N., and T. A. EZHOVA. 1974. Vliyanie ingibitora sinteza DNK 5-aminouratsila na vozniknovenie strukturnykh mutatsii khromosom v semenakh luka-batuna raznogo vozrasta. *Genetika* 10 (12):19–25. [Effect of the DNA synthesis inhibitor 5-aminouracil on the appearance of chromosomal aberrations in *Allium fistulosum* seeds of various ages. *Sov. Genet.* 10:1476–81.]

ORLOVA, N. N., and V. I. NIKITINA. 1968. O momente vozniknoveniya aberratsii khromosom pri starenii semyan. *Genetika* 4 (9):24–32. [The moment of appearance of chromosome aberrations during the aging of seeds. *Sov. Genet.* 4:1153–58.]

ORLOVA, N. N. and N. P. ROGATYKH. 1970. Nekotorye zakonomernosti mutirovaniya v semenakh i prorastkakh luka-batuna. *Genetika* 6 (12):23–26. [Some regularities of mutation in seeds and seedlings of Welsh onions. *Sov. Genet.* 6(12):1565–67.]

ORLOVA, N. N., N. P. ROGATYKH, and G. A. KHARTINA. 1975. Snizhenie mitoticheskogo potentsiala kletok v pokoyashchikhsya semenakh luka-batuna pri khranenii. *Fiziol. Rast.* (Mosc.) 22:734–40. [Decrease in the mitotic potential of cells in dormant seeds of Welsh onion during storage. *Sov. Plant Physiol.* 22:629–35.]

ORLOVA, N. N., and O. P. SOLDATOVA. 1975. Tsitogeneticheskoe izuchenie mutatsionnogo protsessa v semenakh populyatsii i chistykh linii rzhi (*Secale cereale* L.) pri khranenii. *Genetika* 11 (11):9–14. [Cytogenetic study of mutation in stored grain of populations and pure lines of rye. *Sov. Genet.* 11:1362–66.]

ORY, R. L., and A. J. ST. ANGELO. 1982. Effects of lipid oxidation on proteins of oilseeds. In J. P. Cherry, ed., *Food Protein Deterioration*, pp. 55–65. Washington D.C.: American Chemical Society.

OSBORNE, D. J. 1980. Senescence in seeds. In K. V. Thimann, ed., *Senescence in Plants*. Boca Raton: CRC Press.

OSBORNE, D. J. 1982a. Deoxyribonucleic acid integrity and repair in seed germination: the importance in viability and survival. In A. A. Khan, ed., *The Physiology and Biochemistry of Seed Development, Dormancy, and Germination*, pp. 435–63. New York: Elsevier.

OSBORNE, D. J. 1982b. DNA integrity in plant embryos and the importance of DNA repair. In M. M. Burger and R. Weber, eds., *Embryonic Development*, pt. B, pp. 577–92. New York: Liss.

OSBORNE, D. J. 1983. Biochemical control systems operating in the early hours of germination. *Can. J. Bot.* 61:3568–77.

OSBORNE, D. J., M. DOBRZANSKA, and S. SEN. 1977. Factors determining nucleic acid and protein synthesis in the early hours of germination. *Symp. Soc. Exp. Biol.* 31:177–94.

OSBORNE, D. J., B. E. ROBERTS, P. I. PAYNE, and S. SEN. 1974. Protein synthesis and viability in rye embryos. *Bull. R. Soc. N.Z.* 12:805–12.

OSBORNE, D. J., R. SHARON, and R. BEN-ISHAI. 1980/81. Studies on DNA integrity and DNA repair in germinating embryos of rye (*Secale cereale*). *Isr. J. Bot.* 29:259–72.

OSTROMĘCKI, I. 1977. Wpływ warunków przechowywania na zdolność kiełkowania nasion kapusty (*Brassica oleracea* L. var. *capitata*) i próba jego wyjaśnienia za pomocą techniki zymogramowej wybranych enzymów. [Influence of storage conditions on the germination capacity of cabbage seeds (*Brassica oleracea* L. var. *capitata*) and an attempt at its explanation by the zymogram technique applied to selected enzymes.] *Hodowla Rosl. Aklim. Nasienn.* 22:207–22.

OVCHAROV, K. E. 1969. *Fiziologicheskie osnovy vskhozhesti semyan*. Moscow: Nauka. [*Physiological Basis of Seed Germination*. New Delhi: Amerind, 1977.]

OVCHAROV, K. E. 1976. *Fiziologiya formirovaniya i prorastaniya semyan*. Moscow: Kolos.

OVCHAROV, K. E., and K. P. GENKEL'. 1973. Belki semyan raznoi zhiznesposobnosti. [Proteins of seeds with differing viability.] In F. E. Reimers, ed., *Fiziologo-biokhimicheskie problemy semenovedeniya i semenovodstva*, 2:40–45. Irkutsk: Akademiya Nauk SSSR Sibirskoe Otdelenie.

OVCHAROV, K. E., and YU. P. KOSHELEV. 1974. Soderzhanie sakharov v semenakh kukuruzy raznoi zhiznesposobnosti. *Fiziol. Rast.* (Mosc.) 21:969–74. [Sugar content in corn seeds of different viability. *Sov. Plant. Physiol.* 21:805–7.]

OVCHAROV, K. E., YU. P. KOSHELEV, N. D. MURASHOVA, K. P. GENKEL', and D. M. SEDENKO. 1978. Biokhimicheskie izmeneniya pri starenii semyan. [Biochemical changes during seed aging.] *Byull. Vses. Nauchno-Issled. Inst. Rastenievod. im. N. I. Vavilova* (Leningr.) 77:36–39.

OVCHAROV, K. E., and N. D. MURASHOVA. 1973. Aktivnost' degidrogenaz semyan raznoi zhiznesposobnosti. [Activity of dehydrogenases in seeds with differing viability.] In F. E. Reimers, ed., *Fiziologo-biokhimicheskie problemy semenovedeniya i semenovodstva*, 1:107–10. Irkutsk: Akademiya Nauk SSSR Sibirskoe Otdelenie.

OVCHAROV, K. E., and D. M. SEDENKO. 1976. Vymyvanie inozitfosfatov iz semyan kukuruzy raznoi zhiznesposobnosti. *Fiziol. Rast.* (Mosc.) 23:945–51. [Washing out of inositol phosphates from corn seeds of different viability. *Sov. Plant Physiol.* 23:795–800.]

OVEČKA, V. 1960. Studium vlivu stáří osiva na produkční schopnost polních okurek Znojemských nakládaček (*Cucumis sativa* L.) [A study of the effect of seed age on the productivity of the Znojmo field pickling cucumber.] *Sb. Cesk. Akad. Zemed. Ved Rostl. Vyroba* 6:1261–74.

OVERAA, P. 1984. Distinguishing between dormant and inviable seeds. In J. B. Dickie, S. Linington, and J. T. Williams, eds., *Seed Management Techniques for Genebanks*, pp. 182–96. Rome: International Board for Plant Genetic Resources.

OXLEY, T. A., and J. D. JONES. 1944. Apparent respiration of wheat grains and its relation to a fungal mycelium beneath the epidermis. *Nature* (Lond.) 154:826–27.

PACK, D. A., and F. V. OWEN. 1950. Viability of sugar beet seed held in cold storage for 22 years. *Proc. Am. Soc. Sugar Beet* 6:127–29.

PAINTER, E. P., and L. L. NESBITT. 1943. The stability of linseed oil during storage of flaxseed. *N. D. Agric. Exp. Stn. Bimonth. Bull.* 5 (6):36–40.

PALLADIUS, R. T. AE. 1843. *De re rustica*, ed. Cabaret-Dupaty. Paris: Panckouke.

PAMMENTER, N. W., J. H. ADAMSON, and P. BERJAK. 1974. Viability of stored seed: extension by cathodic protection. *Science* (Wash., D.C.) 186:1123–24.

PAN, D., and R. N. BASU. 1985. Mid-storage and pre-sowing seed treatments for lettuce and carrot. *Sci. Hortic.* (Amsterdam) 25:11–19.

PAN, D., B. PUNJABI, and R. N. BASU. 1981. A note on the involvement of protein synthesis in hydration–dehydration treatment of lettuce seed. *Seed Res.* (New Delhi) 9:202–5.

PARICHA, P. C., A. M. RATH, and J. K. SABOO. 1977. Studies on the hygroscopic equilibrium and viability of rice stored under various relative humidities. *Seed Res.* (New Delhi) 5:1–5.

PARRISH, D. J., and C. C. BAHLER. 1983. Maintaining vigor of soybean seeds with lipid antioxidants. *Proc. Plant Growth Regul. Soc. Am.* 10:165–70.

PARRISH, D. J., and A. C. LEOPOLD. 1977. Transient changes during soybean imbibition. *Plant Physiol.* 59:1111–15.

PARRISH, D. J., and A. C. LEOPOLD. 1978. On the mechanism of aging in soybean seeds. *Plant Physiol.* 61:365–68.

PARRISH, D. J., A. C. LEOPOLD, and M. A. HANNA. 1982. Turgor changes with accelerated aging of soybeans. *Crop Sci.* 22:666–69.

PATHAK, G., and R. N. BASU. 1980. Control of seed deterioration in sunflower (*Helianthus annuus* L.). *Curr. Sci.* (Bangalore) 49:67–69.

PATIL, V. N., and C. H. ANDREWS. 1985. Cotton seeds resistant to water absorption and seed deterioration. *Seed Sci. Technol.* 13:193–99.

PATTEE, H. E., J. L. PEARSON, C. T. YOUNG, and F. G. GIESBRECHT. 1982a. Changes in roasted peanut flavor and other quality factors with seed size and storage time. *J. Food Sci.* 47:455–60.

PATTEE, H. E., D. K. SALUNKHE, S. K. SATHE, and N. R. REDDY. 1982b. Legume lipids. *CRC Crit. Rev. Food Sci. Nutn.* 17:97–139.

PAUL, A. K., and S. MUKHERJI. 1972. Change in respiration rate of rice seedlings as affected by storage and viability, and its possible relations with catalase and peroxidase activities during germination. *Biol. Plant.* (Prague) 14:414–19.

PAUL, A. K., S. MUKHERJI, and S. M. SIRCAR. 1970. Metabolic changes in rice seeds during storage. *Indian J. Agric. Sci.* 40:1031–36.

PAULS, K. P., and J. E. THOMPSON. 1981. Effects of *in vitro* treatment with ozone on the physical properties of membranes. *Physiol. Plant.* 53:255–62.

PAULSEN, M. R., W. R. NAVE, T. L. MOUNTS, and L. E. GRAY. 1981. Storability of harvest-damaged soybeans. *Trans. Am. Soc. Agric. Eng.* 24:1583–89.

PAULSON, R. E., and L. M. SRIVASTA. 1968. The fine structure of the embryo of *Lactuca sativa*. I. Dry embryo. *Can. J. Bot* 46:1437–45.

PEARCE, R. S., and I. M. ABDEL SAMAD. 1980. Changes in fatty acid content of polar lipids during ageing of seeds of peanut (*Arachis hypogea* L.). *J. Exp. Bot.* 31:1283–90.

PEHAP, A. 1972. Seed eluates on the germination blotter—a germinability test? *Stud. For. Suec.* 101:1–21.

PENNEY, C. 1842. Vitality of seeds. *Gard. Chron.*, p. 471.

PERCIVAL, J. 1921. *The Wheat Plant*. London: Duckworth.

PERCIVAL, J. 1936. Cereals of ancient Egypt and Mesopotamia. *Nature* (Lond.) 138:270–73.

PERDOK, E. A. 1970. The influence of gradual deterioration of polypoid beet seed on the ploidy pattern. *Proc. Int. Seed Test. Assoc.* 35:813–14.

PERELBERG, T., E. BINI-DA-SILVA, V. CONE, and A. BOVERIS. 1981. Chemiluminescence of soybean seeds upon spontaneous and accelerated aging. *Braz. J. Med. Biol. Res.* 14:217.

PERL, M. 1980. An ATP-synthesizing system in seeds. *Planta* (Berl.) 149:1–6.

PERL, M., I. LURIA, and H. GELMOND. 1978. Biochemical changes in sorghum seeds affected by accelerated aging. *J. Exp. Bot.* 29:497–509.

PERL, M., Z. YANIV, and Z. FEDER. In press. The effect of natural and accelerated aging on the lipid content and on the fatty acid composition in seeds. *Acta Hortic.*

PERNER, E. 1965. Elektronenmikroskopische Untersuchungen an Zellen von Embryonen im Zustand völliger Samenruhe. I. Mitteilung. Die Zelluläre Strukturordnung in der Radikula lufttrockener Samen von *Pisum sativum*. *Planta* (Berl.) 65:334–57.

PERRY, D. A. 1978. Report of the Vigour Test Committee, 1974–1977. *Seed Sci. Technol.* 6:159–81.

PERRY, D. A. 1981. *Handbook of Vigor Test Methods*. Zurich: International Seed Testing Association.

PERRY, D. A., and J. G. HARRISON. 1977. Effects of seed deterioration and seed-bed environment on emergence and yield of spring-sown barley. *Ann. Appl. Biol.* 86:291–300.

PESIS, E., and T. J. NG. 1983. Viability, vigor, and electrolytic leakage of muskmelon seeds subjected to accelerated aging. *HortScience* 18:242–44.

PETER, A. 1893. Culturversuche mit "ruhenden" Samen. *Nachr. K. Ges. Wiss. Georg-Aug. Univ. Goettingen*, pp. 673–91.

PETER, A. 1894. Culturversuche mit "ruhenden Samen." II. Mittheilung. *Nachr. Ges. Wiss. Goettingen, Math. Phys. Kl.*, pp. 373–93.

PETERSEN, H. I., and S. LUND. 1944. Undersøgelser over spireforhold hos frø af nogle ondartede danske ukrudtsplanter. [Investigations into the germination characteristics of seeds of some harmful Danish weeds.] *Tidsskr. Landokonomi*, pp. 425–38.

PETERSON, R. F. 1965. *Wheat: Botany, Cultivation, and Utilization*. New York: Interscience.

PETO, F. H. 1933. The effect of aging and heat on the chromosomal mutation rates in maize and barley. *Can. J. Res.* 9:261–64.

PETRIE, W. M. F. 1914. Mummy wheat. *Ancient Egypt* 1 (2):78–79.

PETRUZZELLI, L., and G. CARELLA. 1983. The effect of ageing conditions on loss of viability in wheat (*T. durum*). *J. Exp. Bot.* 34:221–25.

PETRUZZELLI, L., L. LIOI, G. CARELLO, S. MORGUTTI, and S. COCUCCI. 1982. The effect of fusicoccin and monovalent cations on the viability of wheat seeds. *J. Exp. Bot.* 33:118–24.

PETRUZZELLI, L., and G. TARANTO. 1984. Phospholipid changes in wheat embryos aged under different storage conditions. *J. Exp. Bot.* 35:517–20.

PETRUZZELLI, L., and G. TARANTO. 1985. Effects of permeation with plant growth regulators via acetone on seed viability during accelerated ageing. *Seed Sci. Technol.* 13:183–91.

PEUMANS, W. J., and A. R. CARLIER. 1981. Loss of protein synthesis activity in ageing wheat grains: lesions in the initiation process and in RNA degradation. *Biochem. Physiol. Pflanz. [BPP]* 176:384–95.

PFEIFFER, N. E. 1934. Morphology of the seed of *Symphoricarpos racemosus* and the relation of fungal invasion of the coat to germination capacity. *Contrib. Boyce Thompson Inst.* 6:103–22.

PHILLIS, E., and T. G. MASON. 1945. The effect of extreme desiccation on the viability of cotton seed. *Ann. Bot.* (Lond.) 9:353–59.

PIDOTTI, O. A. 1952. Vliyanie srokov khraneniya semyan travyanistykh mnogoletnikh rastenii na vskhozhest'. [The effect of storage period on germinability of seeds of herbaceous perennial plants.] *Tr. Bot. Inst. Akad. Nauk SSSR*, ser. 6, 2:361–67.

PIECH, J., and S. SUPRYN. 1979. Effect of chromosome deficiencies on seed viability in wheat *Triticum aestivum* L. *Ann. Bot.* (Lond.) 43:115–18.

PIERRET, P. 1875. *Dictionnaire d'archéologie égyptienne*. Paris: Imprimerie Nationale.

PITEL, J. A. 1982. Accelerated aging studies of seeds of jackpine (*Pinus banksiana* Lamb.) and red oak (*Quercus rubra* L.). In B. S. P. Wang and J. A. Pitel, eds., *Proceedings of the International Symposium on Forest Tree Seed Storage, Chalk River, 1980*, pp. 40–54. Ottawa: Canadian Forestry Service.

PIXTON, S. W. 1980. Changes in the quality of wheat during 18 years storage. In J. Shejbal, ed., *Controlled Atmosphere Storage of Grains*, pp. 301–10. New York: Elsevier.

PIXTON, S. W. and S. HENDERSON. 1981. The moisture content–equilibrium relative humidity relationships of five varieties of Canadian wheat and of Candle rapeseed at different temperatures. *J. Stored Prod. Res.* 17:187–90.

PIXTON, S. W., and S. WARBURTON. 1971. Moisture content–relative humidity equilibrium, at different temperatures, of some oilseeds of economic importance. *J. Stored Prod. Res.* 7:261–69.

PIXTON, S. W., S. WARBURTON, and S. T. HILL. 1975. Long-term storage of wheat. III: Some changes in the quality of wheat observed during 16 years of storage. *J. Stored Prod. Res.* 11:177–85

POCSAI, K., and L. SZABÓ. 1982. Metanol-stressz hatása különbözö korú lóbabmagvak imbibíciós légzésére és csírázására. [The effect of methanol-stress during imbibition on respiration and germination of broad bean seeds.] *Novenytermeles* 31:135–39.

POISSON, J. 1903. Observations sur la durée de la vitalité des graines. *Bull. Soc. Bot. Fr.* 50:337–54.

PONCELET [P.] 1779. *Histoire naturelle du froment*. Paris: Desprez.

PORSILD, A. E., C. R. HARINGTON, and G. A. MULLIGAN. 1967. *Lupinus arcticus* Wats. grown from seeds of Pleistocene age. *Science* (Wash., D.C.) 158:113–14.

PORTER, N. G., and P. F. WAREING. 1974. The role of the oxygen permeability of the seed coat in the dormancy of seed of *Xanthium pennsylvanicum* Wallr. *J. Exp. Bot.* 25:583–94.

POTTHAST, K. 1978. Influence of water activity on enzymic activity in biochemical systems. In J. H. Crowe and J. S. Clegg, eds., *Dry Biological Systems*, pp. 323–42. New York: Academic Press.

POTTS, H. C., J. DUANGPATRA, W. G. HAIRSTON, and J. C. DELOUCHE. 1978. Some influences of hardseededness on soybean seed quality. *Crop Sci.* 18:221–24.

POUCHET. F.-A. 1866. Expériences comparées sur la résistance vitale de certains embryons végétaux. *C. R. Hebd. Seances Acad. Sci.* (Paris) 63:939–41.
POWELL, A. A., and G. E. HARMAN. In press. Absence of a consistent association of changes in membranal lipids with the ageing of pea seeds. *Seed Sci. Technol.*
POWELL, A. A., and S. MATTHEWS. 1977. Deteriorative changes in pea seeds (*Pisum sativum* L.) stored in humid or dry conditions. *J. Exp. Bot.* 28:225–34.
POWELL, A. A., and S. MATTHEWS. 1978. The damaging effect of water on dry pea embryos during imbibition. *J. Exp. Bot.* 29:1215–29.
POWELL, A. A., and S. MATTHEWS. 1981a. A physical explanation for solute leakage from dry pea embryos during imbibition. *J. Exp. Bot.* 32:1045–50.
POWELL, A. A., and S. MATTHEWS. 1981b. Association of phospholipid changes with early stages of seed aging. *Ann. Bot.* (Lond.) 47:709–12.
POWELL, A. D., D. W. M. LEUNG, and J. D. BEWLEY. 1983. Long-term storage of dormant Grand Rapids lettuce seeds in the imbibed state: physiological and metabolic changes. *Planta* (Berl.) 159:182–88.
PRADET, A. 1982. Oxidative phosphorylation in seeds during the initial phases of germination. In A. A. Khan, ed., *The Physiology and Biochemistry of Seed Development, Dormancy, and Germination*, pp. 347–69. New York: Elsevier.
PRESLEY, J. T. 1958. Relation of protoplast permeability to cotton seed viability and predisposition to seedling disease. *Plant Dis. Rep.* 42:852.
PRICE, L. W. 1972. *The Periglacial Environment, Permafrost, and Man.* Washington, D.C.: Association of American Geographers.
PRIESTLEY, D. A. 1985. Hugo de Vries and the development of seed aging theory. *Ann. Bot.* (Lond.). 56:267–70.
PRIESTLEY, D. A., and B. DE KRUIJFF. 1982. Phospholipid motional characteristics in a dry biological system: a ^{31}P-nuclear magnetic resonance study of hydrating *Typha latifolia* pollen. *Plant Physiol.* 70:1075–78.
PRIESTLEY, D. A., W. C. GALINAT, and A. C. LEOPOLD. 1981. Preservation of polyunsaturated fatty acid in ancient Anasazi maize seed. *Nature* (Lond.) 292:146–48.
PRIESTLEY, D. A., and A. C. LEOPOLD. 1979. Absence of lipid oxidation during accelerated aging of soybean seeds. *Plant Physiol.* 63:726–29.
PRIESTLEY, D. A., and A. C. LEOPOLD. 1983. Lipid changes during natural aging of soybean seeds. *Physiol. Plant.* 59:467–70.
PRIESTLEY, D. A., M. B. MCBRIDE, and A. C. LEOPOLD. 1980. Tocopherol and organic free radical levels in soybean seeds during natural and accelerated aging. *Plant Physiol.* 66:715–19.
PRIESTLEY, D. A., and M. A. POSTHUMUS. 1982. Extreme longevity of lotus seeds from Pulantien. *Nature* (Lond.) 299:148–49.
PRIESTLEY, D. A., B. G. WERNER, and A. C. LEOPOLD. 1985a. The susceptibility of soybean lipids to artificially enhanced atmospheric oxidation. *J. Exp. Bot.* 36:1653–59.
PRIESTLEY, D. A., B. G. WERNER, A. C. LEOPOLD, and M. B. MCBRIDE. 1985b. Organic free radical levels in seeds and pollen: the effects of hydration and aging. *Physiol. Plant.* 64:88–94.
PRITCHARD, E. W. 1933. How long do seeds retain their germinating power? *S. Aust. J. Agric.* 36:645–46.
PROTOPOPOVA, E. M., A. M. BAGROVA, and G. A. GRIGOR'EVA. 1979. Issledovanie tsitogeneticheskogo effekta kisloroda v dlitel'no khranivshikhsya semenakh *Crepis capillaris*. [A study of the cytogenetic effect of oxygen on *Crepis capillaris* seeds subjected to prolonged storage.] *Genetika* 15:254–60.
PROTOPOPOVA, E. M., A. M. BAGROVA, and N. I. SHAPIRO. 1974. Vozmozhnost' vyyav-

leniya potentsial'nykh narushenii struktury khromosom v semenakh *Crepis capillaris* i *Crepis tectorum*. *Genetika* 10 (2):48–52. [The possibility of the detection of potential damages to the chromosome structure in seeds of *Crepis capillaris* and *Crepis tectorum*. *Sov. Genet.* 10:174–77.]

PROTOPOPOVA, E. M., V. V. SHEVCHENKO, and G. A. GRIGOR'EVA. 1970. Vliyanie vozrasta semyan na kharakter tsitogeneticheskogo deistviya mutagenov s zaderzhannym effektom. *Genetika* 6 (1):29–36. [Influence of the age of the seeds on the nature of the cytogenetic action of mutagens with a delayed effect. *Sov. Genet.* 6:19–23.]

PRYOR, W. A. 1976. The role of free radical reactions in biological systems. In W. A. Pryor, ed., *Free Radicals in Biology*, 1:1–49. New York: Academic Press.

PUGSLEY, H. W. 1928. The longevity of seeds. *J. Bot.* (Lond.) 66:203–4.

PUKACKA, S. 1983. Phospholipid changes and loss of viability in Norway Maple (*Acer platanoides* L.) seeds. *Z. Pflanzenphysiol.* 112:199–205.

PULS, E. E., and V. N. LAMBETH. 1974. Chemical stimulation of germination rate in aged tomato seeds. *J. Am. Soc. Hortic. Sci.* 99:9–12.

PUNJABI, B., and R. N. BASU. 1982. Control of age- and irradiation-induced seed deterioration in lettuce (*Lactuca sativa* L.) by hydration–dehydration treatments. *Proc. Indian Natl. Sci. Acad.*, pt. B, *Biol. Sci.* 48:242–50.

PUNJABI, B., A. K. MANDAL, and R. N. BASU. 1982. Maintenance of vigour, viability, and productivity of stored barley seed. *Seed Res.* (New Delhi) 10:69–71.

PURKAR, J. K., and S. K. BANERJEE. 1979. Genetic changes in relation to seed ageing under tropical storage conditions. *Seed Res.* (New Delhi) 7:190–96.

PURKAR, J. K., and S. K. BANERJEE. 1983. Seed viability as an index of cytogenetical damage. *Seed Res.* (New Delhi) 11:112–24.

PURKAR, J. K., R. B. MEHRA, and S. K. BANERJEE. 1979. Effect of seed ageing on associations between the quantitative characters. *Seed Res.* (New Delhi) 7:197–213.

PURKAR, J. K., R. B. MEHRA, and S. K. BANERJEE. 1981. Quantitative genetical changes in wheat induced through seed ageing. *Seed Res.* (New Delhi) 9:172–87.

PURKAR, J. K., R. B. MEHRA, and S. K. BANERJEE. 1982. Quantitative genetical changes in peas induced through seed ageing. *Seed Res.* (New Delhi) 10:32–40.

PURKAR, J. K., and H. C. S. NEGI. 1982. Initiation of seed deterioration and its localization in peas and wheat. *Seed Res.* (New Delhi) 10:196–200.

QUAGLIA, G., R. CAVAIOLI, P. CATANI, J. SHEJBAL, and M. LOMBARDI. 1980. Preservation of chemical parameters in cereal grains stored in nitrogen. In J. Shejbal, ed., *Controlled Atmosphere Storage of Grains*, pp. 319–33. New York: Elsevier.

QUARLES, R. H., and R. M. C. DAWSON. 1969. The distribution of phospholipase D in developing and mature plants. *Biochem J.* 112:787–94.

QUICK, C. R. 1961. How long can a seed remain alive? *U.S. Dep. Agric. Yearb. Agric.*, pp. 94–99.

RABINOWITCH, H. D., and I. FRIDOVICH. 1983. Superoxide radicals, superoxide dismutases and oxygen toxicity in plants. *Photochem. Photobiol.* 37:679–90.

RADHAKRISHNAN, T. C., V. P. SUKUMARA DEV, P. A. VAR KEY, and R. GOPDLAKRISHNAN. 1976. Storage life of rice seeds of some high yielding varieties in a high humid region. *Agric. Res. J. Kerala* 14:83–84.

RAJBHANDARY, K. L. 1971/76. Yield potentialities of aged wheat seed. *Nepal. J. Agric.* 6/11:165–69.

RAMPTON, H. H., and T. M. CHING. 1966. Longevity and dormancy in seeds of several cool-season grasses and legumes buried in soil. *Agron J.* 58:220–22.

RAMPTON, H. H., and T. M. CHING. 1970. Persistence of crop seeds in soil. *Agron J.* 62:272–77.

RAMSTAD, P. E., and W. F. GEDDES. 1942. The respiration and storage behavior of soybeans. *Minn. Agric. Exp. Stn. Tech. Bull.* 156:1–54.
RANDOLPH, M. L., J. A. HEDDLE, and J. L. HOSSZU. 1968. Dependence of ESR signals in seeds on moisture content. *Radiat. Bot.* 8:339–43.
RAO, A. P., and A. A. FLEMING. 1979. Cytoplasmic-genotype influences on seed viability in a maize inbred. *Can. J. Plant Sci.* 59:241–42.
RAO, A. S., and D. S. WAGLE. 1981. Beta-amylase activity in artificially aged soybean seeds. *Biol. Plant.* (Prague) 23:24–27.
RAO, A. S., and D. S. WAGLE. 1983. Effect of accelerated aging of soyabean seeds on isocitrate lyase and malate synthetase activities. *Seed Res.* (New Delhi) 11:82–86.
RATKOVIĆ, S., G. BAČIČ, Č. RADENOVIĆ, and Ž. VUČINIĆ. 1982. Water in plants: a review of some recent NMR studies concerning the state and transport of water in leaf, root and seed. *Stud. Biophys.* 91:9–18.
RAVALO, E. J., E. D. RODDA, F. D. TENNE, and J. B. SINCLAIR. 1980. Soybean seed storage under controlled and ambient conditions in tropical environments. In F. T. Corbin, ed., *World Soybean Research Conference II, Raleigh, 1979: Proceedings*, pp. 519–32. Boulder: Westview; New York: Granada.
RAY, J. [RAIUS, J.] 1670. *Catalogus plantarum angliae et insularum adjacentium.* London: Martyn.
RAY, M. B., and K. GUPTA. 1979. Effect of storage of rice seeds on solute leakage and nucleic acid synthesis. *Indian J. Agric. Sci.* 49:715–19.
RAY, S. R. 1982. Maintenance of vigour, viability, and yield potential of stored wheat seed. *Seed Res.* (New Delhi) 10:139–42.
RAYMOND, P., A. HOURMANT, J. -M. LEBLANC, A. AL-ANI, and A. PRADET. 1982. Mécanismes régénérateurs d'ATP au cours des premières phases de la germination. *Bull. Soc. Bot. Fr., Actual. Bot.* 129 (2):91–97.
REES, B. 1911. Longevity of seeds and structure and nature of seed coat. *Proc. R. Soc. Victoria* 23:393–414.
REIMERS, F. E. 1979. Dolgozhitel'stvo i pokoi semyan. [Prolonged viability and quiescence of seeds.] In F. E. Reimers and I.E. Illi, eds., *Biokhimicheskie i fiziologicheskie issledovaniya semyan,* pp. 7–17. Irkutsk: Akademiya Nauk SSSR Sibirskoe Otdelenie.
REISS, U., and A. L. TAPPEL. 1973. Fluorescent product formation and changes in structure of DNA reacted with peroxidizing arachidonic acid. *Lipids.* 8:199–202.
REISS, W., and A. STUBEL. 1887. *The Necropolis of Ancon in Peru.* Vol. 3. Trans. A. H. Keane. Berlin: Asher.
RENFREW, J. M. 1973. *Palaeoethnobotany.* London: Methuen.
RENFREW, J. M., M. MONK, and P. MURPHY. [1976.] *First Aid for Seeds.* Hertford: Rescue.
ŘETOVSKÝ, R. 1934. Azotate d'uranyle et l'énergie de la germination de la sémence de la vieille orge. *Bull. Int. Ceska Akad. Ved Umeni (Cl. Math. Nat. Med.)* 35:71–74.
RIGGIO-BEVILACQUA, L. 1979. On alcohol dehydrogenase activity in aged pea seeds. *Boll. Soc. Ital. Biol. Sper.* 55:2401–5.
RINCKER, C. M. 1981. Long-term subfreezing storage of forage crop seeds. *Crop Sci.* 21:424–27.
RINCKER, C. M. 1983. Germination of forage crop seeds after 20 years of subfreezing storage. *Crop Sci.* 23:229–31.
RINCKER, C. M., and J. D. MAGUIRE. 1979. Effect of seed storage on germination and forage production of seven grass cultivars. *Crop Sci.* 19:857–60.
ROBERTS, B. E., and D. J. OSBORNE. 1973a. Protein synthesis and loss of viability in

rye embryos: the lability of transferase enzymes during senescence. *Biochem J.* 135:405–10.

ROBERTS, B. E., and D. J. OSBORNE. 1973b. Protein synthesis and viability in rye grains. In W. Heydecker, ed., *Seed Ecology*, pp. 99–114. State College, Pa.: Pennsylvania University Press.

ROBERTS, B. E., P. I. PAYNE, and D. J. OSBORNE. 1973. Protein synthesis and the viability of rye grains. Loss of activity of protein-synthesizing systems in vitro associated with a loss of viability. *Biochem J.* 131:275–86.

ROBERTS, E. H. 1960. The viability of cereal seed in relation to temperature and moisture. *Ann. Bot.* (Lond.) 24:12–31.

ROBERTS, E. H. 1961. The viability of rice seed in relation to temperature, moisture content, and gaseous environment. *Ann. Bot.* (Lond.) 25:381–90.

ROBERTS, E. H. 1972a. Storage environment and the control of viability. In E. H. Roberts, ed., *Viability of Seeds*, pp. 14–58. Syracuse, N.Y.: Syracuse University Press.

ROBERTS, E. H. 1972b. Cytological, genetical, and metabolic changes associated with loss of viability. In E. H. Roberts, ed., *Viability of Seeds*, pp. 253–306. Syracuse, N.Y.: Syracuse University Press.

ROBERTS, E. H. 1972c. Loss of viability and crop yields. In E. H. Roberts, ed., *Viability of Seeds*, pp. 307–20. Syracuse, N.Y.: Syracuse University Press.

ROBERTS, E. H. 1972d. Dormancy: a factor affecting seed survival in the soil. In E. H. Roberts, ed., *Viability of Seeds*, pp. 321–59. Syracuse, N.Y.: Syracuse University Press.

ROBERTS, E. H. 1973a. Loss of seed viability: chromosomal and genetical aspects. *Seed Sci. Technol.* 1:515–27.

ROBERTS, E. H. 1973b. Loss of viability: ultrastructural and physiological aspects. *Seed Sci. Technol.* 1:529–45.

ROBERTS, E. H. 1978. Mutations during seed storage. *Acta Hortic.* (The Hague) 83:279–82.

ROBERTS, E. H. 1979. Seed deterioration and loss of viability. *Adv. Res. Technol. Seeds* 4:25–42.

ROBERTS, E. H. 1981. Physiology of ageing and its application to drying and storage. *Seed Sci. Technol.* 9:359–72.

ROBERTS, E. H. 1983. Loss of seed viability during storage. *Adv. Res. Technol. Seeds.* 8:9–34.

ROBERTS, E. H., and F. H. ABDALLA. 1968. The influence of temperature, moisture, and oxygen on period of seed viability in barley, broad beans, and peas. *Ann. Bot.* (Lond.) 32:97–117.

ROBERTS, E. H., F. H. ABDALLA, and R. J. OWEN. 1967. Nuclear damage and the ageing of seeds, with a model for seed survival curves. *Symp. Soc. Exp. Biol.* 21:65–99.

ROBERTS, E. H., and R. H. ELLIS. 1977. Prediction of seed longevity at sub-zero termperatures and genetic resources conservation. *Nature* (Lond.) 268:431–33.

ROBERTS, E. H., and R. H. ELLIS. 1982. Physiological, ultrastructural, and metabolic aspects of seed viability. In A. A. Khan, ed., *The Physiology and Biochemistry of Seed Development, Dormancy, and Germination*, pp. 465–85. New York: Elsevier.

ROBERTS, E. H., and R. H. ELLIS. 1984. The implications of the deterioration of orthodox seeds during storage for genetic resources conservation. In J. H. W. Holden and J. T. Williams, eds., *Crop Genetic Resources: Conservation and Evaluation*, pp. 18–37. London and Boston: George Allen & Unwin.

ROBERTS, E. H., M. W. KING, and R. H. ELLIS. 1984. Recalcitrant seeds: their recognition and storage. In J. H. W. Holden and J. T. Williams, eds., *Crop Genetic Resources: Conservation and Evaluation*, pp. 38–52. London and Boston: George Allen & Unwin.

ROBERTS, E. H., and D. L. ROBERTS. 1972. Viability nomographs. In E. H. Roberts, ed., *Viability of Seeds*, pp. 417–23. Syracuse, N.Y.: Syracuse University Press.

ROBERTS, H. A. 1981. Seed banks in soils. *Adv. Appl. Biol.* 6:1–55.

ROBERTS, H. A., and P. M. FEAST. 1973a. Emergence and longevity of seeds of annual weeds in cultivated and undisturbed soil. *J. Appl. Ecol.* 10:133–43.

ROBERTS, H. A., and P. M. FEAST. 1973b. Changes in the numbers of viable seeds in soil under different regimes. *Weed Res.* 13:298–303.

ROBERTSON, D. W., A. M. LUTE, and H. KROEGER. 1943. Germination of 20-year-old wheat, oats, barley, corn, rye, sorghum, and soybeans. *Agron. J.* 35:786–95.

ROBERTSON, F. R., and J. G. CAMPBELL. 1933. Some observations on the increase of free fatty acid in cottonseed. *Oil Soap* 10:146–47.

ROBOCKER, W. C., M. C. WILLIAMS, R. A. EVANS, and P. J. TORELL. 1969. Effects of age, burial, and region on germination and viability of halogeton seed. *Weed Sci.* 17:63–65.

ROCKLAND, L. B. 1969. Water activity and storage stability. *Food Technol.* 23:1241–51.

RODRIGO, P. A. 1939. Study on the vitality of old and new seeds of mungo (*Phaseolus aureus* Roxb). *Phillipp. J. Agric.* 10:285–91.

RODRIGO, P. A., and A. L. TECSON. 1940. Storing some vegetable seeds. *Philipp. J Argric.* 11:383–95.

ROHMEDER, E. 1972. *Das Saatgut in der Forstwirtschaft*. Hamburg: Parey.

ROLSTON, M. P. 1978. Water impermeable dormany. *Bot. Rev.* 44:365–96.

ROMANOV, V. P. 1980. Realizatsiya potentsial'nykh izmenenii v strukturnye mutatsii khromosom pri estestvennom mutageneze v kletkakh *Crepis capillaris*. [Conversion of potential changes into structural chromosome mutations during spontaneous mutagenesis in *Crepis capillaris*.] *Genetika* 16:867–73.

Roos, E. E. 1980. Physiological, biochemical, and genetic changes in seed quality during storage. *HortScience* 15:781–84.

Roos, E. E. 1982. Induced genetic changes in seed germplasm during storage. In A. A. Khan, ed., *The Physiology and Biochemistry of Seed Development, Dormancy, and Germination*, pp. 409–34. New York: Elsevier.

Roos, E. E. 1984. Genetic shifts in mixed bean populations. I. Storage effects. *Crop Sci.* 24:240–44.

Roos, E. E., and C. M. RINCKER. 1982. Genetic stability in 'Pennlate' orchardgrass seed following artificial aging. *Crop Sci.* 22:611–13.

Roos, E. E., S. SOWA, and G. W. BURTON. 1978. Accelerated aging studies of normal and segregating chlorophyll deficient isolines of pearl millet. *Crop Sci.* 18:231–33.

ROSEN, R. 1978. Feedforwards and global system failure: a general mechanism for senescence. *J. Theor. Biol.* 74:579–90.

ROUBAL, W. T. 1970. Trapped radicals in dry lipid-protein systems undergoing oxidation. *J. Am. Oil Chem. Soc.* 47:141–44.

ROWELL, T. A., S. M. WALTERS, and H. J. HARVEY. 1982. The rediscovery of the fen violet, *Viola persicifolia* Schreber, at Wicken Fen, Cambridgeshire. *Watsonia* 14:183–84.

ROWNTREE, L. 1930. Longevity of seeds in the desert. *Horticulture* 8:270.

RUDENKO, S. I. 1953. *Kultura naseleniya Gornogo Altaya v skifskoe vremya*. Moscow: Akademiya Nauk SSSR. [*Frozen Tombs of Siberia*. Berkeley: University of California Press, 1970.]

RUDRAPAL, A. B., and R. N. BASU. 1979. Physiology of hydration-dehydration treatment in the maintenance of seed viability in wheat *Triticum aestivum* L. *Indian J. Exp. Biol.* 17:768–71.

RUDRAPAL, A. B., and R. N. BASU. 1980. Iodine treatment of mungbean seeds for the maintenance of vigour and viability. *Curr. Sci.* (Bangalore) 49:319–20.

RUDRAPAL, A. B., and R. N. BASU. 1981. Use of chlorine and bromine in controlling mustard seed deterioration. *Seed Res.* (New Delhi) 9:188–91.

RUDRAPAL, A. B., and R. N. BASU. 1982. Lipid peroxidation and membrane damage in deteriorating wheat and mustard seeds. *Indian J. Exp. Biol.* 20:465–70.

RUGE, U. 1947. Untersuchungen über keimungsfördernde Wirkstoffe. *Planta* (Berl.) 35:297–318.

RUGE, U. 1952. Über die Steigerung der Keimfähigkeit alten Saatgutes mit Hilfe von Äthylenchlorohydrin. *Angew. Bot.* 26:162–65.

RUPLEY, J. A., E. GRATTON, and G. CARERI. 1983. Water and globular proteins. *Trends Biochem. Sci.* 8:18–22.

RUSSOM, Z. 1983. Effect of storing period and type of storage on seed germination of American, Indonesian soybean lines and their crosses. *Oleagineux* 38:439–43.

RYYNANEN, M. 1980. X-ray radiography of ageing Scots pine seeds. *Silva Fenn.* 14:106–10.

SAHA, R., and R. N. BASU. 1981. Maintenance of soybean seed viability by hydration–dehydration treatments. *Indian Agric.* 25:275–78.

SAHA, R., and R. N. BASU. 1982. Preconditioning soybean seed for viability maintenance by soaking–drying treatment. *Seed Res.* (New Delhi) 10:183–87.

SAHA, R., and R. N. BASU. 1984. Invigoration of soybean seed for the alleviation of soaking injury and ageing damage on germinability. *Seed Sci. Technol.* 12:613–22.

SAHADEVAN, P. C., and M. B. V. NARASINGA RAO. 1947. Note on the deterioration in germination capacity of a paddy strain in Malabar and South Kanara. *Curr. Sci.* (Bangalore) 16:319–20.

ST. ANGELO, A. J., J. C. KUCK, and R. L. ORY. 1979. Role of lipoxygenase and lipid oxidation in quality of oil seeds. *J. Agric. Food Chem.* 27:229–34.

ST. ANGELO, A.J., and R. L. ORY. 1983. Lipid degradation during seed deterioration. *Phytopathology* 73:315–17.

SAIO, K., I. NIKKUNI, Y. ANDO, M. OTSURU, Y. TERAUCHI, and M. KITO. 1980. Soybean quality changes during model storage studies. *Cereal Chem.* 57:77–82.

SALISBURY, E. 1964. *Weeds and Aliens* 2d ed. London: Collins.

SALMONSON, B. J. 1977. Increase in radiosensitivity with increase in age of *Populus tremuloides* seed. In J. Zavitkovski, ed., *The Enterprise, Wisconsin, Radiation Forest,* pp. 185–88. Springfield, V.: Energy Research and Development Administration.

SALTER, J. 1857. On the vitality of seeds after prolonged submersion in the sea. *J. Linn. Soc. Lond., Bot.* 1:140–42.

SALZMANN, R. 1954. Untersuchungen über die Lebensdauer von Unkrautsamen im Boden. *Mitt. Schweiz. Landwirtsch.* 2:170–76.

SAMEK, J. 1888. Ueber die Dauer der Keimkraft der landwirthschaftlichen Samereien. *Tirol. Landwirtsch. Blatt.* 7 (1):1–3.

SAMIMY, C., and A. G. TAYLOR. 1983. Influence of seed quality on ethylene production of germinating snapbeans. *J. Am. Soc. Hortic. Sci.* 108:767–69.

SAMPIETRO, G. 1931. Per prolungare la longevità nei semi di riso. *Risicoltura* 21:1–5.

SAMSHERY, R., and D. BANERJI. 1979. Some biochemical changes accompanying loss of seed viability. *Plant Biochem. J.* 6:54–63.

SAMUILOV, F. D., V. I. NIKIFOROVA, and E. A. NIKIFOROV. 1976a. Issledovanie yadernoi magnitnoi relaksatsii protonov v rastitel'noi tkani vo vrashchayushcheisya i laboratornoi sistemakh koordinat. *Dokl. Adad. Nauk. SSSR* 229:994–97. [Investigation of proton nuclear magnetic relaxation in plant tissue in the rotating and laboratory frames. *Dokl. Adad. Nauk. SSSR, Biophys. Sect.* 229:112–15.]

SAMUILOV, F. D., V. I. NIKIFOROVA, and E. A. NIKIFOROV. 1976b. Ispol'zovanie metoda yadernogo magnitnogo resonansa dlya izucheniya sostoyaniya vody v proprastayushchikh semenakh. *Fiziol. Rast.* (Mosc.) 23:567–72. [Use of nuclear magnetic resonance to study the state of water in germinating seeds. *Sov. Plant Physiol.* 23:480–84.]

SÁNCHEZ, R. A., and L. C. DE MIGUEL. 1983. Ageing of *Datura ferox* seed embryos during dry storage and its reversal during imbibition. *Z. Pflanzenphysiol.* 110:319–29.

SANKARA RAO, D. S., and K. T. ACHAYA. 1969. Occurrence and possible protective function of carbon dioxide in oil seeds. *J. Sci. Food Agric.* 20:531–34.

SASAKI, S. 1977. The physiology, storage, and germination of timber seeds. In H. F. Chin, I. C. Enoch, and R. M. Raja Harun, eds., *Seed Technology in the Tropics*, pp. 111–15. Serdang: Universiti Pertanian Malaysia.

SAVINO, G., A. DELL'AQUILA, and P. DE LEO. 1976. Effetto dell'idratazione e della disidratazione sulla sintesi proteica e respirazione *in vivo* in embrioni e semi di grano. *Boll. Soc. Ital. Biol. Sper.* 52:1187–92.

SAVINO, G., P. M. HAIGH, and P. DE LEO. 1979. Effects of presoaking upon seed vigour and viability during storage. *Seed Sci. Technol.*. 7:57–64.

SAX, K., and H. J. Sax. 1961. The effect of age of seed on the frequency of spontaneous and gamma ray induced chromosome aberrations. *Rad. Bot.* 1:80–83.

SAX, K. and H. J. SAX. 1962. Effects of X-rays on the ageing of seeds. *Nature* (Lond.) 194:459–60.

SAXENA, O. P., and D. C. MAHESHWARI. 1980. Biochemical aspects of viability in soybean. *Acta Bot. Indica* 8:229–34.

SAYRE, J. D. 1940. Storage tests with seed corn. *Ohio J. Sci.* 40:181–85.

SCARASCIA, G. T. 1957. Effetti morfologici e mutativi dell'invecchiamento del seme in *Nicotiana tabacum* L. *Tabacco* (Rome) 61:270–90.

SCARASCIA, G. T., and C. DI GUGLIELMO. 1953. Mutazioni cromosomiche spontanee in *Soja hispida*. *Ann. Sper. Agrar.* 9:1269–73.

SCARASCIA, G. T., and M. E. SCARASCIA-VENEZIAN. 1954. Effetti citologici e caratteristiche biochimiche degli estratti acquosi di una serie di semi di *Soja hispida* M. di diversa età. *Caryologia* 6:247–70.

SCARASCIA, G. T., and M. E. VENEZIAN. 1953. Mutabilità cromosomica spontanea e stato enzimatico in semi di *Nicotiana rustica* di diverse annate. *Tabacco* (Rome) 57:272–98.

SCHAEFER, J., and E. O. STEJSKAL. 1975. Carbon-13 nuclear magnetic resonance analysis of intact oilseeds. *J. Am. Oil Chem. Soc.* 52:366–69.

SCHAFER, D. E., and D. O. CHILCOTE. 1969. Factors influencing persistence and depletion in buried seed populations. I. A model for analysis of parameters of buried seed persistence and depletion. *Crop Sci.* 9:417–19.

SCHAICH, K. M. 1980. Free radical initiation in proteins and amino acids by ionizing and ultraviolet radiations and lipid oxidation. Pt. III. Free radical transfer from oxidizing lipids. *CRC Crit. Rev. Food Sci. Nut.* 13:189–244.

SCHIMPFF, G., H. MULLER, and H. FOLLMAN. 1978. Age-dependent DNA labeling and deoxyribonucleotide synthesis in wheat seeds. *Biochim. Biophys. Acta* 520:70–81.

SCHJELDERUP-EBBE, T. 1935. Über die Lebensfähigkeit alter Samen. *Skr. Nor. Vidensk. Adad. Oslo, Mat. Natvidensk. Kl.* 13:1–178.

SCHKWARNIKOV, P. K., and M. S. NAWASCHIN. [SHKVARNIKOV, P. K., AND M. S.

Navashin.] 1934. Über die Beschleunigung des Mutationsvorganges in ruhenden Samen unter dem Einfluss von Temperaturerhöhung. *Planta* (Berl.) 22:720–36.

SCHLIPF, J. A. 1844. *Manuel populaire d'agriculture*. Trans. N. Nicklès. Paris: Roret.

SCHMID, E., L. HILTNER, L. RICHTER, and F. NOBBE. 1888. Über den Einfluss der Keimungs-energie des Samen auf die Entwickelung der Pflanze. *Landwirtsch. Vers Stn.* 35:137–48.

SCHNEIDER, A. S. 1981. Hydration of biological membranes. In L. B. Rockland and G. F. Stewart, eds., *Water Activity: Influences on Food Quality*, pp. 377–405. New York: Academic Press.

SCHNEIDER, J., and K. WIĄZECKA. 1977. Wartość siewna w czasie długotrwałego przechowywania nasion rzepaku ozimego zebranych w różnych terminach. [Influence of the degree of ripeness of winter rape seeds on their preservation of seeding value in long storage.] *Hodowla Rosl. Aklim. Nasienn.* 21:343–64.

SCHNEIDER, M. J. T., and A. S. SCHNEIDER. 1972. Water in biological membranes: adsorption isotherms and circular dichroism as a function of hydration. *J. Membr. Biol.* 9:127–40.

SCHOETTLE, A. W., and A. C. LEOPOLD. 1984. Solute leakage from artifically aged soybean seeds after imbibition. *Crop Sci.* 24:835–38.

SCHÖNBORN, A. VON. 1964. *Die Aufbewahrung des Saatgutes der Waldbäume*. Munich: Bayerischer Landwirtsverlag.

SCHUCH, L. O. B., and S. S. LIN. 1982. Efeito do envelhecimento rápido sobre o desempenho de sementes e de plantas de trigo. *Pesqui. Agropecu. Bras.* 17:1163–70.

SCHWEMMLE, J. 1940. Keimversuche mit alten Samen. *Z. Bot.* 36:225–61.

SCHWEMMLE, J. 1952. Samenalter und Entwicklung. *Z. Naturforsch., Teil B Chem. Biochem. Biophys. Biol.* 7:255.

SCOTT, G. E. 1981. Improvement for accelerated aging response of seed in maize populations. *Crop Sci.* 21:41–43.

SEDENKO, D. M. 1975. Fosfornye soedineniya i fitaznaya aktivnost' u semyan kukuruzy, khranivshikhsya s raznoi iskhodnoi vlazhnost'yu. *Fiziol. Rast.* (Mosc.) 22:236–40. [Phosphorus compounds and phytase activity in corn seeds stored at different values of initial moisture content. *Sov. Plant Physiol.* 22:193–96.]

SEDENKO, D. M., and K. E. OVCHAROV. 1969. Fosfornyi obmen i dykhanie semyan kukuruzy raznoi zhiznesposobnosti. *Fiziol. Rast.* (Mosc.) 16:795–99. [Phosphorus metabolism and respiration of corn seeds of various vitality. *Sov. Plant Physiol.* 16:661–64.]

SEEWALDT, V., D. A. PRIESTLEY, A. C. LEOPOLD, G. W. FEIGENSON, and F. GOODSAID-ZALDUONDO. 1981. Membrane organization in soybean seeds during hydration. *Planta* (Berl.) 152:19–23.

SEMERDZHYAN, S. P., I. G. IOR-AREVYAN, and D. O. OGANESYAN. 1973. O roli tiolov v opredelenii radiochuvstvitel'nosti semyan pshenitsy. *Radiobiologiya* 13:303–6. [The role of thiols in the determination of the radiosensitivity of wheat seeds. *Radiobiology* 13 (2):183–88.]

SEMERDZHYAN, S. P., D. O. OGANESYAN, and N. V. SIMONYAN. 1969. Radiobiologicheskii effekt u semyan pshenitsy v zavisimosti ot ikh vozrasta. [The radiobiological effect on wheat seeds in relation to their age.] *Biol. Zh. Arm.* 22 (9):47–53.

SEN, S. 1977. Cellular lesions leading to non-viability of seeds. *Seed Res.* (New Delhi) 5:79–92.

SEN, S., and D. J. OSBORNE. 1977. Decline in ribonucleic acid and protein synthesis with loss of viability during the early hours of imbibition of rye (*Secale cereale* L.) embryos. *Biochem. J.* 166:33–38.

Sevov, A., K. Khristov, and P. Khristova. 1973. Biologiya i tsitogenetichni izmeneniya v stari semena ot tsarevitsa, s''khranyavani pri obiknoveni usloviya. [Biological and cytogenetic changes in old maize seeds stored under normal conditions.] *Gen. Sel.* 6:107–15.

Shaidaee, G., B. E. Dahl, and R. M. Hansen. 1969. Germination and emergence of different age seeds of six grasses. *J. Range Manage.* 22:240–43.

Sharma, K. D. 1977. Biochemical changes in stored oil seeds. *Indian J. Agric. Res.* 11:137–41.

Sharrock, R. 1672. *The History of the Propagation and Improvement of Vegetables by the Occurrence of Art and Nature.* 2d ed. Oxford: Davis.

Shaw, J. M., and T. E. Thompson. 1982. Effect of phospholipid oxidation products on transbilayer movement of phospholipids in single lamellar vesicles. *Biochemistry* (Wash., D.C.) 21:920–27.

Shen-Miller, J., J. W. Schopf, and R. Berger. 1983. Germination of a ca. 700-year-old lotus seed from China: evidence of exceptional longevity of seed viability. *Am. J. Bot.* 70 (5, pt. 2):78.

Shenstone, J. C. 1923. The vitality and distribution of seeds. *J. Bot.* (Lond.) 61:297–305.

Sherf, A. F. 1953. Correlation of germination data of corn and soybean seed lots under laboratory, greenhouse, and field conditions. *Proc. Assoc. Off. Seed Anal.* 43:127–30.

Sherman, H. 1921. Respiration of dormant seeds. *Bot. Gaz.* 72:1–30.

Shewry, P. R., M. A. Kirkman, S. R. Burgess, G. N. Festenstein, and B. J. Miflin. 1982. A comparison of the protein and amino acid composition of old and recent barley grain. *New Phytol.* 90:455–66.

Shi, S.-X. 1981. Xiaomai zhongzi sheng huoli yu xibao se su yanghua mei tonggong mei, guo yanghua wu mei tonggong meide guanxi. [The relationship between the vigor of wheat seeds and the activity of cytochrome oxidase and peroxidase isozymes.] *Zhongguo Nongye Kexue [Chung Kuo Nung Yeh K'o Hsueh: Chin. Agric. Sci.]* No. 5, pp. 52–56.

Shi, S.-X., C.-Y. Jiang, and Y. Tian. 1982. Shuidao zhongzi sheng huolide yanjiu. [A study on the viability of rice seed.] *Zhiwu Xuebao [Chih Wu Hsueh Pao: Acta Bot. Sin.]* 24:285–88.

Shih S.-H. [Shi, S.-H.] 1959. *On 'Fan Sheng-Chih Shu,' an Agriculturistic Book of China written by Fan Sheng-Chih in the First Century B.C.* Peking: Science Press.

Shkvarnikov, P. K. 1939. Mutatsionnaya izmenchivost' v semenakh i ee znachenie dlya semenovodstva i selektsii. [Mutational variability in seeds and its significance for seed production and breeding.] *Izv. Akad. Nauk. SSSR Otd. Biol. Nauk,* pp. 1009–54.

Shmidt, P. Yu. 1955. *Anabioz.* 4th ed. Moscow: Akademiya Nauk SSSR.

Short, G.E., and Lacy, M. L. 1976. Carbohydrate exudation from pea seeds: effect of cultivar, seed age, seed color, and temperature. *Phytopathology* 66:182–87.

Shutova, E. A., Z.I. Ballod, A. I. Aprod, and V. L. Kretovich. 1973. Izmenenie belkov i izofermentov peroksidazy pri potere vskhozhesti semyan risa. *Prikl. Biokhim. Mikrobiol.* 9:269–72. [Changes in proteins and peroxidase enzymes with loss of germinating power in rice seeds. *Appl Biochem. Microbiol.* 9:222–25.]

Shvetsova, V. A., and N. I. Sosedov. 1958. Biokhimicheskie izmeneniya pri dlitel'nom khranenii pshenitsy v germeticheskikh usloviyakh. [Biochemical changes during prolonged hermetic storage of wheat.] *Biokhim. Zerna* 4:229–40.

Sifton, H. B. 1920. Longevity of the seeds of cereals, clover, and timothy. *Am. J. Bot.* 7:243–51.

SIJBRING, P. H. 1963. Results of some experiments on the moisture relationship of seeds. *Proc. Int. Seed Test Assoc.* 28:837–43.

SIMAK, M. 1957. The x-ray contrast method for seed testing Scots pine, *Pinus silvestris*. *Medd. Statens. Skogsforskningsinst.* (Swed.) 47 (4):1–22.

SIMAK, M. 1976. Germination improvement of Scots pine seeds from circumpolar regions using polyethylene glycol. In K. Hatano, S. Asakawa, M. Katsuta, S. Sasaki, and T. Yoloyama, eds., *Proceedings of the Second International Symposium on Physiology of Seed Germination, Fuji, 1976*, pp. 145–53. Tokyo: Government Forest Experiment Station.

SIMAK, M., Å. GUSTAFSSON, and G. GRANSTRÖM. 1957. Die Röntgendiagnose in der Samenkontrolle. *Proc. Int. Seed Test. Assoc.* 22:330–43.

SIMANCIK, F. 1968. Germination of presoaked and redried seeds of European larch (*Larix decidua* Mill.) during a 3-year period of storage. *Acta Univ. Agric. Fac. Silvic.* (Brno) 37:42–57.

SIMOLA, L. K. 1974. Ultrastructural changes in the seeds of *Pinus sylvestris* L. during senescence. *Stud. For. Suec.* 119:1–22.

SIMOLA, L. K. 1976. Ultrastructure of non-viable seeds of *Picea abies*. *Z. Pflanenphysiol.* 78:245–52.

SIMON, E. W. 1974. Phospholipids and plant membrane permeability. *New Phytol.* 73:377–420.

SIMON, E. W. 1978. Membranes in dry and imbibing seeds. In J. H. Crowe and J. S. Clegg, eds., *Dry Biological Systems*, pp. 205–24. New York: Academic Press.

SIMON, E. W., and L. K. MILLS. 1983. Imbibition, leakage, and membranes. In C. Nozzolillo, P. J. Lea, and F. A. Loewus, eds., *Mobilization of Reserves in Germination*, pp. 9–27. New York: Plenum.

SIMON, E. W., and R. M. RAJA HARUN. 1972. Leakage during seed imbibition. *J. Exp. Bot.* 23:1076–85.

SIMON, U. 1958. Zur Keimfähigkeit alter Samen. *Z. Acker. Pflanzenb.* 106:108–18.

SIMPSON, D. M. 1946. The longevity of cottonseed as affected by climate and seed treatments. *Agron. J.* 38:32–45.

SIMPSON, D. M. 1953. Cottonseed storage in various gases under controlled temperature and moisture. *Tenn. Agric. Exp. Stn. Bull.* 228:1–16.

SINGH, B. B. 1976. Breeding soybean varieties for the tropics. *INTSOY Ser.* 10:11–17.

SINGH, B. N., P. B. MATHUR, and M. L. MEHTA. 1938. Determination of catalase ratio as a rapid method of seed testing. *Trop. Agric.* 15:260–61.

SINGH, J., B. A. BLACKWELL, R. W. MILLER, and J.D. BEWLEY. 1984. Membrane organization of the desiccation-tolerant moss *Tortula ruralis* in dehydrated states. *Plant Physiol.* 75:1075–79.

SINGH, J. N., and R. K. SETIA. 1974. The germination of different qualities of soybean seeds under varying storage conditions. *Bull. Grain Technol.* 12:3–10.

SIRCAR, P. K., and S. M. SIRCAR. 1971. Role of endogenous ferulic acid and sinapic acids in the viability and germination of rice (*Oryza sativa* L.) seed. *Indian J. Agric. Sci.* 41:584–90.

SIRCAR, S. M. 1970. The physiology of ageing in rice seeds. *J. Indian Bot. Soc.* 49:1–7.

SIRCAR, S. M., and M. BISWAS. 1960. Viability and germination inhibitor of the seed of rice. *Nature* (Lond.) 187:620–21.

SIRCAR, S. M., and B. DEY. 1967. Dormancy and viability of the seed of rice (*Oryza Sativa* L.). In H. Borris, ed., *Physiologie, Ökologie und Biochemie der Keimung*, pp. 969–73. Greifswald: Ernst-Moritz-Arndt Universitat.

SIVORI, E. M., and F. NAKAYAMA. 1973. Planta de "achira" (*Canna* sp.). In E. M. Cigliano, ed., *Tastil, una ciudad preincaica argentina* pp. 547–58. Buenos Aires: Cabargon.

SIVORI, E. M., F. NAKAYAMA, and E. CIGLIANO. 1968. Germination of achira seed (*Canna* sp.) approximately 550 years old. *Nature* (Lond.) 219:1269–70.

SIZOVA, L. I. 1976. Vliyanie stareniya semyan na strukturnye mutatsii khromosom, indutsirovannye gamma-luchami, u khlorofil'nykh mutantov podsolnechnika. *Genetika* 12 (7):24–30. [Effect of seed aging on structural chromosome mutations induced by gamma rays in sunflower chlorophyll mutants. *Sov. Genet.* 12:793–97.]

SKRABKA, H. 1964. Współdziałanie wolnych auksyn z kwasem askorbinowym i glutationem w procesie kiełkowania ziarn pszenicy. Cz. I. Badania ilościowe wolnych auksyn, kwasu askorbinowego i glutationu w kiełkujących ziarach pszenicy a zróżnicowanej żywotności. [Interaction of free auxins with ascorbic acid and glutathione in the process of wheat germination. Pt. I. Quantitative investigations of free auxins, ascorbic acid, and glutathione in germinating wheat grains with various viability.] *Acta Soc. Bot. Pol.* 33:689–704.

SKRABKA, H. 1965. Współdziałanie wolnych auksyn z kwasem askorbinowym i glutationem w procesie kiełkowania ziarn pszenicy. Cz. III. Wpływ kwasu 3-indolilooctowego, kwasu askorbinowego i glutationu na kiełkowanie ziarn pszenicy o różnej żywotności. [Interaction of free auxins with ascorbic acid and glutathione in the process of wheat germination. Pt. III. Effect of 3-indolylacetic acid, ascorbic acid, and glutathione on the germination of wheat seeds of varying viability.] *Acta Soc. Bot. Pol.* 34:713–18.

SKRABKA, H., and Z. SZUWALSKA. 1969. Wpływ związków chelatujących na zdolność kiełkowania nasion o obniżonej żywotności. [The influence of chelating compounds on germinating power of seeds with lowered viability.] *Zesz. Nauk. Wyzsz. Szk. Roln. Wroclawiu, Roln.* 27:17–25.

SKUJINS, J. J., and A. D. MCLAREN. 1967. Enzyme reaction rates at limited water activities. *Science* (Wash., D.C.) 158:1569–70.

SLESARAVICHYUS, A. K. [SLIESARAVIČIUS, A.] 1969a. Izmenenie chisla khromosomnykh aberratsii i nekotorykh biokhimicheskikh protsessov v semenakh raznogo vozrasta. 1. Khromosomnye aberratsii i vskhozhest' semyan viki. [Change in the number of chromosomal abberrations and some biochemical processes in seeds of different age. 1. Chromosomal abberrations and germinability of vetch seeds.] *Liet. TSR Mokslu Akad. Darb.*, Ser. C *Biol. Mokslai [Tr. Akad. Nauk. Lit. SSR*, Ser. B *Biol. Nauk.]*, No. 1, pp. 167–74.

SLESARAVICHYUS, A. K. [SLIESARAVIČIUS, A.] 1969b. Izmenenie chisla khromosomnykh aberratsii i nekotorykh biokhimicheskikh protsessov v semenakh raznogo vozrasta. 2. Izmenenie vital'nosti i chisla khromosomnykh aberratsii v khode stareniya semyan rzhi. [Change in the number of chromosomal aberrations and some biochemical processes in seeds of different age. 2. Change in vitality and number of chromosomal aberrations during aging of rye seeds.] *Liet. TSR Mokslu Akad. Darb.*, Ser. C. *Biol. Mokslai [Tr. Akad. Nauk. Lit. SSR*, Ser. B *Biol. Nauk.]*, no. 1, pp. 175–82.

SLESARAVICHYUS, A. K. [SLIESARAVIČIUS, A.] 1969c. Izmenenie chisla khromosomnykh aberratsii i nekotorykh biokhimicheskikh protsessov v semenakh raznogo vozrasta. 3. Chastota khromosomnykh aberratsii v semenakh rzhi pri raznykh usloviyakh khraneniya. [Change in the number of chromosomal aberrations and some biochemical processes in seeds of different age. 3. Frequency of chromosomal aberrations in rye seeds under different conditions of storage.] *Liet. TSR Mokslu Akad. Darb.*, Ser. C *Biol. Mokslai [Tr. Akad. Nauk. Lit. SSR*, Ser. B *Biol. Nauk.]*, no. 1, pp. 183–86.

SLESARAVICHYUS, A. K. [SLIESARAVIČIUS, A.] 1971. Izmenenie chisla khromosomnykh aberratsii i nekotorykh biokhimicheskikh protessov v semenakh raznogo vozrasta. 4. Rol' zarodysha v natural'nom mutageneze semyan rzhi. [Change in the number of chromosomal aberrations and some biochemical processes in seeds of different age. 4. Role of the embryo in natural mutagenesis of rye seed.] *Liet. TSR Mokslu Akad. Darb.*, Ser. C *Biol. Mokslai [Tr. Akad. Nauk. Lit. SSR*, Ser. B *Biol. Nauk.]*, No. 1, pp. 167–71.

ŠMERDA, V., and S. TICHÝ. 1957. Vliv prohřívání a stáři semene na výnos plodů okurek. [Effect of heating and seed age on yield of cucumbers.] *Sb. Cesk. Akad. Zemed. Ved Rostl. Vyroba* 3:223–30.

SMITH, C. A. D., and C. M. BRAY. 1982. Intracellular levels of polyadenylated RNA and loss of vigour in germinating wheat embryos. *Planta* (Berl.) 156:413–20.

SMITH, C. A. D., and C. M. BRAY. 1984. Polyadenylated RNA levels and macromolecular synthesis during loss of seed vigour. *Plant Sci. Lett.* 34:335–43.

SMITH, M. T. 1978. Cytological changes in artificially aged seeds during imbibition. *Proc. Electron Microsc. Soc. South. Afr. [Elektronmikroskopiever. Suidelike Afr. Verrigt.]* 8:105–6.

SMITH, M. T. 1979. An evaluation of some methods for the fixation of dry seed tissues. *Proc. Electron Microsc. Soc. South. Afr. [Elektronmikroskopiever. Suidelike Afr. Verrigt.]* 9:141–42.

SMITH, M. T. 1980. Studies on the fixation of membrane phospholipids. *Proc. Electron Microsc. Soc. South Afr. [Elektronmikroskopiever. Suidelike Afr. Verrigt.]* 10:81–82.

SMITH, M. T. 1983. Cotyledonary necrosis in aged lettuce seeds. *Proc. Electron Microsc. Soc. South. Afr. [Elektronmikroskopiever. Suidelike Afr. Verrigt.]* 13:129–30.

SMITH, R. D. 1984. The influence of collecting, harvesting and processing on viability of seed. In J. B. Dickie, S. Linington, and J. T. Williams, eds., *Seed Management Techniques for Genebanks*, pp. 42–82. Rome: International Board for Plant Genetic Resources.

SMITH, W. 1851. The raspberry case. *Gard. Chron.*, p. 549.

SMOLEN, J. E., and S. B. SHOHET. 1974. Permeability changes induced by peroxidation in liposomes prepared from human erythrocyte lipids. *J. Lipid Res.* 15:273–80.

SNELL, K. 1912. Über das Vorkommen von keimfähigen Unkrautsamen im Boden. *Landwirtsch. Jahrb.* 43:323–47.

SNIGIREVSKAYA, N. S. 1964. Materialy k morfologii i sistematike roda *Nelumbo* Adans. [Contributions to the morphology and taxonomy of the genus *Nelumbo* Adans.] *Tr. Bot. Inst. Akad. Nauk SSSR*, ser. 1, 13:104–72.

SNYDER, H. 1904. Composition of an ancient Egyptian wheat. *Minn. Agric. Exp. Stn. Bull.* 85:211–12.

SOBIERAJ, B., and K. KULKA. 1983. Aktywność proteaz i inhibitorów trypsyny w starzejącym się ziarnie żyta o różnyn stopniu dojrzałości. [Activity of proteases and of trypsin inhibitors in aging rye seed with different degrees of ripeness.] *Zesz. Probl. Postepow Nauk Roln.* 258:143–51.

SOJKA, E., and R. ZAIDAN. 1983. Effect of winter wheat (cv. Jana) seed ageing on its viability and vigour. *Acta Soc. Bot. Pol.* 52:53–60.

SOLDATOVA, O. P. 1976. Spontannyi mutatsionnyi protsess i vskhozhest' semyan pri dlitel'nom khranenii. [The spontaneous mutation process and the germinability of seeds undergoing long-term storage.] *Byull. Vses. Nauchno-Issled. Inst. Rastenievod. im. N. I. Vavilova* (Leningr.) 60:20–22.

SONAVNE, K. M. 1934. Longevity of crop seeds. Pt. II. *Agric. Livest. India* 4:287–92.

SORGER-DOMENIGG, H., L. S. CUENDET, C. M. CHRISTENSEN, and W. F. GEDDES. 1955.

Grain storage studies. XVII. Effect of mold growth during temporary exposure of wheat to high moisture contents upon the development of germ damage and other indices of deterioration during subsequent storage. *Cereal Chem.* 32:270–85.

SORIANO, A., R. A. SANCHEZ, and B. A. DE EILBERG. 1964. Factors and processes in the germination of *Datura ferox* L. *Can. J. Bot.* 42:1189–203.

SPECTOR, W. S. 1956. *Handbook of Biological Data.* Philadelphia: Saunders.

SPENCER, G. E. L. 1931. The use of cool storage in retaining the germinating power of some oily seeds. *Trop. Agric.* 8:333.

SPENCER, G. F., F. R. EARLE, I. A. WOLFF, and W. H. TALLENT. 1973. Oxygenation of unsaturated fatty acids in seeds during storage. *Chem. Phys. Lipids* 10:191–202.

SPIEGEL, S., and A. MARCUS. 1975. Polyribosome formation in early wheat embryo germination independent of either transcription or polyadenylation. *Nature* (Lond.) 256:228–30.

SPIRA, T. P., and L. K. WAGNER. 1983. Viability of seeds up to 211 years old extracted from adobe brick buildings of California and Northern Mexico. *Am. J. Bot.* 70:303–7.

SREERAMULU, N. 1983a. Leakage during imbibition by seeds of bambarra groundnut (*Voandzeia subterranea* [L.] Thouars) at different stages of loss of viability. *Trop. Agric.* 60:265–68.

SREERAMULU, N. 1983b. Germination and food reserves in bambarra groundnut seeds (*Voandzeia subterranea* Thouars) after different periods of storage. *J. Exp. Bot.* 34:27–33.

SREERAMULU, N. 1983c. Auxins, inhibitors, and phenolics in bambarranut seeds (*Voandzeia subterranea* Thouars) in relation to loss of viability during storage. *Ann. Bot.* (Lond.) 51:209–16.

SRIVASTAVA, A. K., and M. K. GILL. 1975. Physiology and biochemistry of seed deterioration in soyabean. Pt. III. Seedling growth and seed leachate analysis. *Ind. J. Exp. Biol.* 13:481–85.

SRIVASTAVA, A. K., and K. SAREEN. 1974. Physiology and biochemistry of deterioration in soybean seeds during storage. I. Mobilization efficiency and nitrogen metabolism. *Seed Res.* (New Delhi) 2:26–32.

STANDARD, S. A., D. PERRET, and C. M. BRAY. 1983. Nucleotide levels and loss of vigour and viability in germinating wheat embryos. *J. Exp. Bot.* 34:1047–54.

STANWOOD, P. C. 1985. Cryopreservation of seed germplasm for genetic conservation. In K. K. Kartha, ed., *Cryopreservation of Plant Cells and Organs*, pp. 199–226. Boca Raton: CRC Press.

STANWOOD, P. C., and L. N. BASS. 1981. Seed germplasm preservation using liquid nitrogen. *Seed Sci. Technol.* 9:423–37.

STANWOOD, P. C., and E. E. ROOS. 1979. Seed storage of several horticultural species in liquid nitrogen ($-196°C$). *HortScience* 14:628–30.

STARZINGER, E. K., S. H. WEST, and K. HINSON. 1982. An observation on the relationship of soybean seed coat colour to viability maintenance. *Seed Sci. Technol.* 10:301–5.

STEERE, W. C., W. C. LEVENGOOD, and J. M. BONDIE. 1981. An electronic analyser for evaluating seed germination and vigour. *Seed Sci. Technol.* 9:567–76.

STEFANOV, B. ZH., and A. B. DENCHEVA. 1984. Fiziologo-biokhimichni osnovi na zhiznenostta na semenata. II. S''d''rzhanie na sulfkhidrilni grupi v reservnite belt''tsi na semena ot tsarevitsa v''v vr''zka s promenite na tykhnata zhiznenost pod vliyanie na usloviyata na s''khranenie. [Physiological and biochemical basis of seed viability. II. Content of sulfhydryl groups in maize seed reserve proteins in relation to changes in viability induced by storage conditions.] *Fiziol. Rast.* (Sofia) 10 (3):67–72.

STERNBERG, C. VON. 1836. Über die Keimung einiger aus ägyptischen Mumien-Grabern erhaltener Waizenkörner. *Isis von Oken*, no. 3, pp. 231–233

STEVENS, E., and L. STEVENS. 1977. Glucose-6-phosphate dehydrogenase activity under conditions of water limitation: a possible model system for enzyme reactions in unimbibed resting seeds and its relevance to seed viability. *J. Exp. Bot.* 28:292–303.

STEVENS, O. A. 1935. Germination studies on aged and injured seeds. *J. Agric. Res.* (Wash., D.C.) 51:1093–106.

STEWART, R. R. C., and J. D. BEWLEY. 1980. Lipid peroxidation associated with accelerated aging of soybean axes. *Plant Physiol.* 65:245–48.

STONE, A. L. 1930/33. Results of ten years of germination tests on the same samples of vegetable seeds. *Proc. Assoc. Off. Seed Anal.* 23/26:76–80.

STONE, E. C., and G. JUHREN. 1951. The effect of fire on the germination of the seed of *Rhus ovata* Wats. *Am. J. Bot.* 38:368–72.

STONE, G. E., and R. E. SMITH. 1901. Influence of chemical solutions upon the germination of seeds. *Annu. Rep. Miss. Agric. Exp. Stn.* 12:74–83.

STUBBE, H. 1935. Samenalter und Genmutabilität bei *Antirrhinum majus* L. (Nebst einigen Beobachtungen über den Zeitpunkt des Mutierens während der Entwicklung). *Biol. Zentralbl.* 55:209–15.

STURTEVANT, P. 1951. Genealogy of a pea—back to King Tut. *Bull. Gard. Club. Amer.* 39 (1):33.

STYER, R. C., and D. J. CANTLIFFE. 1977. Effect of growth regulators on storage life of onion seed. *Proc. Fla State Hortic. Soc.* 90:415–18.

STYER, R. C., D. J. CANTLIFFE, and C. B. HALL. 1980. The relationship of ATP concentration to germination and seedling vigor of vegetable seeds stored under various conditions. *J. Am. Soc. Hortic. Sci.* 105:298–303.

SUSZKA, B., and T. TYLKOWSKI. 1982. Storage of acorns of the northern red oak *(Quercus borealis* Michx. = *Q. rubra* L.) over 1–5 winters. *Arbor. Kornickie* 26:253–306.

SUTAMIHARDJA, T. M., Y. NAGATA, and K. HAYASHI. 1969. Komugi-shushi roka no seirikagaku. VI. Hatsuga-ritsu to amylase, alcohol dehydrogenase kassei ni tsuite. [Physiology of seed aging. VI. Germinability, amylase and alcohol dehydrogenase activity.] *Gifu Daigaku Nogakubu Kenkyu Hokoku [Res. Bull. Fac. Agric. Gifu Univ.]* 28:122–30.

SUTAMIHARDJA, T. M., Y. NAGATA, and K. HAYASHI. 1970. Physiological decrepitude of seeds. VIII. On the germinability and the contents of vitamins of B group. *Gifu Daigaku Nogakubu Kenkyu Hokoku [Res. Bull. Fac. Agric. Gifu Univ.]* 29:154–62.

SUTULOV, A. N. 1965. Rol' okisleniya v protsesse stareniya i smerti semyan. [The role of oxidation in the aging and death of seeds.] *Byull. Gl. Bot. Sada* (Mosc.) 57:53–60.

SUZUKI, Y. 1969. Studies on the maturity and longevity of solanaceous plant seeds. *Jpn. J. Breed.* 19:149–58.

SWIFT, J. G., and M. S. BUTTROSE. 1972. Freeze-etch studies of protein bodies in wheat scutellum. *J. Ultrastruct. Res.* 40:378–90.

SWIFT, J. G., and M. S. BUTTROSE. 1973. Protein bodies, lipid layers, and amyloplasts in freeze-etched pea cotyledons. *Planta* (Berl.) 109:61–72.

SWIFT, J. G., and T. P. O'BRIEN. 1972. The fine struture of the wheat scutellum before germination. *Aust. J. Biol. Sci.* 25:9–22.

SZABÓ, L., and S. VIRÁNYI. 1971. Változó raktári körülmények között tárolt kultúrnövénymagok csírázási vizsgálata. [Germination data on seeds of different crops after storage in an unconditioned environment.] *Agrobotanika*, pp. 1215–20.

TÄCKHOLM, V. and G. TÄCKHOLM. 1941. *Flora of Egypt*. Vol. 1. Cairo: Fouad I University.

TAGLIASACCHI, A., and R. VOCATURO. 1977. Effect of the seed ageing of *Triticum durum* cv. Cappelli on the length of mitotic cycle as measured by ³H-thymidine incorporation in the root meristem. *Caryologia* 30:225–30.

TAKAHASHI, N., and Y. SUZUKI. 1975. Factors affecting the viability of seeds with special references to maturity and longevity. In T. Matsuo. ed., *Gene Conservation*, pp. 63–69, 107–8. Tokyo: University of Tokyo Press.

TAKAYANAGI, K. 1975. Discernment of seed viability without germination testing. In T. Matsuo, ed., *Gene Conservation*, pp. 69–75, 108–9. Tokyo: University of Tokyo Press.

TAKAYANAGI, K. 1977. Shushi no katsuryoku kentei ni kansuru kenkyū, tokuni shushi yōshitsubutsu-ho ni tsuite. [An examination of seed vitality, with special reference to a method using seed exudate.] *Nogyo Gijutsu Kenkyusho Hokoku D. Seiri Iden Sakumotsu Ippan [Bull. Natl. Inst. Agric. Sci., Ser. D Plant Physiol. Genet. Crops. Gen.]* 28:1–87.

TAKAYANAGI, K. 1980. Seed storage and viability tests. In S. Tsunoda, K. Hinata, and C. Gomez-Campo, eds., *Brassica Crops and Wild Allies*, pp. 303–21. Tokyo: Japanese Scientific Societies Press.

TAKAYANAGI, K., and J. F. HARRINGTON. 1971. Enhancement of germination rate of aged seeds by ethylene. *Plant Physiol.* 47:521–24.

TAKAYANAGI, K., and K. MURAKAMI. 1968. Rapid germinability test with exudates from seed. *Nature* (Lond.) 218:493–94.

TAKAYANAGI, K., and K. MURAKAMI. 1969. New method of seed viability test with exudates from seed. *Proc. Int. Seed Test. Assoc.* 34:243–52.

TANG, P.-S., G.-H. ZHENG, C.-M. GE, C.-J. LIU, and Y.-J. ZHANG. 1964. Biding hegan suan diu yang shu zhongzi qu qing mei huo xing xiaoshide huifu zuoyong. [Pyridine nucleotides restore the dehydrogenase activity of poplar seeds as viability is lost.] *Kexue Tongbao [K'o Hsueh T'ung Pao: Sci. Bull.]* 9:535–37.

TAO, K.-L. J. 1978. Factors causing variation in the conductivity test for soybean seeds. *J. Seed Technol.* 3 (1):10–18.

TARASENKO, N. D., G. D. BERDYSHEV, and V. YU. LOPUSHONOK. 1965. Svobodnye radikaly v obluchennykh semenakh kartofelya s razlichnym srokom khraneniya. *Biofizika* 10:893–95. [Free radicals in irradiated potato seeds stored for different periods. *Biophysics* 10:989–91.]

TAUER, C. G. 1979. Seed tree, vacuum, and temperature effects on eastern cottonwood seed viability during extended storage. *For. Sci.* 25:112–14.

TÄUFEL, K., and R. POHLOUDEK-FABINI. 1954. Beziehungen zwischen Keimfähigkeit und Gehalt an freien Fettsäuren bei Raps- und Rübsensamen. *Pharmazie* 9:511–14.

TÄUFEL, K., and R. POHLOUDEK-FABINI. 1955. Keimfähigkeit und Gehalt an Citronensäure bei gelagerten Pflanzensamen. *Biochem. Z.* 326:317–21.

TAYLOR, G. B. 1984. Effect of burial on the softening of hard seeds of subterranean clover. *Aust. J. Agric. Res.* 35:201–10.

TAYLORSON, R. B. 1970. Changes in dormancy and viability of weed seed in soils. *Weed Sci.* 18:265–69.

TEIXEIRA, J. P. F., M. T. R. DA SILVA, H. A. A. MASCARENHAS, and J. A. MAEDA. 1980. Variação da composição química de sementes de três cultivares de soja, durante o armazenamento. *Bragantia* 39:21–25.

TELLEZ, R., and F. CIFERRI. 1954. *Trigos arqueológicos de España*. Madrid: Instituto Nacional de Investigaciones Agronómicas.

TERÄSVUORI, K. 1930. Iän vaikutus apilaan siemenen kylvösiemenarvoon. [The influence of age on the seeding value of clover.] *Suom. Maataloustiet. Seuran Julk.* (Finland) 20:81–108.

TEWARI, M. N., and P. C. GUPTA. 1981. Effect of genotype, seed grade, and environment on viability and vigour of sunflower seed in storage. *Seed Res.* (New Delhi) 9:126–31.

THEOPHRASTUS. 1916. *Enquiry into Plants. [Peri phytōn historias.]* Vol. 2. Ed. A. Hort. Cambridge: Harvard University Press.

THEOPHRASTUS. 1976. *De causis plantarum. [Peri phytōn aitiōn.].* Vol. 1. Ed. B. Einarson and G. K. K. Link. Cambridge: Harvard University Press.

THISELTON-DYER, T. F. 1889. *The Folk-lore of Plants.* New York: Appleton.

THOMAS, W. 1980. Seed viability in relation to storage conditions in varieties of soybean (*Glycine max* [L] Merrill). *Mysore J. Agric. Sci.* 14:271.

THOMSON, A. 1896. Zum Verhalten alter Samen gegen Fermentlösungen. *Gartenflora* 45:344–45.

THOMSON. W. W. 1979. Ultrastructure of dry seed tissue after a non-aqueous primary fixation. *New Phytol.* 82:207–12.

THOMSON, W. W., and K. PLATT-ALOIA. 1982. Ultrastructure and membrane permeability in cowpea seeds. *Plant Cell Environ.* 5:367–73.

THORNTON, M. L. 1963. Dormancy and longevity of sorghum seed (*Sorghum vulgare* Pers.). *Proc. Assoc. Off. Seed Anal.* 53:107–11.

THOUIN, A. 1845. *Cours de culture.* Vol. 2. Ed. O. Leclerc. Paris: Bouchard-Huzard.

TILDEN, R. L., and S. H. WEST. 1985. Reversal of the effects of aging in soybean seeds. *Plant Physiol.* 77:584–86.

TILLOTSON, C. R. 1921. Storage of coniferous tree seed. *J. Agric. Res.* (Wash., D.C.) 22:479—510.

TŁUCZKIEWICZ, J. 1980. Ageing-induced changes of respiratory activity in tissues and mitochondria form germinating rye grains. *Acta Physiol. Plant.* 2:9–17.

TODA, J., S. MORITAKA, H. NAKATANI, and S. WADA. 1966. Shokuhin-chū no fosufatāze ni kansuru kenkyū. (Dai 4 pō) Komugiko-chū no 5'-ribonukureochido no anteisei. [Phosphatase in food. (Pt. 4) Stability of 5'-ribonucleotides in wheat flour.] *Eiyo to Shokuryo [Food. Nutr.]* 19:171–74.

TOIVIO-KINNUCAN, M. A., and C. STUSHNOFF. 1981. Lipid participation in intracellular freezing avoidance mechanisms of lettuce seed. *Cryobiology* 18:72–78.

TOMPSETT, P. B. 1983. The influence of gaseous environment on the storage life of *Araucaria hunsteinii* seed. *Ann. Bot.* (Lond.) 52:229–37.

TOOLE, E. H., and E. BROWN. 1946. Final results of the Duvel buried seed experiment. *J. Agric. Res.* (Wash., D.C.) 72:201–10.

TOOLE, E. H., and V. K. TOOLE. 1953a. Relation of storage conditions to germination and to abnormal seedlings of bean. *Proc. Int. Seed Test. Assoc.* 18:123–29.

TOOLE, E. H., and V. K. TOOLE. 1953b. Seed dormancy in relation to seed longevity. *Proc. Int. Seed Test. Assoc.* 18:325–28.

TOOLE, E. H., and V. K. TOOLE. 1960. Viability of stored snapbean seed as affected by threshing and processing injury. *U.S. Dept. Agric. Tech. Bull.* 1213:1–9.

TOOLE, E. H., V. K. TOOLE, and H. A. BORTHWICK. 1957. Growth and production of snap beans stored under favorable and unfavorable conditions. *Proc. Int. Seed Test. Assoc.* 22:418–23.

TOOLE, E. H., V. K. TOOLE, and E. A. GORMAN. 1948. Vegetable-seed storage as affected by temperature and relative humidity. *U.S. Dept. Agric. Tech. Bull.* 972:1–24.

TORRIE, J. H. 1958. Comparison of different generations of soybean crosses grown in bulk. *Agron. J.* 50:265–67.

TORTORA, P., G. M. HANOZET, A. GUERRITOE, M. T. VINCENZINI, and P. VANNI. 1978.

Selective denaturation of several yeast enzymes by free fatty acids. *Biochim. Biophys. Acta* 522:297–306.

TOUZARD, J. 1961. Influence de diverse conditions constantes de température et d'humidité sur la longévité des graines de quelques espèces cultivées. In J.-G. Garnaud, ed., *Proceedings of the 15th International Horticultural Congress, Nice, 1958*, 1:339–47. New York: Pergamon.

TOUZARD, J. 1975. Théorie et pratique de la conservation des semences. In R. Chaussat and Y. Le Deunff, eds., *La Germination des semences*, pp. 157–70. Paris: Bordas.

TOYODA, K. 1958. Analysis of gas contained in the fruit of Indian lotus plant. *Bot. Mag.* (Tokyo) 71:371–77.

TOYODA, K. 1960. On the ascorbic acid in the plumule of Indian lotus seed. *Bot. Mag.* (Tokyo) 73:98–103.

TOYODA, K. 1965. Glutathione in the seed of *Nelumbo nucifera*. *Bot. Mag.* (Tokyo) 78:443–51.

TOYODA, K. [TOYOTA, K.] 1974. Gyoda shutsudo no hasu no mi no shizenhatsuga to kaika. [Natural germination and flowering of lotus fruits from Gyoda.] *Shokubtsu Kenkyu Zasshi [Jpn. J. Bot.]* 49:206–13.

TOYODA, K. [TOYOTA, K.] 1980. Hasu. [Lotus.] In M. Hotta, ed., *Shokubutsu Seikatsu Shi*, pp. 149–55. Tokyo: Heibon-Sha.

TRIOLO, E., G. LORENZINI, and G. FAVILLI MANNERUCCI. 1975. Alterazioni virus-simili non ereditarie in pisello della cv. S. Cipriano, connesse con l'invecchiamento del seme. *Riv. Patol. Veg.*, ser. 4, 11:75–88.

TROOP, J. 1914. *Melon Culture*. New York: Orange Judd.

TSOUNTAS, C., and J. I. MANNATT. 1897. *The Mycenaean Age*. Boston: Houghton Mifflin.

TSVETKOV, R., K. KAMBUROVA, and M. LUCHEVA. 1975. Prouchvane v''rkhu s''khranyavane posevnite kachestva na semenata ot nyakoi eterichno-masleni rasteniya. [A study on the preservation of the sowing qualities of seeds of some essential oil-bearing plants.] *Rasteniev''d. Nauki* 12 (5):46–56.

TUPPER, M. F. 1840. Extraordinary vitality of seeds. *Times* (Lond.), October 9, p. 7.

TUPPER, M. F. 1886. *My Life as an Author*. London: Sampson Low, Marston, Searle & Rivington.

TURNER, J. H. 1933. The viability of seeds. *Kew Bull.*, pp. 257–69.

TURRILL, W. B. 1933. A study of variation in *Glaucium flavum*. *Kew Bull.* pp. 174–84.

UEMATSU, T., and K. ISHII. 1981. Rakkasei no hinshitsu ni oyobosu chozo-joken no eikyo. [Effects of storage conditions on the quality of peanuts.] *Nihon Daigaku Nojuigakubu Gakujutsu Kenkyu Hokoku [Bull. Coll. Agric. Vet. Med. Nihon Univ.]* 38:125–34.

UGARYNKO, A. 1967/68. Wpływ warunków przechowywania nasion cebuli na ich wartość produkcyjną. [Effect of storage conditions on the productivity of onion seeds.] *Biul. Warzywniczy* 9:337–47.

ULLMAN, W. 1949. Über die Keimfähigkeitsdauer (Lebensdauer) von landwirtschaftlichen und gartenbaulichen Samen. *Saatgutwirtschaft* 1:174–75, 195–96.

UNGER, F. 1846. *Grundzüge der Anatomie und Physiologie der Pflanzen*. Vienna: Gerold.

UNGER, F. 1859. Botanische Streifzüge auf dem Gebiete der Culturgeschichte. IV. Die Pflanzen des alten Ägyptens. *Sitzungsber. K. Akad. Wiss. Wien, Math. Naturwiss. Kl.* 38:69–140.

UNGER, F. 1866. Botanische Streifzüge auf dem Gebiete der Culturgeschichte. VII. Ein Ziegel der Dashurpyramide in Ägypten: nach seinem Inhalte an organischen Einschlüssen. *Sitzungsber. K. Akad. Wiss. Wien, Math. Naturwiss. Kl.* 54:33–62.

UNITED STATES FOREST SERVICE. 1974. Seeds of woody plants in the United States. *U.S. Dep. Agric. Handb.* 450:1–883.

USBERTI, R. 1982. Relações entre teste de envelhecimento acelerado, potencial de armazenamento e tamanho de sementes em lotes de amendoim. *Rev. Bras. Semen.* 4:31–44.

VAINAGII, I. V. 1971. Dinamika skhozhosti i zhittyezdatnosti nasinnya deyakikh trav'yanikh roslin Karpat. [Dynamics of germinability and viability of seeds of some herbaceous plants from the Carpathians.] *Ukr. Bot. Zh.* 28:449–55.

VAINAGII, I. V. 1973. Rezul'tati dal'shikh doslidzhen' dinamiki skhozhosti ta zhittyezdatnosti nasinnya trav'yanistikh roslin Karpat. [Results of further investigations into the dynamics of germinability and viability of seeds of herbaceous plants from the Carpathians.] *Ukr. Bot. Zh.* 30:104–10.

VAN DER MAESEN, L. J. G. 1984. Seed storage, viability, and rejuvenation. In J. R. Witcombe and W. Erskine, eds., *Genetic Resources and Their Exploitation: Chickpeas, Faba Beans, and Lentils*, pp. 13–22. The Hague: Nijhoff-Junk.

VAN DER MEY, J. A. M., R. A. KILPATRICK, and I. B. J. SMITH. 1982. The germination of wheat and oat seed stored at Bethlehem, Republic of South Africa, 1956–1981. *Cereal Res. Commun.* 10:159–64.

VAN ONCKELEN, H. A., R. VERBEEK, and A. A. KHAN. 1974. Relationship of ribonucleic acid metabolism in embryo and aleurone to α-amylase synthesis in barley. *Plant Physiol.* 53:562–68.

VAN STADEN, J. 1978. Seed viability in *Protea neriifolia*. II. The effects of different storage conditions on seed longevity. *Agroplantae* 10:69–72.

VAN STADEN, J., J. E. DAVEY, and L. M. DU PLESSIS. 1976. Lipid utilization in viable and non-viable *Protea compacta* embryos during germination. *Z. Pflanzenphysiol.* 77:113–19.

VAN STADEN, J., M. G. GILLILAND, and N. A. C. BROWN. 1975. Ultrastructure of dry viable and non-viable *Protea compata* embryos. *Z. Pflanzenphysiol.* 76:28–35.

VAN STADEN, J., M. G. GILLILAND, and D. L. DIX. 1981. Long-term storage of *Protea neriifolia* seed. *S. Afr. J. Sci. [S. Afr. Tydskr. Wes.]* 77:140–41.

VAN TIEGHEM, P. 1884. *Traité de botanique*. Paris: Savy.

VAN ZUTPHEN, H., and D. G. CORNWELL. 1973. Some studies on lipid peroxidation in monomolecular and bimolecular lipid films. *J. Membr. Biol.* 13:79–88.

VARRO, M. T. 1934. *De re rustica*. In W. D. Hooper and H. B. Ash, eds., *Marcus Porcius Cato: On Agriculture: Marcus Terentius Varro: On Agriculture*. Cambridge: Harvard University Press.

VASIL'EVA, V. T., and M. I. LAZUKOV. 1970. Osobennosti nabukhaniya raznovozrastnykh semyan redisa i belokochannoi kapusty. [Characteristics of swelling in radish and cabbage seeds of different ages.] *Dokl. Timiryazevsk. S'kh. Akad.* (Mosc.) 158:43–48.

VAUGHAN, C. E., and J. C. DELOUCHE. 1960. Relation of rate of swelling to viability in small seeded legumes. *Proc. Assoc. Off. Seed Anal.* 50:109–11.

VAUGHAN, C. E., and J. C. DELOUCHE. 1968. Physical properties of seeds associated with viability in small-seeded legumes. *Proc. Assoc. Off. Seed Anal.* 58:128–41.

VAUGHAN, C. E., and R. P. MOORE. 1970. Tetrazolium evaluation of the nature and progress of deterioration of peanut (*Arachis hypogaea* L.) seed in storage. *Proc. Assoc. Off. Seed Anal.* 60:104–17.

VENATOR, C. R. 1972. Effects of gibberellic acid on germination of low-vigor Honduras pine seeds. *For. Sci.* 18:331.

VERMA, R. S., and P. C. GUPTA. 1975. Storage behaviour of soybean varieties vastly differing in size. *Seed Res.* (New Delhi) 3:39–44.
VERTUCCI, C. W., and A. C. LEOPOLD. 1984. Bound water in soybean seed and its relation to respiration and imbibitional damage. *Plant Physiol.* 75:114–17.
VIEIRA, C. 1966. Effect of seed age on germination and yield of field bean (*Phaseolus vulgaris* L). *Turrialba* 16:396–98.
VIGIL, E. L., R. L. STEERE, W. P. WERGIN, and M. N. CHRISTIANSEN. 1984. Tissue preparation and fine structure of the radicle apex from cotton seeds. *Am. J. Bot.* 71:645–59.
VILLIERS, T. A. 1972. Cytological studies in dormancy. II. Pathological ageing changes during prolonged dormancy and recovery upon dormancy release. *New Phytol.* 71:145–52.
VILLIERS, T. A. 1973. Ageing and the longevity of seeds in field conditions. In W. Heydecker, ed., *Seed Ecology*, pp. 265–88. State College, Pa.: Pennsylvania State University Press.
VILLIERS, T. A. 1974. Seed aging: chromosome stability and extended viability of seeds stored fully imbibed. *Plant Physiol.* 53:875–78.
VILLIERS, T. A. 1975. Genetic maintenance of seeds in imbibed storage. In O. H. Frankel and J. G. Hawkes, eds., *Crop Genetic Resources for Today and Tomorrow*, pp. 297–315. New York: Cambridge University Press.
VILLIERS, T. A. 1980. Ultrastructural changes in seed dormancy and senescence. In K. V. Thimann, ed., *Senescence in Plants*, pp. 39–66. Boca Raton: CRC Press.
VILLIERS, T. A., and D. J. EDGCUMBE. 1975. On the cause of seed deterioration in dry storage. *Seed Sci. Technol.* 3:761–74.
VINCENT, G. 1929. Stárnutí semen jehličnaů. [The aging of conifer seeds.] *Sb. Cesk. Akad. Zemed.,* Rada A. 4:453–92.
VINCENT, G., and A. FREUDL. 1931. Časnásklizeň šišek jehličnaů a jakost jejich semen. [The effects of early harvesting of coniferous cones on seed quality.] *Lesn. Pr.* 10:248–56.
VISHNYAKOVA, I. A., N. P. KRASNOOK, and Z. T. BUKHTOYAROVA. 1979. Elektronnomikroskopicheskoe izuchenie strukturnoi organizatsii zapasnykh veshchestv v zarodyshe zernovki risa pri potere zhiznesposobnosti. *Fiziol. Rast.* (Mosc.) 26:750–55. [Electron-microscopic study of the structural organization of storage substances in the rice caryopsis embryo with loss of viability. *Sov. Plant Physiol.* 26:602–7.]
VISHNYAKOVA, I. A., N. P. KRASNOOK, R. I. POVAROVA, E. A. MORGUNOVA, and Z. T. BUKHTOYAROVA. 1976. Ul'trastruktura kletok zarodyshei zhiznesposobnykh i nezhiznesposobnykh semyan risa pri nabukhanii. *Fiziol. Rast.* (Mosc.) 23:361–64. [Ultrastructure of cells of the embryos of viable and unviable rice seeds in the course of swelling. *Sov. Plant Physiol.* 23:307–11.]
VLADIMIROV, YU. A., V. I. OLENOV, T. B. SUSLOVA, and Z. P. CHEREMISINA. 1980. Lipid peroxidation in mitochondrial membrane. *Adv. Lipid Res.* 17:173–249.
VOLYNETS, A. P., and L. A. PAL'CHENKO. 1980. Sostav i soderzhanie svobodnykh fenol'nykh soedinenii pri prorastanii semyan raznoi zhiznesposobnosti. [Content and composition of free phenolic compounds during germination of seeds of different viability.] *Fiziol. Biokhim. Kul't. Rast.* 12:511–15.
VOLYNETS, A. P., and L. A. PAL'CHENKO. 1982. Sostav i soderzhanie fenol'nykh konyugatov v protsesse prorastaniya semyan raznoi zhiznesposobnosti. [Content and composition of phenol conjugates during germination of seeds of different viability.] *Fiziol. Biokhim. Kul't. Rast.* 14:225–31.

WAHAB, A. H., and J. S. BURRIS. 1971. Physiological and chemical differences in low- and high-quality soybean seeds. *Proc. Ass. Off. Seed Anal.* 61:58–67.
WAHLEN, F. T. 1929. Hardseededness and longevity in clover seeds. *Proc. Int. Seed Test. Assoc.* 2 (9/10):34–39.
WALKER, M. J. 1973. T-tests on prehistoric and modern charred grain measurements. *Sci. Archaeol*, no. 10, pp. 11–32.
WALLEN, V. R., M. A. WALLACE, and W. BELL. 1955. Response of aged vegetable seed to seed treatment. *Plant Dis. Rep.* 39:115–17.
WALLER, A. D. 1901. An attempt to estimate the vitality of seeds by an electrical method. *Proc. R. Soc. Lond.* 68:79–93.
WALTERS, S. M. 1974. The rediscovery of *Senecio paludosus* L. in Britain. *Watsonia* 10:49–54.
WANG, B. S. P. 1974. Tree-seed storage. *Can. For. Serv. Publ.* 1335:1–30.
WANG, B. S. P. 1982. Long-term storage of *Abies, Betula, Larix, Picea, Pinus*, and *Populus* seeds. In B. S. P. Wang and J. A. Pitel, eds., *Proceedings of the International Symposium on Forest Tree Seed Storage, Chalk River, 1980*, pp. 212–18. Ottawa: Canadian Forestry Service.
WARD, F. H., and A. A. POWELL. 1983. Evidence for repair processes in onion seeds during storage at high seed moisture contents. *J. Exp. Bot.* 34:277–82.
WAREING, P. F. 1966. Ecological aspects of seed dormancy and germination. In J. G. Hawkes, ed., *Reproductive Biology and Taxonomy of Vascular Plants*, pp. 103–21. Oxford: Pergamon.
WAUGH, F. A. 1897. The enzymic ferments in plant physiology. *Science* (Wash., D.C.) 6:950–52.
WEBSTER, B. D., and A. C. LEOPOLD. 1977. The ultrastructure of dry and imbibed cotyledons of soybean. *Am. J. Bot.* 64:1286–93.
WEIDNER, S., and K. ZALEWSKI. 1982. Ribonucleic acids and ribosomal proteins synthesis during germination of unripe and aged wheat caryopses. *Acta Soc. Bot. Pol.* 51:291–300.
WEISS, M. G., and J. B. WENTZ. 1937. Effect of *luteus* genes on longevity of seed in maize. *Agron. J.* 29:63–75.
WENT, F. W. 1957. *The Experimental Control of Plant Growth*. Waltham, Mass.: Chronica Botanica.
WENT, F. W. 1969. A long-term test of seed longevity. II. *Aliso* 7:1–12.
WENT, F. W. 1974. Reflections and speculations. *Annu. Rev. Plant Physiol.* 25:1–26.
WENT, F. W., and P. A. MUNZ. 1949. A long-term test of seed longevity. *Aliso* 2:63–75.
WENZEL, G. 1922. Beobachtungen über die Langlebigkeit von Pflanzensamen. *Ber. Naturwiss. Ver. Bielefeld Umgegend* 4:246–48.
WESSON, G., and P. F. WAREING. 1969a. The role of light in the germination of naturally occurring populations of buried weed seeds. *J. Exp. Bot.* 20:402–13.
WESSON, G., and P. F. WAREING. 1969b. The induction of light sensitivity in weed seeds by burial. *J. Exp. Bot.* 20:414–25.
WEST, S. H., and H. C. HARRIS. 1963. Seedcoat colors associated with physiological changes in alfalfa and crimson and white clovers. *Crop Sci.* 3:190–93.
WESTER, H. V. 1973. Further evidence of age of ancient viable lotus seeds from Pulantien deposit, Manchuria. *HortScience* 8:371–77.
WHITE, J. W. 1882/84. Flora of the Avon-bed. *Proc. Bristol Nat. Soc.* 4:107–15.
WHITE, O. E. 1946. The germination of peas in Florida and King Tut's tomb. *Turtox News* 24:6–8.

WHITEHEAD, P. 1981. *The British Museum (Natural History)*. London: Wilson.
WHITMORE, T. C. 1983. Secondary succession from seed in tropical rain forests. *For. Abstr.* 44:767–79.
WHITTERN, C. C., E. E. MILLER, and D. E. PRATT. 1984. Cottonseed flavonoids as lipid antioxidants. *J. Am. Oil Chem. Soc.* 61:1075–78.
WHYMPER, R. 1913. The influence of age on the vitality and chemical composition of the wheat berry. *Knowledge* (Lond.) 36:85–90, 135–38.
WIEN, H. C., and E. A. KUENEMAN. 1981. Soybean seed deterioration in the tropics. II. Varietal differences and techniques for screening. *Field Crops Res.* 4:123–32.
WIESNER, J. R. VON. 1913. *Biologie der Pflanzen*. 3d ed. Vienna: Holder.
WIŁKOJĆ, A. BISKUPSKI, J. KIERSNOWSKI, W. LONC, and M. NARKIEWICZ-JODKO. 1983. Wpłwyw czasu i warunków przechowywania ziarna siewnego pszenicy, żyta i jęczmienia na jego właśćiwości biologiczne i technologiczne. [The influence of length and conditions of seed storage on the biological and technological properties of wheat, rye, and barley.] *Hodowla Rosl. Aklim. Nasienn.* 27:277–96.
WILSON, A. T., M. VICKERS, and L. R. B. MANN. 1979. Metabolism in dry pollen—a novel technique for studying anhydrobiosis. *Naturwissenschaften* 66:53–54.
WILSON, D. O., and M. B. MCDONALD. In press. The lipid peroxidation model of seed aging. *Seed Sci Technol.*
WILTON, A. C., C. E. TOWNSEND, R. J. LORENZ, and G. A. ROGLER. 1978. Longevity of alfalfa seed. *Crop Sci.* 18:1091–93.
WINKLER, A. 1891. Die Keimfähigkeit des Samens der *Malva moschata* L. *Dtsch Bot. Monatsschr.* 9:4–5.
WIŚNIEWSKI, K., and K. KULKA. 1979. Ribonucleic acids in the embryos of rye stored at various air humidities. *Bull. Acad. Pol. Sci.*, ser. *Sci. Biol.* 27:387–91.
WITTMACK, L. 1873. *Gras- und Kleesamen*. Berlin: Wiegandt & Hempel.
WITTMACK, [L.] 1888. Über einen Roggen aus dem dreissigjährigen Kriege. *Jahrb. Dtsch Landwirtsch. Ges.* 3:69–76.
WITTMACK, L. 1903. Über die in Pompeji gefundenen pflanzlichen Reste. *Bot. Jahrb., Beibl.* 73:38–66.
WITTMACK, L. 1922. *Landwirtschaftliche Samenkunde*. Berlin: Parey.
WITTMER, G. 1958. Prime osservazioni sull'attività citologica dell'olio estratto da semi di *N. tabacum* e *N. rustica*. *Tabacco* (Rome) 62:384–405.
WITTMER, G. 1960. Attività citologica dell'olio estratto da semi vecchi e freschi di *Nicotiana rustica*. *Tabacco* (Rome) 64:172–93.
WITTMER, G. 1961a. Attività citologica dell'olio estratto da semi vecchi e freschi di *Nicotiana tabacum* L. *Tabacco* (Rome) 65:229–37.
WITTMER, G. 1961b. Attività citologica di alcuni acidi grassi componenti gli olii estratti da semi vecchi e freschi di Nicotianae. *Tabacco* (Rome) 65:434–40.
WOODSTOCK, L. W. 1973. Physiological and biochemical tests for seed vigor. *Seed Sci. Technol.* 1:127–57.
WOODSTOCK, L. W., and J. FEELEY. 1965. Early seedling growth and initial respiration rates as potential indicators of seed vigor in corn. *Proc. Assoc. Off. Seed Anal.* 55:131–39.
WOODSTOCK, L. W., K. FURMAN, and T. SOLOMOS. 1984. Changes in respiratory metabolism during aging in seeds and isolated axes of soybean. *Plant Cell Physiol.* 25:15–26.
WOODSTOCK, L. W., and D. F. GRABE. 1967. Relationships between seed respiration during imbibition and subsequent seedling growth in *Zea mays* L. *Plant Physiol.* 42:1071–76.

WOODSTOCK, L. W., S. MAXON, K. FAUL, and L. BASS. 1983. Use of freeze-drying and acetone impregnation with natural and synthetic anti-oxidants to improve storability of onion, pepper, and parsley seeds. *J. Am. Soc. Hortic. Sci.* 108:692–96.

WOODSTOCK, L. W., J. SIMKIN, and E. SCHROEDER. 1976. Freeze-drying to improve seed storability. *Seed Sci. Technol.* 4:301–11.

WOODSTOCK, L. W., and K. L. J. TAO. 1981. Prevention of imbibitional injury in low vigor soybean embryonic axes by osmotic control of water uptake. *Physiol. Plant.* 51:133–39.

WOODSTOCK, L. W., and R. B. TAYLORSON. 1981a. Ethanol and acetaldehyde in imbibing soybean seeds in relation to deterioration. *Plant Physiol.* 67:424–28.

WOODSTOCK, L. W., and R. B. TAYLORSON. 1981b. Soaking injury and its reversal with polyethylene glycol in relation to respiratory metabolism in high and low vigor soybean seeds. *Physiol. Plant.* 53:263–68.

WU, G., R. A. STEIN, and J. F. MEAD. 1979. Autoxidation of fatty acid monolayers adsorbed on silica gel. IV. Effect of antioxidants. *Lipids* 14:644–50.

WYTTENBACH, E. 1955. Der Einfluss verschiedener Lagerungsfaktoren auf die Haltbarkeit von Feldsämereien (Luzerne, Rotklee und Gemeinem Schotenklee) bei länger dauernder Aufbewahrung. *Landwirtsch. Jahrb. Schweiz* 69:161–96.

YAMAGUCHI, H., T. NAITO, and A. TATARA. 1978. Decreased activity of DNA polymerase in seeds of barley during storage. *Jpn. J. Gen.* 53:133–35.

YAMAGUCHI, T., M. NAKATANI, and J. F. SUTCLIFFE. 1983. Promotion of growth in aged tobacco seeds by pre-treatment with gibberellic acid. *Ann. Bot.* (Lond.) 51:157–59.

YAMAGUCHI, T., T. WAKIZUKA, K. OKUI, and E. OHTA. 1982. Promotion of germination in aged rice and bean seeds *in vitro*. In A. Fujiwara, ed., *Proceedings of the 5th International Congress on Plant Tissue and Cell Culture, Tokyo, 1982*, pp. 785–86. Tokyo: Japanese Association for Plant Tissue Culture.

YANAGISAWA, T. 1965. Shinyōju no tane no hatsuga to jumyō ni oyobosu kyūka no seijukudo no eikyō. [The effect of cone maturity on the viability and longevity of coniferous seeds.] *Ringyo Shikenjo Kenkyu Hokoku [Bull. For. Prod. Res. Inst.]* 172:45–94.

YANG, S. F., and Y.-B. YU. 1982. Lipid peroxidation in relation to aging and loss of seed viability. *Search (Am. Seed Res. Foundn.)* 16 (1):2–7.

YARCHUK, T. A. 1966. Dolgoletie semyan kukuruzy razlichnoi konsistentsii. [Longevity of maize seed of differing consistency.] *Tr. Prikl. Bot. Genet. Sel.* 38 (1):157–59.

YARCHUK, T. A. 1974. Prodolzhitel'nost zhizni semyan kukuruzy v zavisimosti ot srokov i sposobov khraneniya. [Longevity of maize seed in relation to the length and conditions of storage.] *Byull. Vses. Nauchno-Issled. Inst. Rastenievod. im. N. I. Vavilova* (Leningr.) 43:16–20.

YARCHUK, T. A., and I. V. LEIZERSON. 1971. Zhiznesposobnost' semyan kukuruzy v zavisimosti ot sroka khraneniya. [Viability of maize seed as a function of their period of storage.] *Tr. Prikl. Bot. Genet. Sel.* 44:215–19.

YARCHUK, T. A., and I. V. LEIZERSON. 1972. Zavisimost' kachestva semyan kukuruzy ot dlitel'nosti ikh khraneniya. [The relationship between maize seed quality and length of storage.] *Sel. Semenovod.* 37 (1):59–60.

YAROSH, N. P., and O. G. ANTONOVA. 1978. Izmenenie sostava legkorastvorimykh belkov i nekotorykh oksidoreduktaz pri dlitel'nom khranenii semyan yachmenya, risa, grechikhi. [Change in the constitution of readily soluble proteins and various oxidoreductases during prolonged storage of barley, rice, and buckwheat.] *Byull. Vses. Nauchno-Issled. Inst. Rast. im. N. I. Vavilova* (Leningr.) 77:30–36.

YATSU, L. Y. 1965. The ultrastructure of cotyledonary tissue from *Gossypium hirsutum* L. seeds. *J. Cell Biol.* 25:193–99.
YONEYAMA, T., I. SUZUKI, and M. MUROHASHI. 1970. Natural maturing of wheat flour. I. Changes in some chemical components and in farinograph and extensigraph properties. *Cereal Chem.*. 47:19–26.
YOO, B. Y. 1970. Ultrastructural changes in cells of pea embryo radicles during germination. *J. Cell Biol.* 45:158–71.
YOUNGMAN, B. J. 1951. Germination of old seeds. *Kew Bull.* 6:423–26.
YU, Y.-B. 1981. Zhongzi zhi lao hua sheng li. [Physiology of seed senescence.] *Zhongguo Yuanyi [Chung Kuo Yuen Yi: Chin. Hortic.* (Taipei)*]* 27:151–59.
ZALEWSKI, K. 1982. Formowanie polirybosomów w zardodkach kiełkującego ziarna pszenicy o różnej żywotności. [Polyribosome formation in embryos of germinating wheat grain of varying viability.] *Hodowla Rosl. Aklim. Nasienn.* 26:1–9.
ZALEWSKI, K., and S. WEIDNER. 1982. Changes in ribosomal proteins of wheat embryos during accelerated ageing of the grain. *Acta Soc. Bot. Pol.* 51:301–8.
ZASADA, J. C., and R. A. DENSMORE. 1977. Changes in seed viability during storage for selected Alaskan Salicaceae *Seed Sci. Technol.* 5:509–18.
ZELENCHUK, T. K., and S. O. GELEMEI. 1965. Trivalist' zberigannya skhozhosti nasinnya luchnikh zlakiv i bobovikh u laboratornikh umovakh. [Longevity of pasture grass and legume seeds under laboratory conditions.] *Ukr. Bot. Zhurn.* 22 (3):44–51.
ZELENOV, A. N. 1968. Tsitologicheskoe izuchenie estestvennogo mutatsionnogo protsessa v strarykh i nezrelykh semenakh gorokha. [Cytological change associated with the natural mutation process in aged and unripe pea seeds.] *Dokl. Timiryazevsk. S'kh. Akad.* (Mosc.) 136:61–69.
ZELENSKII, G. V. 1982. Aktivnost' proteoliticheskikh fermentov v khranyashchikhsya semenakh soi. [The activity of proteolytic enzymes in stored soybean seeds.] *Byull. Vses Nauchno-Issled. Inst. Rastenievod. im. N. I. Vavilova* (Leningr.) 118:38–40.
ZELENSKII, G. V., and T. A. ZELENSKAYA. 1983. Aktivnost' proteoliticheskikh fermentov pri potere zhiznesposobnosti semyan soi. *Dokl. Vses. Akad. S'kh. Nauk im. V. I. Lenina* (Mosc.), No. 8, pp. 16–19. [Activity of proteolytic enzymes in relation to loss of viability in soya seeds. *Soviet Agric. Sci.*, No. 8, pp. 24–28.]
ZELENY, L., and D. A. COLEMAN. 1938. Acidity in cereals and cereal products, its determination and significance. *Cereal Chem.* 15:580–95.
ZELENY, L., and D. A. COLEMAN. 1939. The chemical determination of soundness in corn. *U.S. Dept. Agric. Tech. Bull.* 644:1–23.
ZENTSCH, W. 1970. The nitrogen, fat, and sugar content of *Picea abies* Karst. and *Pinus silvestris* L. seeds during storage under different conditions and in a water-free state. In S. Białobok and B. Suszka, eds., *Proceedings of the International Symposium on Seed Physiology of Woody Plants, Kórnik, 1968*, pp. 73–78. Warsaw: Państwowe Wydannictwo Naukowe.
ZEVEN, A. C., G. J. DOEKES, and M. KISLEV. 1975. Proteins in old grains of *Triticum* sp. *J. Archaeol. Sci.* 2:209–13.
ZHENG, G.-H., C.-M. GE, C.-J. LIU, Y.-J. ZHANG, and P.-S. TANG. 1964. Biding hegan suan dui yang shu zhongzi qu qing mei huo xing xiaoshide huifu zuoyong. [Pyridine nucleotides restore the dehydrogenase activity of poplar seeds as viability is lost.] *Zhiwu Xuebao [Chih Wu Hsueh Pao: Acta Bot. Sin.]* 12:325–32.
ZHENG, G.-H., and Q.-S. YAN. 1964. Yong yingguang fa kuaisu ceding zhongzi sheng huolide shiyan. [The use of fluorimetry as a rapid test of vigor.] *Zhiwu Shenglixue Tongxun [Chih Wu Sheng Li Hsueh T'ung Hsun: Bull. Plant Physiol.]* No. 3, pp. 21–25.

ZHENG, X.-Y., Z.-H. GU, B.-M. XU, and G.-H. ZHENG. 1982. Ji zhong shu rong zhongzide huoli shiyan. [Experiments on the seed vigor of some vegetable crops.] *Zhongguo Nongye Kexue [Chung Kuo Nung Yeh K'o Hsueh: Chin. Agric. Sci.]* No. 4, pp. 58–64.

ZHUKOVA, G. YA., and M. S. YAKOVLEV. 1976. Khloroplasty pochechki zarodysha iskopaemogo plodika lotosa (elektronnomikroskopicheskoe issledovanie). [Chloroplasts of the embryonic plumule of a fossilized lotus fruit: a study with the electron microscope.] *Bot. Zh.* (Leningr.) 61:869–72.

ZORNER, P.S., R. L. ZIMDAHL, and E. E. SCHWEIZER. 1984. Sources of viable seed loss in buried dormant and non-dormant populations of wild oat (*Avena fatua* L.) seed in colorado. *Weed Res.* 24:143–50.

Species Index

Abelmoschus esculentus (okra), 66, 74, 148–50, 197
   manihot, 74
Abutilon hybridum, 197
   indicum, 197
   theophrasti, 66
Acacia, 81, 103
Acanthus latifolius, 197
   mollis, 197
Acer platanoides (Norway maple), 187
   saccharinum (silver maple), 104
   saccharum (sugar maple), 104
Achillea millefolium, 197, 206
Aconitum napellus, 197
Adelocaryum coelestinum, 204
Adlumia fungosa, 197
Agathis macrophylla, 52n
Ageratum houstonianum, 197, 204, 205
Agropyron intermedium, 64
Agrostemma githago (corn cockle), 93, 96
Agrostis, 201
   gigantea, 206
   nebulosa, 197
Albizia, 81
   distachya, 197
   julibrissin, 86, 87
Alcea rosea, 197, 203
Alfalfa. See *Medicago sativa*
Allium, 74, 127
   ampeloprasum, 197, 206
   cepa, 14, 49, 53, 55, 60, 61, 108, 123, 127, 138, 147, 149, 170, 174, 184, 187–88, 191, 197, 201–3, 206
   fistulosum, 174
   schoenoprasum, 197
Alopecurus, 201
   myosuroides, 102
   pratensis, 64, 197, 202, 206
Althaea officinalis, 197
Amaranthus, 92
   caudatus, 197
   retroflexus, 93
Ambrosia artemisiifolia, 93
Ammobium alatum, 204
Ampelopsis, 197
Anchusa capensis, 197, 204
Anemone coronaria, 197
Anethum graveolens, 197
Angelica archangelica, 197
Anthemis cotula, 93
Anthriscus cerefolium, 197, 206
Anthyllis vulneraria, 82, 197, 206
Antirrhinum majus, 197, 204–5
Apium graveolens (celery), 51, 60, 92, 184, 197, 206
Aquilegia, 197
   caerulea, 204
   vulgaris, 204
Arabidopsis thaliana, 45
Arabis alpina, 197, 204
Arachis hypogaea (groundnut or peanut), 24, 52, 61, 65, 129, 132, 138, 144,

Arachis hypogaea (cont.)
   146–49, 157, 186, 203
Arctotis stoechadifolia, 197, 203, 205
Argemone polyanthemos, 197
Armeria juncea, 204
   pseudarmeria, 197⁻
Armoracia rusticana, 197
Arnica montana, 197
Arrenatherum elatius, 64, 198, 202, 206
Artemisia absinthium, 198
   dracunculus, 198
   vulgaris, 198
Asarina, 198
Asclepias syriaca, 198
Ash, European (*Fraxinus excelsior*), 187
Asparagus densiflorus, 198
   officinalis, 198, 206
   setaceus, 198
Aspen (*Populus*), 60
Aspergillus, 19, 71, 130
Asperula orientalis, 204
Asphodeline liburnica, 198
Asphodelus microcarpus, 198
Aster, 198
   alpinus, 205
Astragalus cicer, 205
   sinaicus, 66
Atriplex hortensis, 198
Aurinia saxatilis, 204
Avena fatua (wild oats), 32, 100–101, 148
   sativa (cultivated oats), 24, 31, 32, 47, 63, 64, 68, 86, 87, 146, 161, 183, 198, 201–3, 206

Bambara groundnut. See *Voandzeia subterranea*
Baptisia australis, 198
   tinctoria, 198
Barley. See *Hordeum vulgare*
Bean, broad. See *Vicia faba*
   fava. See *Vicia faba*
   kidney. See *Phaseolus vulgaris*
   lima. See *Phaseolus lunatus*
   mung. See *Vigna radiata*
   snap. See *Phaseolus vulgaris*
Beech (*Fagus sylvatica*), 91
Beet. See *Beta vulgaris*
Begonia cucullata, 198, 204
   gracilis, 198
   rex, 198
Bellis perennis, 198, 204–5
Bergenia crassifolia, 205
Beta vulgaris (beet or mangel), 49, 74, 155, 170, 198, 202, 206
Bindweed. See *Convolvulus arvensis*
Borago officinalis, 198

Brachycome iberidifolia, 198
Brassica hirta (white mustard), 42, 198, 203, 206
   juncea (brown mustard), 150
   kaber (charlock or field mustard), 32, 49, 94, 101–2
   napus (rape), 14, 23n, 24, 64, 136, 138, 165, 169–70, 185, 198, 202, 206
   nigra, 93, 198
   oleracea (cabbage or cauliflower), 14, 60, 64, 65, 137, 159–60, 170–71, 191, 198, 201–2, 206
   rapa (turnip), 198, 202, 206
Bromus arvensis, 198
   erectus, 201
   inermis, 64, 206
   mollis, 206
   secalinus (chess), 93, 96
Browallia speciosa, 198

Cabbage. See *Brassica oleracea*
Cacao (*Theobroma cacao*), 21
Calceolaria tripartita, 198
Calendula, 203
   officinalis, 198, 204–5
Callistephus chinensis, 52, 198, 203–4
Calluna vulgaris, 91
Caltha palustris, 198
Campanula medium, 198, 204–5
   persicifolia, 198, 205
Canavalia ensiformis (jack bean), 29
Cannabis sativa, 198, 201, 206
Canna compacta, 113
   indica, 198
Capsella bursa-pastoris, 93, 97
Capsicum annuum (sweet pepper), 52, 66, 147, 149, 184, 198, 203
   frutescens (tabasco pepper), 46, 51
Cardaria draba (hoary cress), 101
Carrot. See *Daucus carota*
Carthamus tinctorius (safflower), 190
Carum carvi, 198, 206
Cassia bicapsularis, 86
   multijuga, 79, 84, 86
Castor bean. See *Ricinus communis*
Catharanthus roseus, 198
Cauliflower. See *Brassica oleracea*
Ceanothus sanguineus, 103
Celery. See *Apium graveolens*
Celosia cristata, 198, 204–5
Centaurea americana, 204
   cyanus, 198, 203
   gymnocarpa, 203
   moschata, 198, 203
Centaurium erythraea, 198
Centranthus ruber, 198
Cephalaria gigantea, 205

Species Index

Cerastium tomentosum, 198, 204
Chaerophyllum bulbosum, 198
Chamaemelum nobile, 198
Charlock. See *Brassica kaber*
Chasmanthium latifolium, 198
Cheiranthus allionii, 204
   cheiri, 198
Chenopodium album (lambs-quarters), 94, 97, 98, 101–3, 117–18
   hybridum, 103
Chess *(Bromus secalinus)*, 93, 96
Chickpea. See *Cicer arietinum*
Chrysanthemum carinatum, 203
   coccineum, 198
   coronarium, 198, 204
   leucanthemum, 203
   maximum, 204
   parthenifolium, 198
   parthenium, 204
   segetum, 198, 203–4
Cicer arietinum (chickpea), 53, 55, 72, 73, 123, 140, 160, 169, 201, 206
Cichorium endiva, 198
   intybus, 149, 198, 206
Citrullus lanatus (watermelon), 50, 71, 198, 201
Clarkia amoena, 204
   unguiculata, 198, 204
Clematis paniculata, 198
Cleome, 198
Clover. See *Trifolium*
   crimson. See *Trifolium incarnatum*
   red. See *Trifolium pratense*
   subterranean. See *Trifolium subterraneum*
   white. See *Trifolium repens*
Cobaea scandens, 198
Coffea arabica (coffee), 21
Coix lacryma-jobi, 198
Coleus ×*hybridus*, 198
Consolida, 198
   regalis, 204
Convolvulus arvensis (bindweed), 96, 98, 99, 111
   tricolor, 198, 204
Corchorus capsularis and *C. olitorius* (jute), 49, 71, 155, 186, 193
Cordyline indivisa, 198
Coreopsis, 203
   basalis, 198
   coronata, 204
   grandiflora, 198, 205
   lanceolata, 198
   tinctoria, 204
Coriandrum sativum, 198
Corn cockle (*Agrostemma githago*), 93, 96

Cortaderia selloana, 198
Cosmos, 198, 203
Cotton. See *Gossypium hirsutum*
Cowpea. See *Vigna unguiculata*
Crepis, 149
   capillaris, 174
   tectorum, 172
Cryptostemma calendulaceum, 198
Cucumber. See *Cucumis sativus*
Cucumis melo (melon), 50, 71, 134, 198, 202–3
   sativus (cucumber), 50, 58, 71, 134, 138, 147, 150, 170, 198, 206
Cucurbita, 202
   pepo (squash), 50, 198
Cuphea llavea, 204
Cyclamen persicum, 198
   purpurascens, 198
Cynara cardunculus, 198
   scolymus, 198
Cynosurus cristatus, 198, 202
   echinatus (dogtail), 96
Cyperus alternifolius, 198
   difformis, 91
   fuscus, 91

Dactylis glomerata, 64, 198, 202, 206
Dahlia imperialis, 198
   pinnata, 198, 205
Date palm (*Phoenix dactylifera*), 14
Datura ferox, 190
   stramonium (jimson weed), 175, 203
Daucus carota (carrot), 64, 66, 129, 155, 198, 202, 206
Delphinium, 45
   grandiflorum, 72, 203
   hybridum, 198
Dianthus barbatus, 198, 204–5
   caryophyllus, 198, 203–4
   chinensis, 198, 204
   discolor, 205
   fischeri, 205
   fragrans, 205
   giganteus, 205
   heddewigii, 203
   plumarius, 198, 205
   seguieri, 205
Digitalis ciliata, 205
   grandiflora, 205
   purpurea, 198, 204
Dimorphotheca pluvialis, 198, 205
Dipsacus sylvestris, 206
Dodecatheon meadia, 205
Dogtail (*Cynosurus echinatus*), 96
Dolichos lablab, 198
Doreanthus bellidiformis, 204

Dorotheanthus tricolor, 198
Dracaena draco, 199
Dracocephalum ruyschiana, 205

Echinochloa, 155
Echinocystis lobata, 199
Echinops ritro, 199
Echium lycopsis, 204
Eggplant. See *Solanum melongena*
Elderberry (*Sambucus*), 192
Elm, American (*Ulmus americana*), 60
Emilia, 199
  javanica, 204
Ensete ventricosum, 199
Erechtites hieracifolia, 93
Erica, 199
Eryingium alpinum, 199
Erysimum hieraciifolium, 199
  pulchellum, 199
Eschscholzia californica, 199, 203–4
Eupatorium cannabinum, 199
Euphorbia esula (leafy spurge), 101
  maculata, 93
  marginata, 199

Fagopyrum esculentum, 199, 207
  tartaricum, 199
Fagus sylvatica (beech), 91
Festuca, 202
  elatior, 202, 207
  ovina, 199, 202, 207
  pratensis, 199, 202
  rubra, 199, 202, 207
  tenuifolia, 202
Fir, Douglas (*Pseudotsuga menziesii*), 63
Flax. See *Linum usitatissimum*
Foeniculum vulgare, 199
Fraxinus excelsior (European ash), 187
Fuchsia, 199

Gaillardia ×*grandiflora*, 199
  hybrida, 204
  pulchella, 199
Gentiana lagodechiana, 205
  septemfida, 205
Geranium bohemicum, 103
Gerbera jamesonii, 199
Geum coccineum, 199
Gilia capitata, 203
Glaucium flavum, 117
  serpieri, 117
Gloxinia, 199
Glycine max (soybean), 19, 23n, 24, 25, 26n, 29–31, 33, 35, 48, 51–55, 57, 60, 62, 65, 67, 71, 73–75, 123, 129, 132–39, 141, 144–50, 152–55, 157, 159–61, 167–68, 170, 176, 177n, 190, 199, 202–3, 207
  soja (wild soybean), 65
Gompholobium, 81
Gomphrena globosa, 199
Goodia lotifolia, 81, 86
Gossypium hirsutum (cotton), 23n, 24, 36, 49, 60, 63, 67, 71, 129, 134, 138, 141–42, 146, 155
Grevillea robusta, 199
Ground cherry (*Physalis subglabrata*), 118
Goundnut. See *Arachis hypogaea*
Gypsophyla elegans, 199, 205
  paniculata, 199

Hardenbergia, 81
Haynaldia villosa, 178
Hedysarum alpinum, 205
Helenium autumnale, 199
  hoopesii, 199
Helianthus, 199
  annuus (sunflower), 65, 134, 138, 140, 146, 149, 155, 157, 199, 202, 207
  debilis, 199
Helichrysum, 199
  bracteatum, 203–5
Heliopsis helianthoides, 199
Heliotropium arborescens, 199
  europaeum, 112
Helipterum, 199
  manglesii, 204
  roseum, 199, 204
Hesperis matronalis, 199, 204–5
Heuchera, 199
  americana, 205
Hevea brasiliensis (rubber), 21
Hibiscus cannabinus (kenaf), 25, 153
  trionum, 199
Hippeastrum, 199
Hoary cress (*Cardaria draba*), 101
Holcus lanatus, 207
Hordeum vulgare (barley), 13, 24, 31, 43, 46–48, 53, 55, 57, 64, 65, 86, 87, 107–9, 111, 114–16, 123, 134–6, 158, 160, 170, 174–76, 184, 186, 192, 199, 202–3, 207
Horse nettle (*Solanum carolinense*), 101
Hovea linearis, 81, 86
Hunnemannia fumariifolia, 199
Hyssopus officinalis, 199

Iberis amara, 205
  imperialis, 204
  sempervirens, 199
  umbellata, 199, 203, 205
Impatiens balsamina, 199, 204–5

Impatiens balsamina (cont.)
   wallerana, 199
Indigofera, 81
Inula helenium, 199
Ipomoea nil, 199
   purpurea, 199
Iris ×*germanica*, 199
   kaempferi, 199
Isatis tinctoria, 207

Jack bean (*Canavalia ensiformis*), 29
Jimson weed (*Datura stramonium*), 175, 202
Juncus effusus, 91
Jute. See *Corchorus capsularis* and *C. olitorius*

Kenaf (*Hibiscus cannabinus*), 25, 153
Kennedia, 81
Kentucky bluegrass. See *Poa pratensis*
Kniphofia hybrida, 199
Kochia scoparia, 199, 203

Lactuca sativa (lettuce), 23, 46, 49–51, 52n, 61, 66, 129–30, 132, 155, 161, 170, 174, 184, 187–89, 199, 202–3, 207
Lagenaria siceraria, 199
Lamb's-quarters. See *Chenopodium album*
Lantana camara, 199
Larch. See *Larix decidua*
Larix decidua (larch), 191
Lathyrus latifolius, 199
   odoratus, 199, 203–4
   sativus, 199, 203, 207
Lavandula angustifolia, 199
Lavatera trimestris, 199
Lens culinaris (lentil), 70n, 87, 129, 199, 203, 207
Lentil. See *Lens culinaris*
Leontopodium alpinum, 199
Lepidium sativum, 129, 199, 207
   virginicum, 93
Lettuce. See *Lactuca sativa*
Levisticum offinale, 199
Liatris spicata, 199
Lilium, 199
   regale, 205
Limonium latifolium, 199
   sinuatum, 199, 204
Linaria maroccana, 199, 204
Linseed. See *Linum usitatissimum*
Linum altaicum, 205
   grandiflorum, 199, 204
   perenne, 199

   usitatissimum (flax or linseed), 24, 72, 129, 146, 186, 199, 202, 203, 207
Lobelia cardinalis, 62, 199
   erinus, 199
Lobularia maritima, 199, 203
Lolium, 202
   multiflorum, 199, 202, 207
   perenne (perennial ryegrass), 176, 199, 202, 207
London rocket (*Sisymbrium irio*), 89
Lotus. See *Nelumbo*
Lotus corniculatus, 67, 82, 199, 207
   pedunculatus, 199, 207
   tetragonolobus, 199
   uliginosus, 86
Lunaria annua, 199
Lupine, Arctic. See *Lupinus arcticus*
   yellow. See *Lupinus luteus*
Lupinus (lupine), 129, 184
   albus, 199, 203
   angustifolius, 72, 199, 207
   arcticus (Arctic lupine), 119
   hartwegii, 199
   hirsutus, 199
   luteus (yellow lupine), 66, 100, 103, 158, 186, 199, 203, 207
   mutabilis, 199
   nanus, 199
   polyphyllus, 52n, 199, 205
Lychnis chalcedonica, 199
   coronaria, 199
   viscaria, 200
Lycopersicon esculentum, 200, 202–3, 207
Lythrum flexuosum, 200

Macleaya cordata, 200
Maize. See *Zea mays*
Malope trifida, 204
Malva moschata (musk mallow), 98, 200
   pusilla, 92, 93
Mangel (*Beta vulgaris*), 49
Maple, Norway (*Acer platanoides*), 187
   silver (*Acer saccharinum*), 104
   sugar (*Acer saccharum*), 104
Matricaria, 200
Matthiola, 45, 203
   annua, 204
   incana, 205
   longipetala, 200
Medicago (medick), 21, 81
   lupulina (black medick), 42, 112, 200, 202, 207
   polymorpha, 113
   sativa (alfalfa), 48, 67, 71, 82, 166, 200, 202, 207
Medick. See *Medicago*
   black. See *Medicago lupulina*

Melilotus, 81
   alba, 200, 207
   officinalis, 200
Melon. See *Cucumis melo*
Mentzelia lindleyi, 200
Mesembryanthemum crystallinum, 200
Mesquite (*Prosopis juliflora*), 98
Millet, foxtail. See *Setaria italica*
   pearl (*Pennisetum glaucum*), 73
   prosa. See *Panicum miliaceum*
Mimosa glomerata, 79n
   pudica, 86, 200
Mimulus ×*hybridus*, 200, 204
   moschatus, 200
Mirabilis jalapa, 200
Momordica, 200
Musk mallow (*Malva moschata*), 98, 200
Mustard, brown (*Brassica juncea*), 150
   field. See *Brassica kaber*
   white. See *Brassica hirta*
Myosotis alpestris, 200, 204
   scorpioides, 200
Myrrhis odorata, 200

Nasturtium, 200
   officinale, 200
Nelumbo (lotus), 78n–79n, 87, 104
   luteus (American lotus), 84–86, 122
   nucifera (Asiatic lotus), 84–86, 119–24, 147, 148n
Nemesia grandiflora, 204
   strumosa, 200
Nepeta grandiflora, 205
Nicotiana alata, 200
   glauca, 200
   rustica (tobacco), 142, 174
   sylvestris, 200
   tabacum (tobacco), 49, 64, 92, 142, 146, 174, 176, 182, 184, 193, 200, 203, 207
Nigella damascena, 200, 203
   sativa, 200

Oats, cultivated. See *Avena sativa*
   wild. See *Avena fatua*
Ocimum basilicum, 200
Oenothera, 129, 172n
   berteriana, 74
   biennis, 45, 93, 200
   campylocalyx, 45
   drumondii, 200
   erythrosepala, 147
   missouriensis, 200
   odorata, 74
   rosea, 200
Okra. See *Abelmoschus esculentus*
Onion. See *Allium cepa*
Onobrychis viciifolia, 200, 202, 207
Origanum majorana, 200
Ornithopus sativus, 200, 207
Oryza sativa (rice), 24, 35, 52n, 61, 71, 72, 91, 115, 127, 132, 138, 141, 146, 154–56, 159–62, 168, 170, 178, 181–86, 193, 202–3
Osteospermum, 204
Oxylobium, 81

Paeonia suffruticosa, 200
Panicum dichotomiflorum (panicum), 115, 118
   maximum, 193
   miliaceum (proso millet), 71, 202–3, 207
Papaver (poppy), 203
   dubium, 200, 205
   nudicaule, 200, 205
   orientale, 147, 200
   persicum, 205
   rhoeas (field poppy), 89, 94, 96, 204
   somniferum, 200, 203, 205, 207
Parsley. See *Petroselinum crispum*
Parsnip *(Pastinaca sativa)*, 61, 64
Pastinaca sativa (parsnip), 200, 202, 206
Pea. See *Pisum sativum*
Peach (*Prunus persica*), 111
Peanut. See *Arachis hypogaea*
Pelargonium, 67
   carnosum, 200
   ×*hortorum*, 200
   zonale, 200
Penicillium, 19
Pennisetum glaucum (pearl millet), 73n
   setaceum, 200
   villosum, 200
Penny cress. See *Thlaspi arvense*, 96, 101
Penstemon, 200
   barbatus, 205
   gracilis, 205
Pepper, sweet. See *Capsicum annuum*
   tabasco (*Capsicum frutescens*), 46, 51
Perilla frutescens (perilla), 146, 202, 207
Petrorhagia saxifraga, 200
Petroselinum crispum (parsley), 149, 200, 207
Petunia ×*hybrida*, 200, 203, 204
Phacelia tanacetifolia, 200
Phalaris arundinacea, 200
   canariensis, 200, 207
Phaseolus, 114, 116
   acutifolius, 207
   coccineus, 129, 200, 207
   lunatus (lima bean), 166, 207
   vulgaris (kidney or snap bean), 47, 48,

Species Index

vulgaris (cont.)
    71, 73, 74, 86, 129, 157, 160, 185–86, 200, 202–3, 207
Phleum pratense (timothy), 41, 63, 64, 200, 202, 207
Phlox cuspidata, 200
    drumondii, 200, 203, 204
    paniculata, 200
Phoenix dactylifera (date palm), 14
Physalis alkekengi, 200
    peruviana, 200
    subglabrata (ground cherry), 118
Phytolacca americana, 92–94
Picea abies (Norway spruce), 45, 52, 137
Pimpinella anisum, 200
Pine. See *Pinus*
    Cuban. See *Pinus caribaea*
    jack. See *Pinus banksiana*
    longleaf. See *Pinus palustris*
    Scots. See *Pinus sylvestris*
Pinus (pine), 23, 60, 64, 90, 100, 103, 137, 157
    attenuata, 103
    banksiana (jack pine), 161, 178
    caribaea (Cuban pine), 184
    contorta, 103
    densiflora, 182
    palustris (longleaf pine), 23, 149–50
    pinea, 130, 157, 160
    ponderosa, 31
    serotina, 103
    sylvestris (Scots pine), 63, 130, 137, 190
Pisum sativum (pea), 23n, 35, 48, 50, 52n, 66, 71, 73, 107, 111n, 114, 117, 129–31, 136, 138–39, 147, 149–50, 159, 163, 168–70, 174, 176, 181–82, 185, 190–91, 200, 202–3, 207
Plantago lanceolata, 102
    major, 93
Platycodon, 200
Poa, 129
    annua, 97
    nemoralis, 202, 207
    palustris, 202
    pratensis (Kentucky bluegrass), 51, 200, 202, 207
    trivialis, 200, 202, 207
Polemonium caeruleum, 205
Polygonum hydropiper, 93
Poplar. See *Populus*
    Black. See *Populus nigra*
Poppy, field. See *Papaver rhoeas*
Populus (aspen or poplar), 60, 104, 120
    nigra (black poplar), 163
Portulaca grandiflora, 200, 204

    oleracea, 93, 200
Potato *(Solanum tuberosum)*, 191, 206
Potentilla recta, 205
Poterium sanguisorba, 200
Primula auricula, 200
    ×*kewensis*, 200
    malacoides, 200
    obconica, 52, 200
    sinensis, 200
    veris, 200
    vulgaris, 200
Prosopis juliflora (mesquite), 100
Protea compacta, 130, 147
    neriifolia, 130
Prunella grandiflora, 205
Prunus persica (peach), 111
Pseudotsuga menziesii (Douglas fir), 63
Psylliostachys suworowii, 200
Pueraria lobata, 200

Quercus, 22

Radish. See *Raphanus sativus*
Rape. See *Brassica napus*
Raphanus sativus (radish), 49, 137, 170, 200, 203, 207
Raspberry. See *Rubus idaeus*, 112
Reseda odorata, 200
Rheum, 200, 207
Rhus (sumac), 119n
    ovata, 103
Ribes, 90
Rice. See *Oryza sativa*
Ricinus communis (castor bean), 68, 142, 146, 161, 200, 202, 207
Rosmarinus officinalis, 200
Rubber *(Hevea brasiliensis)*, 21
Rubus idaeus (raspberry), 112
Rudbeckia, 200
    hirta, 204
Rumex crispus, 93, 98, 103
    obtusifolius, 103
    patientia, 200
Russian thistle *(Salsola kali)*, 96
Ruta graveolens, 200
Rye. See *Secale cereale*
Ryegrass, perennial. See *Lolium perenne*

Safflower *(Carthamus tinctorius)*, 190
Saintpaulia ionantha, 200
Salix (willow), 104
Salpiglossis, 203
    sinuata, 200, 204
Salsola kali (Russian thistle), 96
Salvia azurea, 200
    farinacea, 200
    officinalis, 200

Salvia azurea (cont.)
   patens, 201
   sclarea, 201
   splendens, 52, 201, 204
   viridis, 204
Sambucus (elderberry), 192
Sanvitalia procumbens, 204
Saponaria ocymoides, 201
   pyramidata, 201
Satureja hortensis, 201
Saxifraga caespitosa, 205
   rotundifolia, 205
Scabiosa atropurpurea, 201, 203–4
   caucasia, 201
Schizanthus, 201
   ×*wisetonensis*, 203
Scirpus cernuus, 201
Scorzonera hispanica, 201, 208
Secale cereale (rye), 30, 64, 68, 115, 127, 130–32, 137, 146, 154, 159–60, 177–83, 189, 201–2, 208
Sedge (*Cyperus*), 91
Sedum kirilowii, 205
   reflexum, 201
Senecio paludosus, 91
   sylvaticus, 80
Sesame. See *Sesamum indicum*
Sesamum indicum (sesame), 146
Setaria italica (foxtail millet), 71, 201, 203
   pumila, 93
Sinningia speciosa, 201
Sisymbrium irio (London rocket), 89
   orientale, 90
Smilax, 201
Snowberry (*Symphoricarpos albus*), 98
Solanum, 201
   carolinense (horse nettle), 101
   melongena (eggplant), 49, 51, 184, 201
   nigrum, 94
   tuberosum (potato), 191, 208
Sorghum bicolor (sorghum), 24, 46, 60, 63, 83n, 133, 136, 158–59, 190, 202–3
   sudanense, 83
Soybean, cultivated. See *Glycine max*
   Wild (*Glycine soja*), 65
Spergula arvensis, 102, 117–18, 201, 208
Spinach. see *Spinacia oleracea*
Spinacia oleracea (spinach), 192, 201, 203, 208
Spruce, Norway (*Picea abies*), 45, 52, 137
Spurge, leafy (*Euphorbia esula*), 101
Squash (*Cucurbita pepo*), 150
Stellaria media, 93
Stevia serrata, 201
Stipa pennata, 201

Stokesia laevis, 201
Sumac (*Rhus*), 119
Sunflower. See *Helianthus annuus*
Symphoricarpos albus (snowberry), 98

Tagetes, 201
   erecta, 203–4
   patula, 203
Taraxacum officinale, 201
Tellima grandiflora, 205
Tetragonia tetragonioides, 201
Thelesperma burridgeanum, 204
Theobroma cacao (cacao), 21
Thlaspi arvense (penny cress), 96, 101
Thunbergia alata, 201
   coccinea, 201
   fragrans, 201
   grandiflora, 201
Thymus vulgaris, 201
Timothy. See *Phleum pratense*
Tobacco. See *Nicotiana rustica*; *N. tabacum*
Tomato. See *Lycopersicon esculentum*, 23, 68, 71, 132, 134, 138, 155, 185, 190–91
Torenia fournieri, 201
Tortula ruralis, 37
Torula, 29
Tragopogon porrifolius, 201, 208
Trifolium (clover), 64, 66, 82, 96, 128, 202
   alexandrinum, 201
   arvensis, 201
   hybridum, 48, 66, 201–2, 208
   incarnatum (crimson clover), 48, 170, 201–2, 208
   pratense (red clover), 48, 61, 66, 86, 92, 201, 202, 208
   repens (white clover), 48, 61, 91, 93, 201–2, 208
   subterraneum (subterranean clover), 98, 147
Trigonella caerulea, 201
Trisetum flavescens, 201
Triticum aestivum (common wheat), 14, 16, 30, 31, 43, 46, 47, 52n, 62, 65, 70, 71, 73, 87, 89, 96, 107–10, 114, 115, 117, 127, 130–32, 138–40, 146, 148, 150–51, 155, 157–60, 162, 164, 166, 168–69, 172, 175–76, 181–84, 186, 190, 192–93, 201–3, 208
   dicoccon (emmer wheat), 107, 114, 116
   durum (durum wheat), 43, 61, 107, 114, 127–28, 174, 178, 181–82, 184–86, 191, 193
   turgidum, 110–11, 208
Tropaeolum, 201
   majus, 201, 203

Species Index

Turnip (*Brassica rapa*), 49
Typha latifolia, 37

Ulmus americana (American elm), 60
Ursinia anthemoides, 204

Valeriana officinalis, 201
Valerianella locusta, 201, 208
Venidium fastuosum, 204
Verbascum blattaria, 92, 93
 thapsiforme, 93, 117–18, 201
 thapsus, 92
Verbena, 203
 canadensis, 201
 ×*hybrida*, 201, 204
 platensis, 46
 rigida, 201
Veronica gentianoides, 205
 spicata, 201
Vetch. See *Vicia*
Voandzeia subterranea (bambara groundnut), 137–38, 183, 186
Vicia (vetch), 71
 angustifolia, 99
 cracca, 82
 ervilia, 208
 faba (broad or fava bean), 48, 52n, 66, 129, 147, 160, 173, 176, 201, 203, 208
 hirsuta, 94, 98
 narbonensis, 208

 sativa, 201–3, 208
 villosa, 201, 203, 208
Vigna angularis, 208
 radiata (mung bean), 48, 57, 169, 208
 unguiculata (cowpea), 35, 53, 55, 123
Vinca major, 201
 minor, 201
Viola, 203
 cornuta, 201
 odorata, 201
 persicifolia, 91
 tricolor, 201, 204

Watermelon. See *Citrullus lanatus*
Wheat, common. See *Triticum aestivum*
 durum. See *Triticum durum*
 emmer. See *Triticum dicoccon*
 miracle or Mummy. See *Triticum turgidum*
Willow *(Salix)*, 104

Xeranthemum annuum, 201, 204

Zea mays (corn or maize), 23n, 26n, 36, 43, 44, 46, 47, 65, 71–75, 107–8, 110–11, 115–16, 131–32, 139, 147–48, 152, 157–58, 160, 162, 170–72, 174, 176, 183, 188, 191, 193, 201–2, 208
Zinnia, 203
 elegans, 201
 haageana, 201

Subject Index

Abnormalities in seedlings, 17, 45, 171
Abscisic acid, 185–86
Accelerated aging, 29, 44, 57–58, 71, 147, 152, 190, 193
Acetaldehyde, 150
Acid fuschin, 133
Acid violet, 133
Action potentials, 133
Activation energy, 153, 161
Adenosine diphosphate (ADP), 169
Adenosine monophosphate (AMP), 169, 170
Adenosine triphosphate (ATP), 159, 169–71, 180
Adobe, seeds recovered from, 113
Aging, definition of, 18
Aizoaceae, 78
Alanine aminotransferase, 158
Albumins, 157
Alcohol dehydrogenase, 159, 160
Alcohols, 144
Aldehydes, 146
Aleurone, 115, 126–27, 184
Alkaloids, 72
Alkanes, 144
Amaranthaceae, 93
Amino acids, 31, 116, 134, 166
Amino acyl-tRNA, 182
AMP, 169, 170
Amphidiploidy, 74
Amylase, 30, 159, 192

alpha (α), 183–84
beta (β), 158–59
Anaphase, 174, 189
Ancient seeds, 105–24
Annuals, 72
Anoxia during germination, 43
Anthocyanins, 191
Antioxidants, 61, 144–45, 148–49
Aquatic species, 21
Arginine, 116
Argon, 61
Arid climates, 104–5, 113. *See also* Deserts
Aristotle, 14
Arrhenius, S. A., 59
Ascorbate oxidase, 159
Ascorbic acid, 144, 148
Atacama Desert, flora of, 83
ATP, 159, 169–71, 180
Autopolyploidy, 74
Autoxidation, 144, 148
Auxins, 183

Balsaminaceae, 77–78
Barium, 133
Battlefields, new growth from, 89
Beal, W. J., 91–93
Biotin, 162
Bogs, 122
Boiling, germination after, 21
Boraginaceae, 78

299

British Association, 77–79
British Isles, flora of, 83
British Museum, 84, 87, 110
Bromine, 149
Brown, R., 84–85, 110
Buried seeds, 88–104, 117–24
Burning, germination after, 81n, 103. *See also* Carbonization
Butylated hydroxytoluene, 149
Butylhydroquinone, 149
Butyl sulfide, 149

Caffeic acid, 185
Calcium, 66
Cannaceae, 77n
Carbohydrates, 30, 37
Carbon dioxide, 32–33, 59–62, 100, 123, 166–67
Carbonization, 106–8, 114
Carbonyls, 144, 150, 153
Carboxylation, 162
Caryophyllaceae, 78
Cassianus Bassus, 14n
Catalase, 123, 158, 163
Cathodic treatment, 154, 193
Cell cycle, 173–74
Cellulose, 24
Cell walls, 25, 35, 114, 130
Cereals, 46–47, 62, 70, 108, 140, 169
Charring. *See* Carbonization
Chelators, 145, 192
Chemical composition and storability, 68–70
Chemiluminescence, 144, 151, 153–54
Chenopodiaceae, 77n, 82, 95, 103
Chilling stress, 43–44
Chlorine, 149
Chloroethanol, 192
Chloroethylphosphonic acid, 185
Chlorophyll, 73n, 175–76
Chloroplasts, 33, 142
Chromatid aberrations, 173–74
Chromatin, 115, 130, 178
Chromosome aberrations, 172–75, 187, 189
Citrate, 166
Climate and storability, 42, 56–59, 68n. *See also* Temperature and deterioration
Coenzyme A, 162
Coenzymes, 162–63
Color and longevity, 72–73, 108, 114–15, 128–29, 158
Columella, L. J. M., 14
Complex, definition of (*Oenothera* genetics), 74
Compositae, 78, 82–83, 93, 95
Conductivity. *See* Leakage
Constantine VII, Roman emperor, 14n

Convolvulaceae, 77n, 81, 83, 95
Cotyledons, 129
Cruciferae, 78, 82, 93, 95, 103, 133
Cryogenics, 21, 58–59
Cucurbitaceae, 50, 58, 78
Cyanide insensitivity, 168
Cysteine, 146, 157
Cytochrome c oxidase, 159, 168
Cytokinins, 184
Cytosol, 161

Darkening. *See* Color and longevity
Darwin, C. R., 111, 117, 186–87
Decarboxylation, 32
Dehydrogenases, pyridine-linked, 163
Delinting, 63
Denmark, 82, 91, 94, 117, 122
Density. *See* Specific gravity
Deoxyribonuclease, 160, 177, 179
Deoxyribonucleic acid. *See* DNA
Deserts, 83, 148
Desiccation, extreme, 51–25, 134, 145, 174
Deterioration equations:
  soil, 96, 118
  storage, 51–56, 96, 123, 138
Dictyosomes, 35, 131
Dinitrophenol, 168
Diploids, 74
Dispersal, 89
DNA:
  denaturation and fragmentation of, 145, 177–80
  synthesis of, 177–79, 189
DNA polymerase, 191
Dormancy, 42, 44, 59, 77, 104, 187–88, 191, 193
  in buried seeds, 88, 92, 96, 98, 100–102, 186
Drouillard de la Marre, N.-M.-H., 110
Drying, artificial, 63

Earthworms, 118
Ecological succession, 88, 103
Egypt, 14, 105, 107–10
Elastic modulus, 132
Electrolytes. *See* Leakage
Electron microscopy. *See* Ultrastructure
Electron spin resonance, 107, 151–52, 155, 191
Electron transport, 33, 143, 162, 168
Elongation factors, 182
Embryos, 114–15, 126–28
Emergence. *See* Field emergence
Endopeptidase, 160. *See also* Proteinase
Endoplasmic reticulum, 35, 115

Endosperm, 127, 129
Energy charge, 169
Engler, G. H. A., 77–78
Enzymes, 29–30, 156, 158–65, 161
　inhibitors of, 161, 179
Erythrocytes, 37
Esterase, 160–61
Ethanol, 150
Ethylene, 185
Euphorbiaceae, 78, 80, 93
Evans blue, 133
Evolution, 176

Family affiliations. *See* Systematics
Fan Sheng-zhi, 14
Faraday, M., 110
Fat acidity, 114, 140–42, 158
Fats. *See* Lipids
Fatty acids:
　free, 140–43
　hydroxy and oxygenated, 149
　short-chain, 142
Ferulic acid, 186
Field emergence, 16, 40, 43–44, 46–48, 150
Fire, germination after, 81n, 103. *See also* Carbonization
Fixation for electron microscopy, 34–35, 130
Flavin nucleotides, 162
Flavonoids, 145
Flower morphology, 45
Fluorescence, 129–30
Forage species, 48, 82
Forest seed banks, 90, 100
Formazan, 163, 165
Fossils, 107, 114, 117
Free radicals, 144, 151, 191
Freeze fracture, 35
Freezing, 19, 37, 51–52, 62, 98
Fungi, 19, 32, 62–64, 98, 149–50. *See also* Microflora
　in humid storage conditions, 57, 59, 71, 73, 139–41, 157–58, 161
　lipid degradation by, 139–41
Fungicides, 57, 192
Fusicoccin, 193

Genetic complexes, 74
Genetic predisposition to aging, 71–75
　and cytoplasmic effects, 73–75
　and dominance, 73–74
Geographical populations, 98
Geoponika, 14
Geotropism, 113
Geraniaceae, 77n, 80, 82
Germination, 17, 40, 88
　rate of, 44–45

Germ plasm, 58, 175–77
Gibberellins, 183–84
Globulins, 157
Glucose, 133, 136, 168–69
Glucose-6-phosphate dehydrogenase, 29, 31
Glutamate decarboxylase, 158, 184
Glutamate dehydrogenase, 161, 163
Glutathione, 148, 193
Gluten, 115
Glyceraldehyde-phosphate dehydrogenase, 159
Glycine, 116
Glycolysis, 166, 168–69
Golgi bodies, 35, 131
Gramineae, 78, 82–83, 87, 93, 95, 115
Grauballe Man, 122
Growth inhibitors, 185–86
Guanosine triphosphate (GTP), 180
Gymnosperms, 64, 103, 132, 140

Hairs, 129
Half-viability period (P50), 40, 43, 205–8
Hard seeds, 42, 65–67, 71–72, 82, 84, 147
　in soil, 98–99, 119, 186
Harvesting, 63–64
Heating, 43, 107, 123, 151, 153, 172, 177
　in soil, 19, 98
Helium, 61
Hemoproteins, 145
Henslow, J. S., 113
Histidine, 146
Histones, 178–79
Hooker, J. D. H., 113
Hormones, 183–86, 193
Hulling, 63–64
Hydration, 21–38, 56, 59, 72
　and storage environment, 51–60, 72
　and stress during imbibition, 36, 51, 134, 154
Hydrophyllaceae, 78
Hydroxybenzoic acid, 186
Hydroxyl radical, 143

IAA, 183, 185
IAA-oxidase, 160
Ibn al-'Awwām, 15
Imbibition, 137
Incas, 108, 111
Indigo carmine, 133
Indole acetic acid, 183, 185
Injury, mechanical, 62–63, 65, 67
Invertase, 136
Iodine, 149
Iodine value, 146, 148
Iridaceae, 78, 80
Iron Age, 13, 117

Isocitrate lyase, 159, 161
Isozymes, 160, 161

Kaempferol, 185
α-Ketoglutarate, 166
Key cells, 42, 166
Kinetin, 184

Labiatae, 77–78, 80, 82–83
Lactate, 166
Lauremberg, P., 15
Leakage, 36, 64, 126, 132–37, 163, 188, 190–91
   membrane lipids and, 140, 145
Leguminosae, 47–48, 66, 77–78, 80–84, 93, 95, 98, 103, 127–28
Leucine aminopeptidase, 161
Lichen, 31
Light, red, 193
Liliaceae, 78, 95
Limestone, 117
Lindley, J., 110–13
Linoleate, 123, 144, 146–47
Linolenate, 123, 143–44, 146–47
Lipase, 31, 140–42, 159, 161
Lipid bodies, 33, 37, 115, 130, 137
Lipid content, 68–69, 114, 123, 138, 165
Lipid hydroperoxides, 143, 145–46
Lipids:
   peroxidation of, 101, 107, 116, 139, 142–53, 156–57, 179
   and polyunsaturation, 146–48
   and water binding, 23, 25, 38
Lipoxygenase, 31, 142, 144, 153
London, fires of, 89–90
Longevity:
   in soil, 88–104, 117–24
   in storage, 76–87, 197–208
Louis XIV, king of France, 117n
Luteus gene, 73
Lysine, 116, 146, 158
Lysophospholipids, 139
Lysozyme, 29

Macrobiotic species, 80, 83
Maillard reactions, 115, 158, 166
Malachite green, 163
Malate dehydrogenase, 160–61
Malondialdehyde, 149–50
Malvaceae, 66, 77–78, 80–84, 93, 95, 98
Maturity, 64–65
Membranes, 34–38, 130–37
   and fluidity, 144–45
   and non-bilayer and hexagonal phases, 36–37, 145
Mercaptoethanol, 149

Meristems, 171
Mesobiotic species, 80
Metal ions, 143, 145
Methanol, 43
Methionine, 116, 146, 158
Methylene blue, 163
Microbiotic species, 80
Microflora, 19–20, 57, 59, 100–102, 126, 132, 138, 165. *See also* Fungi
Mitochondria, 161–63, 165, 167–68, 187
   lipid-related changes in, 142, 145
   ultrastructure of, 35, 131
Mitosis, 172–75, 177, 188
Mold. *See Fungi*
Monosomics, 175
Moraceae, 78
Moss, 37
Mummy seeds, 107–11
Mutagens, 127, 176, 179–80, 191
Mutations, 171–75, 177
Myrtaceae, 80, 83

Nelumbonaceae, 79n
Niacin, 162
Nicotinamide adenine dinucleotide (phosphate), 162–63
Nicotinic acid, 162
Nitrate, 185, 192–93
Nitrogen, 58, 61–62, 66, 123
Nuclear magnetic resonance:
   ^{1}H-, 27–29
   ^{13}C-, 30
   ^{31}P-, 30, 37
Nucleolus, 115
Nucleoside triphosphates, 170, 180, 182
Nucleotides, 30
Nucleus, 115, 130, 187
Nutrients, 66
Nymphaeaceae, 78, 80

Oil bodies. *See* Lipid bodies
Oil ducts, 129
Oilseeds, 48, 68, 72, 140, 142, 150
Oleate, 144
Olefins, 144
Onagraceae, 78, 93
Orthodox seeds, 21, 59, 104
Oxalacetate, 166
Oxidative phosphorylation, 142, 169
Oxygen, 33, 101, 122–23, 157, 166–68, 179, 187
   in lipid oxidation, 143, 153
   in storage, 59–62

Subject Index

Paddy seed bank, 91
Palladius, R. T. Ae., 14
Pantothenic acid, 162
Papaveraceae, 78
Peat, 122
Pentose phosphate pathway, 168–69
Perennials, 72
Permafrost, 119
Peroxidase, 154, 158, 160–62
Peroxide value, 149–50
Peroxy radicals, 143–44
Phenolase, 162
Phenolics, 185–66
Phenylalanyl-tRNA synthetase, 182
Phosphatase, 30, 132, 158–60
Phosphate, 133
Phosphatidic acid, 139
Phosphatidylcholine, 38
Phosphoenolpyruvate, 170
Phospholipase, 140
  A, 139
  B, 31
  D, 31, 139
Phospholipids, 36–37, 137–38
Phytin, 159
Plantaginaceae, 93, 95
Plasmalemma, 35–36, 115, 130, 132
Plasmolysis, 132, 136
Plastids, 131
Ploidy, 74
Polemoniaceae, 78, 82
Pollen, 31, 36, 167
  abortion of, 175–76
Polyamines, 186
Polyethylene glycol, 190
Polygonaceae, 78, 80, 93, 95, 103
Polymorphism, 98
Polysomes, 131, 180, 187
Pompeii, 111
Portulacaceae, 78, 93
Potassium, 66, 193
Preharvest environment, 62–67
Priming, 189–91
Probits, 40
Propyl gallate, 149
Proteinase, 157–61, 190, 192
Protein bodies, 35
Proteins, 116, 156–58, 187
  content of, 66, 68–69, 114, 123, 157, 165
  denaturation of, 116–17, 145, 156
  synthesis of, 159, 161, 179, 180–82, 189
  and water binding, 24–25, 28
Proton extrusion, 193
Protoplasts, 37
Pulantien, 84, 120–23

Putrescine, 186
Pyridoxine, 162
Pyruvate, 166
Pyruvate decarboxylase, 159

Quercetin, 185

Radiation, 151, 153, 155–56, 179, 191
Radiocarbon dating, 84, 105, 113, 119–21
Radiosensitivity, 191
Rancidity, 143n
Range species, 82, 100
Ranunculaceae, 78
Recalcitrant seeds, 59, 84, 103–4
Reinvigoration, 190–93
Repair, 51, 102, 125, 132, 139, 156, 177, 186–87, 190–91
Respiration, 32–33, 66, 100, 122, 163, 165–68, 187
Riboflavin, 162
Ribonuclease, 158, 160, 182
Ribonucleic acid. See RNA
Ribonucleoside-diphosphate reductase, 160
Ribonucleotides, 30
Ribosomes, 131, 181–82
Ridolfi, C., 110
RNA, 159
  messenger, 145, 178, 180–82
  polyadenylated, 180–81
  ribosomal, 181–82
  synthesis of, 180–81, 187, 189
Root crops, 49
Rosaceae, 78

Sarcoplasmic reticulum, 37
Scarification, 67, 122
Scrophulariaceae, 78, 93
Sea water, 44
Seed banks, 89–91
Seed coat, 33, 62, 65, 70, 79, 128–29
Selection for longevity, 74, 186
Selenium, 163
Senescence, 18
Sick grain, 129, 158
Sirius, 14–15
Size and longevity, 65, 70
Snow, 119
Sod, seeds recovered from, 113n
Sodium thiosulfate, 149
Softening, 98
Soil, longevity in, 88–104
Solanaceae, 77n, 78, 95, 103
Sorption isotherms, 24–27
Sorption zones, 19, 24–25, 29, 39, 58, 125, 145, 153
Species differences in storability, 67–70

Specific gravity, 65–66
Spermidine, 186
Starch, 68–69, 72n, 81, 114–15, 123, 165
  and water binding, 24–25, 28
Starch phosphate, 148
Stem morphology, 45
Sterculiaceae, 77–78, 80, 83
Sternberg, C. von, 108–9
Sterols, 140
Storage, 13, 39–75, 86
Succinate, 163
Succinate dehydrogenase, 159–60
Succinate semialdehyde dehydrogenase, 159
Sucrose, 136
Sugars, 134, 136, 165
  reducing, 158, 165–66
Sulfate, 193
Sulfhydryls, 157–58
Superoxide dismutase, 145
Superoxides, 143–44
Systematics, 76–84, 102–3

Tannins, 70n
Tellurium, 163
Temperature and deterioration, 51–54. See also Climate and storability
Terpenoids, 129
Tetraploids, 74
Tetrazolium salts, 113, 127, 136, 163–65
Theophrastus, 13
Thermodormancy, 187–88
Thaimin, 162
Thiobarbituric acid test, 150
Thiourea, 193
Tiliaceae, 77–78, 80
Tocopherols, 144, 148–49
Trace elements, 191
Transamination, 31, 162
Transcription of mRNA, 177–78, 181
Translation of mRNA into protein, 145, 180–82
Transplantation experiments, 126–27
Tree seeds, 64, 68, 157
Trees, wind-fallen, 90
Triacylglycerols, 137, 140
Tricarboxylic acid cycle, 31, 166

Triploids, 74
Tropical climates, 58, 65, 68, 73–75
Troy, pea seeds recovered at, 117n
Trypsin inhibitor, 159
Tupper, M. F., 109–10
Turgor, 132

Ultrasonics, 193
Ultrastructure, 34–36, 115, 123, 130, 187–88
Ultraviolet radiation, 129, 133, 192
Umbelliferae, 78, 95
Urease, 29
Uridine triphosphate (UTP), 170
Urinalysis papers, 133

Vacuum storage, 60, 62, 87
Valerianaceae, 78
Vanillic acid, 186
Varietal differences in storability, 71–73
Varro, M. T., 14
Verbenaceae, 95
Vigor, 17, 43–45, 58, 127, 136
  testing for, 44, 57, 126
Vital stains, 133
Vitamin B group, 162
Volatiles, 150–51
Volunteering, 96

Water:
  "bound," 25
  "free," 25–29, 31, 33, 153
Waterloo, Battle of, 89
Weathering, 62
Weeds, 88, 96, 98
Weight loss, 165
Wetting-and-drying cycles, 154–56, 187, 189–91
Wilkinson, G., 109
World War I, 89
Wrinkling, 73

X rays, 37, 133

Yeast, 36, 140
Yield, 46–50

Zoroaster, 15n

THE LIBRARY
ST. MARY'S COLLEGE OF MARYLAND
ST. MARY'S CITY, MARYLAND 20686